T0311959

Sustainable Water Engineering

Sustainable Water Engineering

Edited by
Susanne Charlesworth
Colin A. Booth
Kemi Adeyeye

Elsevier
Radarweg 29, PO Box 211, 1000 AE Amsterdam, Netherlands
The Boulevard, Langford Lane, Kidlington, Oxford OX5 1GB, United Kingdom 50 Hampshire Street, 5th Floor, Cambridge, MA 02139, United States

British Library Cataloguing-in-Publication Data
A catalogue record for this book is available from the British Library

Library of Congress Cataloging-in-Publication Data
A catalog record for this book is available from the Library of Congress

ISBN: 978-0-12-816120-3

For Information on all Elsevier publications visit our website at https://www.elsevier.com/books-and-journals

Publisher: Susan Dennis
Acquisition Editor: Anita A. Koch
Editorial Project Manager: Sara Pianavilla
Production Project Manager: Joy Christel Neumarin Honest Thangiah
Cover Designer: Miles Hitchen

Typeset by Aptara, New Delhi, India

Dedication—

This book is dedicated to those in whose hands the future rests:
"The World is not ours to keep. We hold it in trust for future generations"
Kofi Annan.

Contents

Contributors

Colin A. Booth Architecture and the Built Environment, University of the West of England, Bristol, UK; Centre for Architecture and Built Environment Research (CABER), University of the West of England (UWE), Bristol, UK; University of the West of England, Bristol, UK

Susanne M. Charlesworth Coventry University, Coventry, UK; Centre for Agroecology, Water and Resilience, Coventry University, Ryton Gardens, Wolston Lane, Coventry CV8 3LG, UK

Kemi Adeyeye Architecture and Civil Engineering, University of Bath, UK; Department of Architecture and Civil Engineering, University of Bath, Bath, UK

Kate Ward School of Science and Engineering, University of Dundee, UK

Simon D. Smith School of Engineering, University of Edinburgh, UK

Martin Crapper Department of Mechanical and Construction Engineering, Northumbria University, UK

John Griggs FCIPHE, Chartered Institute of Plumbing and Heating Engineering, Hornchurch, UK

Katherine Hyde School of the Built Environment, University of Reading, Reading, Berkshire, UK

Ines Meireles Risco, Department of Civil Engineering, University of Aveiro, Portugal

Jeremy Gibberd Built Environment, CSIR, Pretoria, South Africa

Mynepalli K.C. Sridhar Department of Environmental Health Sciences, Faculty of Public Health, College of Medicine, University of Ibadan, Ibadan, Nigeria

Mumuni Adejumo Department of Environmental Health Sciences, Faculty of Public Health, College of Medicine, University of Ibadan, Ibadan, Nigeria

Omolara Lade Department of Civil Engineering, University of Ibadan, Nigeria

David Oloke Faculty of Science and Engineering, University of Wolverhampton, UK

Akinwale O. Coker Department of Civil Engineering, Faculty of Technology, University of Ibadan, Ibadan, Nigeria

Olalekan I. Shittu Department of Civil Engineering, Faculty of Technology, University of Ibadan, Ibadan, Nigeria

Temitope A. Laniyan Department of Environmental Health Sciences, Faculty of Public Health, College of Medicine, University of Ibadan, Ibadan, Nigeria

Chibueze G. Achi Department of Civil Engineering, Faculty of Technology, University of Ibadan, Ibadan, Nigeria

Alireza Fathollahi Centre for Agroecology, Water and Resilience (CAWR), Coventry University, UK

Stephen J. Coupe Centre for Agroecology, Water and Resilience (CAWR), Coventry University, UK

Luis A. Sañudo-Fontaneda Centre for Agroecology, Water and Resilience (CAWR), Coventry University, UK; INDUROT Research Institute, GICONSIME Research Group, Department of Construction and Manufacturing Engineering, University of Oviedo, Spain
Frank Warwick School of Energy, Construction and Environment, Faculty of Engineering, Environment and Computing, Coventry University, Priory Street, Coventry CV1 5FB, UK
Tom Lavers School of Energy, Construction and Environment, Coventry University, Sir John Laing Building, Much Park Street, Coventry CV1 2LT, UK
Ian Berry School of Architecture and the Built Environment, University of the West of England Frenchay Campus, Coldharbour Lane, Bristol BS16 1QY, UK
Armando Carravetta Department of Civil, Architectural and Environmental, Engineering, Università degli Studi di Napoli "Federico II", Napoli, Italy
Miguel Crespo Chacon Trinity College of Dublin, Department of Civil, Structural and Environmental Engineering, Dublin, Ireland
Oreste Fecarotta Department of Civil, Architectural and Environmental, Engineering, Università degli Studi di Napoli "Federico II", Napoli, Italy
Aonghus Mcnabola Trinity College of Dublin, Department of Civil, Structural and Environmental Engineering, Dublin, Ireland
Helena M. Ramos Instituto Superior Técnico (IST), Departamento de Engenharia Civil, Arquitectura e Georrecursos, Lisbon, Portugal
Carme Machí Castañer São PauloS,P,, São PauloS,P, 05508-080, Brazil
Daniel Jato-Espino GITECO Research Group, Universidad de Cantabria, Av. de los Castros 44, Santander, 39005, Spain
Carly B. Rose Visiting Fellow, CABER, UWE, Bristol, UK

Preface

The purpose of *Sustainable Water Engineering* is threefold. Firstly, it integrates the concept of 'sustainability' with engineering infrastructure and approaches to provide solutions across the water industry. Secondly, it gives real-world examples of the ways in which engineers or engineering principles have been used to apply these concepts by introducing the latest thinking from academic, stakeholder and practitioner perspectives. Thirdly, it investigates how these principles can deliver sustainable water supplies, of sufficient quality, at scale and which principle is cost effective.

Sustainable Water Engineering covers issues around the supply of quality water and its management, as well as challenges around surface water in urban, rural, developed and developing contexts. It tackles flooding, water quality, water supply, environmental quality and the future for sustainable water engineering. In addition, it addresses historical legacies, the implementation of integrated strategies at multiple scales and the policies and governance required to encourage their use. It also includes worldwide case studies whereby these innovative principles have been successfully implemented by water utility operators.

The book includes latest research in the field, presenting real-world applications, to show how engineers, environmental consultancies and international institutions can apply the concepts and strategies presented throughout the book. Practical and accessible, it is a resource across academia, for practitioners and stakeholders working for engineering and water companies, local authorities and water-related consultancies.

CHAPTER 1

From taps to toilets and ponds to pipes—A paradigm shift in sustainable water engineering

Colin A. Booth[a,*], **Susanne M. Charlesworth**[b], **Kemi Adeyeye**[c]

[a]*Architecture and the Built Environment, University of the West of England, Bristol, UK* [b]*Coventry University, Coventry, UK* [c]*Architecture and Civil Engineering, University of Bath, Bath, UK*
**Corresponding author.*

1.1 Introduction

Contemporary water engineering extends beyond the traditional teachings of hydrology and hydraulics to incorporate environmental, technological, managerial, cultural, and societal issues. Current and anticipated water crises can be deemed the intended and/or unintended consequences of long-term changes (i.e. slow evolution) of social norms and values (or more broadly, culture), ideology or political systems, which are not typically anticipated or accounted for in coping with water-related issues [17]. The recognition and integration of these issues allows water engineers to understand, unravel and contribute towards solving both local and global challenges (Fig. 1.1), and make a stepwise change in the delivery of sustainable infrastructure, sustainable buildings, sustainable businesses, and sustainable behaviours.

The United Nations Sustainable Development Goals (SDGs) represent a blueprint towards the achievement of a better and more sustainable future for everyone.[1] The 17 SDGs are an international attempt to address the challenges that societies face, including those related to poverty, inequality, climate change, environmental degradation, peace, and justice. Water is central to most SDGs. The SDG outcomes are interdependent with complex couplings between human, technical and natural systems [23] in which water engineers will play a fundamental role towards their success.

Sustainable Development Goal 6 (SDG–6) focuses on promoting access to water and sanitation for everyone. This SDG aimed at ensuring availability and sustainable management of water and sanitation for all is probably the greatest challenge we face in water resources

[1] www.un.org/sustainabledevelopment.

Sustainable Water Engineering. DOI: 10.1016/C2017-0-04301-X

Fig. 1.1: Dealing with the effects of extreme weather events is amongst the portfolio of activities facing today's water engineer: (A) the dried-out reservoir bed of the dammed Berg River, in Cape Town, South Africa (in 2018); (B) high street and retail premises partly-submerged by floodwaters in Shropshire, England (in 2020).

management [29]. Towards this, the United Nations states "Clean, accessible water for all is an essential part of the world we want to live in and there is sufficient fresh water on the planet to achieve this. However, due to bad economics or poor infrastructure, millions of people including children die every year from diseases associated with inadequate water supply, sanitation, and hygiene. Water scarcity, poor water quality and inadequate sanitation negatively impact food security, livelihood choices and educational opportunities for poor families across the world. At the current time, more than 2 billion people are living with the risk of reduced access to freshwater resources and by 2050, at least one in four people is likely to live in a country affected by chronic or recurring shortages of fresh water. Drought specifically afflict some of the world's poorest countries, worsening hunger, and malnutrition. Fortunately, there has been great progress made in the past decade regarding drinking sources and sanitation, whereby over 90 percent of the world's population now has access to improved sources of drinking water. To improve sanitation and access to drinking water, there needs to be increased investment in management of freshwater ecosystems and sanitation facilities on a local level in several developing countries within Sub–Saharan Africa, Central Asia, Southern Asia, Eastern Asia and South–Eastern Asia" [27].

The other SDGs further highlight the pivotal contribution of water engineering to the attainment of sustainability [28]. For instance, infrastructure either directly or indirectly influence all 17 of the SDGs, including 121 of the 169 targets (72 percent). For 5 of the 17 SDG goals (SDGs 3, 6, 7, 9 and 11), all the targets are influenced by infrastructure; whereas, for 15 of the SDGs more than half of the targets are influenced by infrastructure. Water infrastructure includes wastewater and sanitation services (Fig. 1.2), and infrastructure systems to protect against flooding, as well as water supply, and has therefore, overall, the largest direct influence across all SDG targets [26].

Fig. 1.2: Protecting society, wildlife and the environment from polluting wastewater: (A) a wastewater "primary settlement" treatment tank; and (B) a wastewater "secondary aeration" treatment lane part of a sewage treatment facility in the West Midlands, England.

Meeting the UN SDGs is not straightforward. For example, efforts to achieve the targets for clean water and sanitation can have unintended consequences on food and energy security and can contribute to environmental degradation [17]. The importance of the right sustainable water engineering to achieve each SGD goals becomes clearer. SDG–1, which focuses on ending poverty in all its forms everywhere, highlights several hundred million people still live in extreme poverty and struggle with basic needs of access to water and sanitation. For most, marginalised water access exacerbates levels of poverty. Poor or inequitable water infrastructure can make water unaffordable for the poor; leading to people with already limited resources having to obtain water from far distances or having to pay exorbitantly for someone to transport water to them in order to meet their basic water needs [2]. SDG–2 and SDG–3 target zero hunger and aims to promote good health and well–being. They seek to improve sanitation and hygiene across all ages. The challenge, especially in developing countries, of inadequate access to safe water, improved hygiene, and sanitation facilities on one hand, and increased frequency and intensity of resource stress on the other, impact on the livelihood, productivity, health and well–being of those affected [22]. These SDGs also acknowledge the pressures that climate change is placing on food security through increasing natural disasters linked to drought and flooding. Today, land–use and food systems also account for a quarter of greenhouse–gas emissions, over 90 percent of scarcity–weighted water use, most losses of biodiversity, overexploitation of fisheries, eutrophication through nutrient overload and pollution of water and air [23]. New and innovative water engineering solutions can help mitigate the negative impacts of current practices and support new agricultural practices to help deliver SDG–2. –SDG–4 and SDG–5 pursue quality education and promotes gender equality. This appeals for improvements in water and electricity access to schools. It also recognizes that without safe drinking water, adequate sanitation, and hygiene facilities, it is disproportionately harder for women and girls to lead safe, productive, and healthy lives. SDG–7, which ensures access to affordable and clean energy, intends to progress the use of renewable energy from water, solar and wind power. Developing clean energy can benefit populations in developing countries in the long run, but it can also compromise other SDG goals. However, this SDG may problematise other SDGs, such as those promoting the protection of water related ecosystems (SDG 6.6) [20]. Therefore, engineers have a role in maintaining the delicate balance between natural processes, environmental, economic, and social needs. This includes the engineer's role in achieving SDG–9, which encourages investment in infrastructure crucial to achieving sustainable development and empowering communities across many countries. SDGs related to industrialization, like SDG–9 for industry, need to consider and reconcile with other SDGs (e.g. SDG 6.3 regarding water quality and protecting ecosystems). Engineered solutions that impact on natural water quality and systems; such as SDG–11, which aims for sustainable cities and communities, but acknowledges that rapid urbanization is exerting pressure on fresh water supplies, sewage, the living environment, and public health. SDG–12, which addresses responsible consumption and production, concedes that excessive use of water is contributing to global water stress. Engineering and technological solutions, for example water efficient irrigation at the large sale, or for domestic activities such as showering, have been shown to go a long way to reducing water waste. SDG–13, urges action to combat climate change, accepts sea levels are rising, weather events are becoming more extreme and, consequently, droughts and floods are becoming commonplace. Therefore, the use of engineering solutions, such as Sustainable Drainage Systems (SuDS), combined with recycling and reuse schemes are highly beneficial and need to be more prevalent particularly in urban areas. SDG–14, looks to conserve and sustainably use the oceans, seas and marine resources, accepting that careful management of these resources are a fundamental feature of a sustainable future, because coastal waters are deteriorating due to pollution and eutrophication. Soft engineered flood attenuation (e.g. swales and integrated wastewater treatment and reuse systems) can help to reduce the contamination of water bodies and the impact on associated ecologies. This is particularly useful as SDG–15, which is concerned with continuing life on earth, appreciates water plays a key role in protecting biodiversity and improving land productivity. SDG–16, which promotes peace and justice, knows hydro–politics is important to promote peaceful and inclusive societies. Thus, SDG–17, which seeks to revitalise the role of partnerships becomes important as it identifies that all stakeholders (from individuals, organisations and governments) need to operate as a partnership built upon shared principles, values and vision. This relates to SDG 6.B, which emphasises participatory governance and community consultation related to water management [20].

Humans have significantly influenced and have been influenced by the hydrological regime [31]. Access to water has always been fundamental to the evolution of villages, towns, and cities. Some of the earliest known settlements across all countries tended to locate themselves close to available supplies of freshwater (e.g. nearby a river or a spring) until the pioneering civil engineers of their time realised they could design systems to transfer water over long distances using aqueducts or similar infrastructure. These systems enabled the delivery of potable water to large populations so they could prosper and, moreover, allowed farmers to irrigate their fields so they could grow/supply food for themselves and surrounding settlements (e.g. people of the Byzantine Empire or Inca people of Machu Picchu city) [14].

Engineered civil infrastructures are also needed to remove unwanted wastewater away from settlements. Faecal contaminated water puts people at risk of contracting dysentery, cholera, typhoid, schistosomiasis, trachoma, and intestinal worms [30], so it is a necessity to collect, remove, treat, and dispose of wastewater appropriately. Ever since Dr. John Snow, a public–health worker, investigated a severe cholera outbreak in Soho, London, in 1854, and identified that faecal–contaminated water was linked to a neighbouring cess pit, all modern cities now have a network of underground sewers beneath them. These require a sloping gradient and careful design because the water flows away by gravity and, with it, carries along human waste solids [11].

Hydrological change has shaped how human society responds to water crises, droughts, and floods in multiple ways, formally or not [4,16] . This has necessitated structures constructed to hold back water, including flood defences, such as levees built to limit hurricane–induced flooding of New Orleans (USA) or the Thames Barrier erected to stop London (UK) from inundations of tidal flooding, or less costly designs, such as sea walls fronted by rip-rap (Fig. 1.3), wooden revetments or gabion cages of shingle, which serve to dissipate wave energy and minimise the impacts. Water engineering structures have traditionally provided solutions to control the movement of water for a variety of purposes across a range of landscapes. Most fundamentally, engineered infrastructure provides essential services to people, such as water and energy, and protects them from hazards, such as floods or pathogens in sewage [26]. For instance, during the 18th century, dams were built primarily to store water for canals; moving into the 19th century, water supply was the priority (Fig. 1.4) and by the early 20th century,

Fig. 1.3: Coastal protection: (A) an undermined collapsed seawall and damaged footpath (Gran Canaria, Spain); and (B) a seawall fronted by an amour of modular concrete rip-rap to dissipate wave energy and provide flood protection during times of high water (Scarborough, England).

Fig. 1.4: A stone-built dam (constructed in 1888), standing 45 m high, 37 m wide at its base and 355 m long, creating Lake Vyrnwy reservoir, Powys, Wales, was built to provide reliable supply of freshwater to the City of Liverpool, England (photos show both sides of the same structure).

they were used to supply power to industry [8]. The latter half of the 20th century saw a peak in dam construction due to an increase in population and demand for hydro–electric power [25]. In more recent times advances in desalination have meant water engineers have been able to transform the inexhaustible supply of sea water into potable fresh water [21].

Property–level flood defences, such as door aperture guards and air brick covers, or adaptations made to the outside of properties, such as raised step thresholds or external tanking, have been purposed to restrict floodwaters from entering buildings (Fig. 1.5). Similarly, with the impacts of climate change unlikely to disappear, internal adaptations are becoming commonplace, such as raised electrical sockets and services/appliances, so as to

Fig. 1.5: Property-level flood protection (Gloucestershire, England): (A) a row of terraced houses sheltered behind a floodwall to stop fluvial flooding; and (B) a garden sump pumping rising groundwater over the same floodwall.

minimise the costs of repairs and time spent away from properties during times of post-flood reinstatement [7].

Droughts and water shortage are also commonplace for many populations. Household–level water saving, and water conservation is becoming a routine requirement across both the western world and the low– and middle–income countries. There has been a dramatic increase in water demand since the 1940s. For example, Europeans are using an average of 3550 L per capita per day and this amount is increasing steadily as incomes rise [24]. Per capita water use in England and Wales is 143 L per person per day.[2] However, water use varied depending on household size, with single–person household consuming up to 149 L/d.[3] Thus, the amount of water used can be attributed to multiple factors, including presence or absence of water meters, number of people in households, culture and lifestyles, as well as access to water efficient devices and technologies. Water efficiency can be achieved through a range of technical and operational interventions [1,15]. For instance, changing a toilet from an old cistern (9 L) to new type cistern (4.5 L) can halve the water used per flush and fitting an aerated showerhead can reduce the flow–rate by 28 percent (~3 L/min) [18]. However, technology improvements alone are not the answer.

Without doubt, water is fundamental to everyone's quality of life. Therefore, water could, in some circumstances, be considered more valuable than gold or diamonds. In fact, many wars and conflicts have been fought over rights of access to water e.g. disputes between India and Bangladesh over the River Ganges; disputes between Iraq, Syria and Turkey over the Rivers Euphrates and Tigris; disputes between Israel, Jordan, Lebanon and Palestine over the River Jordan; disputes between Kazakhstan, Kyrgyzstan, Turkmenistan, Tajikistan and Uzbekistan over the Aral Sea, amongst many others [5,6,12]. Thus, effective water solutions need to consider how water resources link different parts of society and how decisions in one sector may affect water users in other areas and sectors, as well as to adopt a participatory and inclusive approach by involving all actors and stakeholders, from all levels, who use and potentially pollute water, so that it is managed equitably and sustainably [29]. Sustainable water engineering provides the necessary basis for a holistic, yet equitable, approach towards water resources management [9,10,13]. This supports the need for a paradigm shift in engineering thinking and practice. Therefore it is important for the modern–day water engineer to know far more than the Darcy–Weisbach equation or the Hagen–Poiseuille equation [19]; they must be trained and experienced with a multifaceted skillset of expertise and knowledge from pressure loss per unit length of pipe to political peace.

A shift in paradigm to sustainable water engineering, which incorporates the necessities of socio–technical aspects, by considering user preferences and requirements is key for the wiser sustainable protection of water sources and supplies, without compromises to health,

[2] https://discoverwater.co.uk/amount–we–use.

[3] https://www.statista.com/statistics/827278/liters–per–day–household–water–usage–united–kingdom–uk/.

wellbeing, and livelihood. At the domestic scale, understanding household attitudes and personal behaviours towards water usage are important challenges for the water industry. Policymakers, alongside the engineering and manufacturing sectors, need to understand public perceptions and the decision–making process behind consumer choices [3]. The delivery of sustainable water strategies goes beyond advancements in engineering and technologically innovative solutions. With projected increases in demands for good quality fresh water, educating society about sustainable personal water use and water quality threats becomes an absolute necessity [24]. Therefore, in proffering innovative solutions for sustainable development and to meet the challenges of climate change, engineers need to better engage with users., and better engage with policy makers. Specialisms and outlooks beyond the status quo are needed for a stepwise change to occur towards achieving many of the sustainable development goals and make a shift towards sustainable water engineering.

Thus, the value of this book is that, on one hand, it showcases innovation in sustainable water engineering solutions from nature to the user; whilst, on the other hand, it argues for a new type of sustainable water engineer – one who is well–rounded in the techniques, measures, and strategies for delivering complete, resilient, and integrated environmental, economic, and social water solutions.

1.2 Structure of this book

This book comprises three sections, which are collated into sixteen chapters.

The first chapters of the book expose the insights and issues of sustainable water engineering through an assemblage of chapters that focus on the collection, storage, transfer, and conservation of water supplies. These include: historical water supply (Chapter 2), potable water supplies (Chapter 3), greywater engineering (Chapter 4), water efficiency in buildings (Chapter 5), water resilient cities (Chapter 6), water, sanitation and hygiene (Chapter 7) and rainwater harvesting (Chapter 8).

The second sets of chapters present chapters that examine wastewater treatment, drainage, flooding, and protection. These include: phytotechnology for wastewater treatment (Chapter 9), highway drainage (Chapter 10), sustainable drainage systems (Chapter 11) and community/property flood protection (Chapter12).

The final chapters include a collection of chapters that reveal insights into examples of varying scales of engineering designs. These include: micropower generation (Chapter 13), soft–water engineering (Chapter 14), the past, the present and the future of canals (Chapter 15) and some closing thoughts on the directions that technology may steer sustainable water engineering (Chapter 16).

1.3 Conclusions

Satisfying the ever–growing potable water demands of an ever–growing global population with an amount of water that remains constant, clearly requires a considered and careful management of water resources. Ensuring water is available and accessible requires future water engineers to think outside the traditional box. Engineers have an important role in an integrated water resources management approach that can enable a sustainable future, with sufficient quantity and adequate quality of water for all. Unfortunately, water is not universally abundant and for many nations the uneven distribution of water and human settlement continues to create growing problems of freshwater availability and accessibility.

Balancing water resources management with engineering ability is entirely essential. Too little versus too much is a challenge. The frequency of drought and flood events are continuing to increase worldwide, which means there is growing need to conserve available water resources, whilst at other times rainfall and runoff are completely undesirable. Most of society tends to follow the belief that new technologies will provide the solutions we need to save us from tragedy and adversity. However, engineers also must be mindful of human attitudes and behaviours. For instance, novel products exist to conserve/reduce water use in buildings and they also exist to restrict floodwater entry into buildings, yet there are some people who blatantly misuse or refuse to install (through personal preference or cost) new products or, in some cases, are wholly unaware of them or the advantages they offer.

If this introductory chapter has wetted your taste buds to know more about sustainable water engineering, then the proceeding chapters in this book will hopefully give you a giant head start in your quest for greater knowledge and understanding, by providing insights, inspiration and innovation. For those interested in making strides towards sustainable water–focused lifestyle changes, some chapters also contain guidance on how to address potable water shortages and minimise the impacts of floodwater excesses... In the meantime, we hope you will carry on reading and enjoy rest of the book!

References

[1] K. Adeyeye, Water Efficiency in Buildings, Wiley–Blackwells, Oxford, 2014, ISBN: 978-1118456576.

[2] K. Adeyeye, J. Gibberd, J. Chakwizira, Water marginality in rural and peri-urban settlements, J. Cleaner Prod. 273 (2020) article No 122594 doi: 10.1016/j.jclepro.2020.122594.

[3] K. Adeyeye, K. She, A. Baïri, Design factors and functionality matching in sustainability products: a study of eco–showerheads, J. Cleaner Prod. 142 (2017) 4214–4229.

[4] W.N. Adger, T. Quinn, I. Lorenzoni, C. Murphy, J. Sweeney, Changing social contracts in climate-change adaptation, Nat. Clim. Chang. 3 (4) (2013) 330–333.

[5] T. Allan, J.A. Allan, The Middle East Water Question: Hydropolitics and the Global Economy, Published by I.B. Tauris, New York, 2000, ISBN: 978-1860645822.

[6] G. Baranyai, European Water Law and Hydropolitics: An Inquiry into the Resilience of Transboundary Water Governance in the European Union, Springer, Switzerland, 2019, ISBN: 978-3030225407.

[7] D.W. Beddoes, C.A. Booth, J.E. Lamond, Towards complete property–level flood protection of domestic buildings in the UK, in: S. Hernández, S. Mambretti, D.G. Proverbs, J. Puertas (Eds.), Urban Water Systems and Floods II, WIT Press, Southampton, 2018, pp. 27–38.
[8] L.S. Blake, Civil Engineer's Reference Book, 4th ed, Butterworth–Heinmann Ltd, Oxford, 1989.
[9] C.A. Booth, S.M. Charlesworth, Water Resources for the Built Environment: Management Issues and Solutions, Wiley–Blackwells, Oxford, 2014, ISBN: 978-0470670910.
[10] C.A. Booth, F.N. Hammond, J.E. Lamond, D.G. Proverbs, Solutions to Climate Change Challenges in the Built Environment, Wiley–Blackwells, Oxford, 2012, ISBN: 978-1405195072.
[11] C.A. Booth, D. Oloke, A. Gooding, S.M. Charlesworth, The necessity for urban wastewater collection, treatment, and disposal, in: S.M. Charlesworth, C.A. Booth (Eds.), Urban Pollution: Science and Management, Wiley–Blackwells, Oxford, 2018, pp. 119–130.
[12] S.M. Charlesworth, C.A. Booth, Water resources challenges – penury and peace, in: C.A. Booth, S.M. Charlesworth (Eds.), Water Resources in the Built Environment: Management Issues and Solutions, Wiley–Blackwells, Oxford, 2014, pp. 403–406.
[13] S.M. Charlesworth, C.A. Booth, Sustainable Surface Water Management: A Handbook for SuDS, Wiley–Blackwells, Oxford, 2017, ISBN: 978-1118897690.
[14] S.M. Charlesworth, L.A. Sañudo–Fontaneda, L.W. Mays, Back to the future? history and contemporary application of sustainable drainage techniques, in: S.M. Charlesworth, C.A. Booth (Eds.), Sustainable Surface Water Management: A Handbook for SuDS, 2017, Wiley–Blackwells, Oxford, 2017, pp. 13–30.
[15] S. Churchill, C.A. Booth, S.M. Charlesworth, Building regulations for water conservation, in: C.A. Booth, S.M. Charlesworth (Eds.), Water Resources in the Built Environment: Management Issues and Solutions, Wiley–Blackwells, Oxford, 2014, pp. 135–150.
[16] Di Baldassarre, G. F. Martinez, Z. Kalantari, A. Viglione, Drought and flood in the Anthropocene: feedback mechanisms in reservoir operation, Earth Syst. Dynam. 8 (2017) 225–233.
[17] G. Di Baldassarre, M. Sivapalan, M. Rusca, C. Cudennec, M. Garcia, H. Kreibich, et al., Sociohydrology: scientific challenges in addressing the sustainable development goals, Water Resour. Res. 55 (8) (2019) 6327–6355.
[18] Environment Agency, Conserving Water in Buildings, Environment Agency Publications, Bristol, 2007.
[19] W. Graebels, Advanced Fluid Mechanics, Academic Press, Elsevier, California, 2007, ISBN: 978-012370885.
[20] V. Herrera, Reconciling global aspirations and local realities: challenges facing the sustainable development goals for water and sanitation, World Dev. 118 (2019) 106–117.
[21] A. Ifelebuegu, S.M. Charlesworth, C.A. Booth, Water, water everywhere and not a drop to drink, in: C.A. Booth, S.M. Charlesworth (Eds.), Water Resources in the Built Environment: Management Issues and Solutions, Wiley–Blackwells, Oxford, 2014, pp. 92–103.
[22] B. Ray, R. Shaw, Defining urban water insecurity: concepts and r, in: B. Ray, R. Shaw (Eds.), Urban Drought. Disaster Risk Reduction (Methods, Approaches and Practices), Springer, Singapore, 2019.
[23] J.D. Sachs, G. Schmidt–Traub, M. Mazzucato, D. Messner, N. Nakicenovic, J. Rockström, Six transformations to achieve the sustainable development goals, Nat. Sustain. 2 (9) (2019) 805–814.
[24] L.M. Seelen, G. Flaim, E. Jennings, L.N.D.S. Domis, Saving water for the future: public awareness of water usage and water quality, J. Environ. Manage. 242 (2019) 246–257.
[25] K. Tannahill, P. Mills, C.A. Booth, Impacts and issues of dams and reservoirs, in: C.A. Booth, S.M. Charlesworth (Eds.), Water Resources in the Built Environment: Management Issues and Solutions, Wiley–Blackwells, Oxford, 2014, pp. 47–60.
[26] S. Thacker, D. Adshead, M. Fay, S. Hallegatte, M. Harvey, H. Meller, N. O'Regan, J. Rozenberg, G. Watkins, J.W. Hall, Infrastructure for sustainable development, Nat. Sustain. 2 (4) (2019) 324–331.
[27] United Nations (n/d) Sustainable Development Goal 6: Ensure Access to Water and Sanitation for All. Available online at: https://www.un.org/sustainabledevelopment/water–and–sanitation/ (accessed January 2020).
[28] United Nations, The Sustainable Development Goals Report 2019, Publication issued by the Department of Economic and Social Affairs, New York, United Nations, 2019 ISBN: 978–92110140374.

[29] UN Water, Sustainable Development Goal 6, Synthesis Report 2018 on Water and Sanitation, Executive Summary, 2018, http://www.unwater.org/publications/executive-summary-sdg-6-synthesis-report-2018-;on-water-and-sanitation (accessed July 2020).
[30] United Nations Educational, Scientific and Cultural Organization (UNESCO), Water: A Shared Responsibility, Berghahn Books, New York, USA, 2006, pp. 203–240.
[31] C.J. Vörösmarty, C. Pahl-Wostl, S.E. Bunn, R. Lawford, Global water, the anthropocene and the transformation of a science, Curr. Opin. Environ. Sustain. 5 (6) (2013) 539–550.

CHAPTER 2

Using the byzantine water supply of Constantinople to examine modern concepts of sustainability

Kate Ward[a], **Simon D. Smith**[b,*], **Martin Crapper**[c]

[a]*School of Science and Engineering, University of Dundee, UK* [b]*School of Engineering, University of Edinburgh, UK* [c]*Department of Mechanical and Construction Engineering, Northumbria University, UK*
**Corresponding author.*

2.1 Introduction

Constantinople was created in the early fourth century in order to replace Rome as the capital of the Roman Empire. After the fall of Rome, it continued, first as the capital of the Eastern Roman (or Byzantine) Empire, then as the capital of the Ottoman Empire from the 15th century to the 20th century. Today, as Istanbul, its population is estimated at over 15 million. The city's location, straddling the meeting point of Europe and Asia and overlooking the Bosphorus, has been hailed as key to its long survival as one of the most important cities in history. However, without its complex and innovative water infrastructure, the small promontory at the end of the Thracian peninsula would have remained in its natural state, largely dry and capable of sustaining no more than a few thousand people as it did in its earliest days as the Greek fishing village of Byzantium.

The water supply infrastructure in Constantinople – three aqueducts, two of unparalleled length, and a complex network of water storage cisterns – was the foundation that allowed the city to grow and prosper for over a millennium. Even after the original water infrastructure eventually fell into disuse, it formed the basis of the Ottoman era water system which lasted until the start of the 20th century.

There is no doubt that Constantinople was a resilient city, successfully surviving both threats and change over many centuries, but was it also a sustainable city? Constantinople's longevity offers a chance to explore sustainability and its numerous facets from a long-term point of view. Purposeful sustainability is relatively new but in this chapter we will explore the sustainability of Constantinople through various lenses of sustainable thought. Applying the sustainability frameworks of Bell [4] and the model of the interconnection of resilience and

Sustainable Water Engineering. DOI: 10.1016/C2017-0-04301-X

sustainability by Elmqvist et al. [15] to a city that sustained its citizens for many centuries can offer insight, particularly for those responsible for the major pieces of civil engineering infrastructure that underpin and enable much of modern life.

2.2 Sustainability

Although sustainability is now a common concept across a wide range of fields, it is a contested term, with a range of interpretations.

Sustainability is an emerging concept and as such lacks a generally agreed definition but is in a state of ferment [12]. However, the degree of conflict and range within the debate on sustainability is perhaps more apparent to those in academic fields than to those trying to implement sustainability in practice [12].

Davison [12], proceeding from the notion of 20th century dualism (that practice is derived from and subservient to established theory) suggests that this is in the process of breaking down and advocates not the development of a new theory to replace the old but rather an emphasis on practical reason from which a new theory in time might emerge. Considering the case of Constantinople, which has proved both resilient and, in some senses, sustainable, and is very much the product of practical reason rather than any grand theory, should shed light on Davison's ideas.

Bell [4], from a survey of the current literature on urban water sustainability, derives five distinct framings which is used as analytical lenses to study specific instances of London's water infrastructure that have given rise to controversy in terms of their sustainability. By turning the same analytical lenses on the historical instance of Constantinople Bell's contemporary example can be complemented. The longevity of Constantinople as a development allows us to appreciate these five framings from a different angle.

Elmqvist et al. [15] draw an important distinction between sustainability and resilience and suggest that in certain circumstances they may conflict. They offer a new framework which addresses the conflict and supports the practical applications necessary for achieving transformation and global sustainability. Constantinople can serve as a model for examining this framework.

The five framings identified by Bell's [4] analysis of literature on urban water sustainability are ecological modernism, socio-technical systems, political ecology, sustainable development and radical ecology. Bell [4] concludes that although ecological modernism is the most dominant approach, the other frameworks identify alternative ways to interpret and explain decisions and developments regarding the sustainability of water infrastructure.

In ecological modernism, also known as ecological pragmatism, the focus is on using human knowledge and technological innovation to meet the needs of humans while reducing environmental impact [3,30]. The socio-technical system framing incorporates a wide range of analysis methods that recognise and emphasise the interactions between society, institutions, technology and the environment to understand the interdependencies between social and political practices and the development of physical infrastructure. Political ecology considers the power relationships surrounding infrastructure investment and governance and examines how these relationships impact society and the environment. Sustainable development is the most familiar framing of sustainability and was established in the United Nation's Brundtland Report. It articulates the need to balance what is required for present needs with avoiding doing harm to the needs of future generations. Radical ecology appears hostile to engineering infrastructure approaches, characterising them as attempts to dominate and control, and places importance on focussing on the needs of non-human nature. However, this implies that engineering approaches that take a cooperative approach and see nature as a partner, rather than a subject, are acceptable.

Although the development of Constantinople's water supply was driven by practical reasoning, it is notable that in studying its development we can recognise principles of sustainability and apply the theory described in five framings identified by Bell [4] and consider the applicability of the framework offered by Elmqvist et al. [15].

2.3 Development of the water supply system

2.3.1 Pre-Constantinople

Only in the earliest stages of Byzantium, the provincial settlement that existed before Constantinople was established, were the local water resources enough to sustain the population: wells fed by groundwater, rainwater collected in small cisterns and, perhaps, water collected from the Lycus, a watercourse that was outside the settlement's defensive wall. However, by the second century, Byzantium had constructed an aqueduct under the auspices of the Emperor Hadrian to augment the meagre local water resources ([23], pp. 18–21). Although little is known of this aqueduct, which is first mentioned in a law code of the late fourth century where it is referred to as the Aqueduct of Hadrian, evidence of Roman structures in Ottoman bridges of the Kırkçeşme water supply system are thought to be part of this Aqueduct (Tursun Bey and Gilles, cited in [11], pp. 242–43). It is assumed that the two systems shared the same water source in the Belgrad Forest north of the city. No physical remains have been discovered of the Aqueduct of Hadrian but it continued to play a key role throughout the Byzantine era of Constantinople ([11], p. 20). The route of the aqueduct within the city has been reconstructed by considering the textual evidence of which structures the aqueduct is reported to have served ([39,40], pp. 190–205).

2.3.2 Constantinople

When Byzantium became Constantinople in the fourth century, the population grew and the occupied area increased beyond the bounds of the valley at the east end of the peninsula, moving into the higher ground to the west. The Aqueduct of Hadrian was initially the new city's only water supply, although it may have been augmented by cisterns during the earliest period of Constantinople. In a letter, dated to around 360–361 to Honoratus, Libanius writes that the city has "abundant reservoirs". However, this dates to before the existence of the Valens Aqueduct (Libanius, Letters 251 [6], pp. 103–105), which is discussed below.

The city was initially bounded by the Constantinian Wall before being extended by the Theodosian Wall constructed in the early fifth century. Built on seven hills, six of them forming a ridge on the north side of the peninsula and a seventh, isolated hill on the southern side, the development of Constantinople included reclamation of land along the coast and the development of harbours on both the north and south coasts [24]. Not only was there an increased demand for water because of population growth, but there was also a need for water to be supplied at a higher elevation in order to reach the new areas of the city. To achieve the increased elevation, a new aqueduct was constructed that exploited springs in the Thracian hinterland at a distance from the city so great that the aqueduct was the longest in the Roman world [9,11,33] and had gradients typically of less than 0.1 percent. The new aqueduct, now known as the Aqueduct of Valens, was a remarkable feat of engineering, collecting water from two spring sources in Danamandıra and Pınarca (over 50 km North West of the city), which is generally agreed to have existed in 373 ([11], p. 10). Bringing water to the higher reaches of the area bounded by the Constantinian Wall enabled expansion into these areas, an important requirement for the growing city. The elevation and route of the aqueduct in the city are indicated by the Bozdoğan Kemeri, which still stands today, the aqueduct bridge that allowed water to cross a valley in the city. The aqueduct followed a route high on the ridge of hills that make up the spine of the city, which would enable water to be delivered to a far greater area than before.

For a time, this aqueduct appears to have satisfied the population's water demand, but in the fifth century a second, parallel, aqueduct was added which exploited springs even more distant [11], creating an aqueduct system over 560 km long ([33], pp. 97–109). This development saw a rapid increase in the number of large public baths, the *Notitia Urbis* identifying seven that were active by about 425 ([27,31], pp. 138–140). These baths would have had a considerable water demand and alongside a growing population, may account for the second phase of aqueduct construction identified by archaeological fieldwork, tapping a source at Pazarlı some 100 km from the city ([11], pp. 29–31). In fact, many Roman aqueducts were constructed to supply public baths ([17], p. 6), for example the restoration of the Aqua Marcia and construction of the Aqua Antoniniana for the Baths of Caracalla ([13], p. 16). Although this second phase of water distribution infrastructure is not verified in ancient texts, it must have brought either increased volume or reliability of water to warrant the vast investment required. There is no clear start or end date for the second phase but given its length, complexity and likely construction duration, planning is likely to have commenced soon after the completion of the first phase in 373.

Although the population continued to grow until the devastating plagues of the sixth century, the municipal authorities did not appear to seek more water sources or construct more aqueducts. Perhaps the second aqueduct brought water to the city to satisfy future population growth; however, while it was being constructed, the authorities also started to construct cisterns, suggesting that water demands continued to grow and a different strategy was needed to store and manage available water. Given that the authorities had already constructed the two longest aqueducts in the Roman world to obtain its water, they may have been reluctant or unable to continue the typical Roman strategy of constructing more and more aqueducts to meet water needs. The cisterns, ranging from tiny room-sized structures to enormous open-air reservoirs, continued to be constructed all over the city and throughout Byzantine Constantinople ([1,11], pp. 125–155; [39], pp. 165–190 and Appendix B).

One of the final major expansions to the water supply came in the early sixth century when the largest of the open-air cisterns, Mokios, was constructed (*Patria* 3.84 cited in [11], p. 231). With the construction of this cistern all areas (except perhaps the highest points adjacent to the Theodosian Wall) could be supplied with water: the Aqueduct of Hadrian was able to provide water to the lower slopes of the peninsula; the Aqueduct of Valens to both slopes of the ridge made up of Hills One to Six; and the Mokios cistern to Hill Seven (see Fig. 2.1). The water source for Mokios is unknown, although Crow et al. [[11], p. 132] suggest that it may have been fed from a branch of the Aqueduct of Valens – recent analysis ([39], pp. 263–276) shows that this was entirely feasible. This branch in the early sixth century may have been the final piece of infrastructure constructed to bring water into the city, but inside the city's walls, investment in infrastructure to store and distribute water continued from the earliest period until the 15th century.

Although cisterns had been a common water supply technology since prehistoric times [28], in Constantinople they reach unprecedented scale and complexity, apparently being combined in a network to manage the city's limited water resources. The first mention of cisterns in Constantinople is in a letter written by Libanius before the construction of the Aqueduct of Valens (Letters 251, trans. [6], pp. 103–105). So at a time when a growing city was having to manage with the water supply of a (presumably much) smaller community, one of the technologies it employed was the cistern. Next, and still before the existence of the Aqueduct of Valens, the Modestus cistern was constructed in 363 (PLRE 1, [21], p. 606).

The construction of the second phase of the Aqueduct of Valens also marked the beginning of a series of major cistern-building periods. In the early to mid-fifth century five large cisterns are known to have been constructed, including, on the periphery of the city,

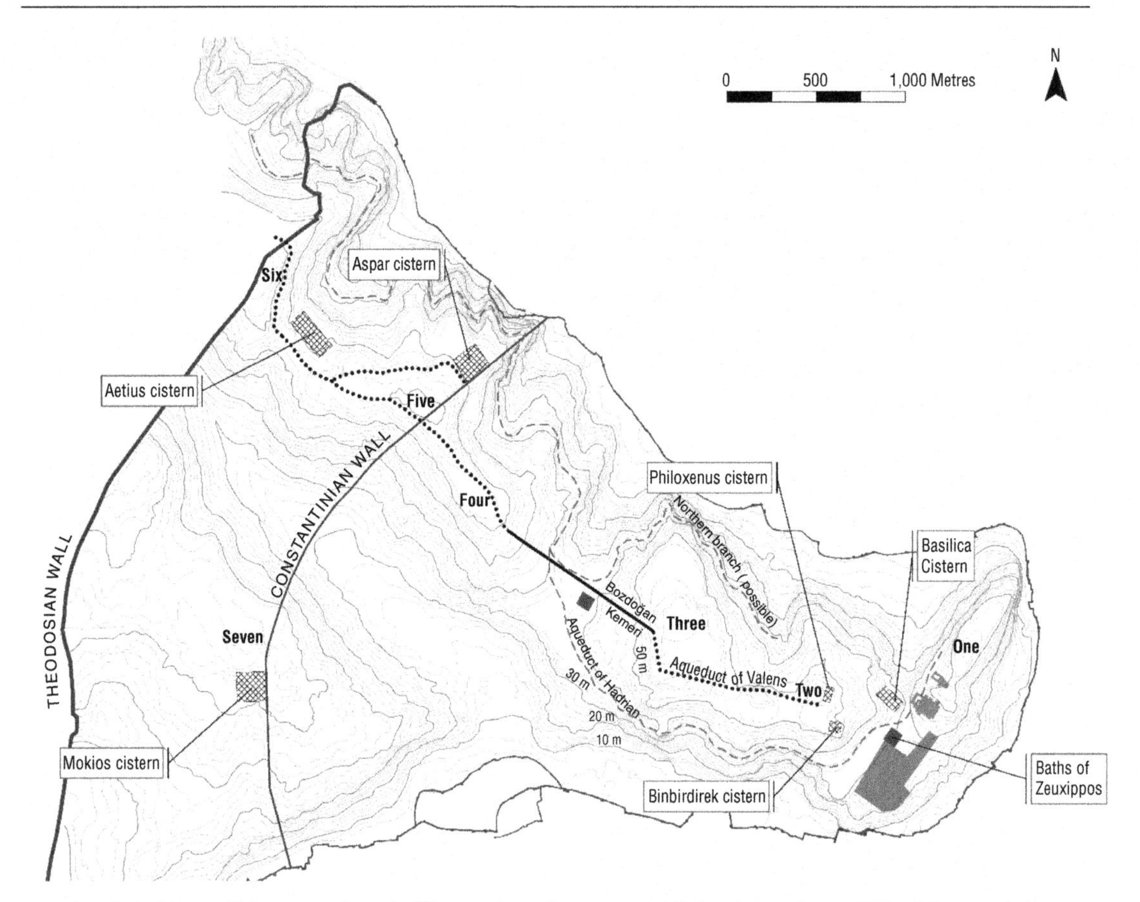

Fig. 2.1: Map of Constantinople illustrating the routes of the Aqueduct of Hadrian and the Aqueduct of Valens (after [39]).

two of the largest: Aspar and Aetius. Since the more central Theodosian and Arcadiaca cisterns are recorded in the *Notitia Urbis*, they were presumably large and perhaps also open-air ([27], p.89 & 94). A second wave of cistern-building occurred in the early sixth century, including the largest open-air cistern, Mokios ([11], p. 132), and the largest covered cistern, the Basilica cistern, in 527 and also possibly the Binbirdirek cistern in the same period ([11], p. 123). The Justinianic plague devastated the population in the mid to late sixth century but was followed by another wave of major cistern construction in the late sixth and early seventh century, although the precise location of many of these cisterns is no longer known ([11], p. 128).

The Basilica cistern is the largest covered cistern known in Constantinople. It has a capacity of over 80,000 m³ and, having been "rediscovered" by Pierre Gilles in the 16th century, is a significant tourist attraction in Istanbul today (see Fig. 2.2). There is no definitive answer as to why cisterns played such a key role in Constantinople's water

Fig. 2.2: The Basilica cistern, the most well-known of the 211 cisterns of Byzantine Constantinople is now a major tourist attraction in Istanbul (Photo: Ward, 2014).

supply. The rationale for constructing the Basilica cistern is given by Procopius as addressing summer shortages:

> In the summer season the imperial city used to suffer from scarcity of water as a general thing, though at the other seasons it enjoyed a sufficiency...the Emperor Justinian made a suitable storage reservoir for the summer season, to contain the water which had been wasted because of its very abundance during the other seasons...*Procopius Buildings 1.xi.10–15 ([14], 91)*

This may therefore be a reason for all the cisterns in Constantinople, although it is worth noting that at the time of the construction of the Basilica cistern, there was already a considerable amount of water storage in the city, largely in the Mokios, Aetius and Aspar reservoirs (along with the unknown but significant volumes of the Modestus, Arcadiaca and Theodosius cisterns). It is possible that the water in open-air cisterns was not used for drinking (although in times of scarcity, all available water sources are likely to be exploited) or merely that the location of these cisterns, on the periphery of the city, made the water less accessible.

Security is a possible driver for the reliance on cisterns, as this would allow the city to continue in the event of the external water supply being cut off, making it more resilient during a siege. There are two recorded instances of the water supply being tampered with – firstly for a brief period in the late fifth century by Theodoric Strabo, who was familiar with Constantinople's infrastructure (Malalas, Chronicle 15.9 trans. [20]), and secondly during the Avar siege of the city in the early seventh century. This time the interruption to the water supply was considerably longer than a few days: the Aqueduct of Valens is reported to have not been repaired until 765, 140 years after the Avar siege (Theophanes *Chronicle* AM 6258, trans. [25]). Since life continued in the city, the Avars clearly did not cut both the aqueducts (see discussion in [11], pp. 19–20). The population was able to survive well enough on the water available from the Aqueduct of Hadrian although only having water at this lower elevation would have curtailed activities in many areas of the city.

2.3.3 The end of the Byzantine water supply and the start of the Ottoman water supply

In the later stages of Constantinople, parts of the water supply, particularly the long aqueducts, were damaged beyond repair (*Deeds of John and Manuel Comnenus* 6.8 trans. [7]) and adaptations to use closer, less plentiful sources may have been made. Visitors to the city reported a functioning water system up to the mid-12th century ([11], pp. 238–239) but around this time some of the bridges outside the city appear to have deteriorated beyond repair – by the late 12th century the "old arcades which conveyed water to Byzantion were long since collapsed" (Kinnamos, trans. [7], pp. 205–206). Although this may have marked the end of the city being supplied by the Roman world's longest aqueduct, the system in the city appears to have still been in use – water is reported to cross the Bozdoğan Kemeri in the 15th century (Clavijo trans. [22], pp. 88) – so an alternative water source closer to the city may have been used and the Aqueduct of Hadrian may still have been in utilised.

Following the Ottoman conquest in 1453, the city, including the water supply, underwent significant changes. Initially the Ottomans probably used much of the same water supply infrastructure, but this was gradually replaced by aqueducts from two main sources – the Kırkçeşme system, which took water from the Belgrad Forest, and the Halkalı system that comprised multiple channels collecting from many small dispersed springs in the Halkalı area some 12 km from the city ([8], Plan 9). This marked the end of the water supply of Byzantine Constantinople although elements of it, including the Bozdoğan Kemeri and Basilica cistern, continued to serve the city as part of the Ottoman water supply.

2.4 Was Constantinople a sustainable city?

What can the story of Constantinople's water supply tell us about current concepts of sustainability and resilience? In this section, we will consider the five analytical lenses identified by Bell [4] and how they relate to Constantinople and its water supply.

2.4.1 Ecological modernism

The ecological modernist perspective is that technological innovation will provide solutions to current environmental problems and enable us to continue our current lifestyles in a sustainable way. The success of Constantinople is in accordance with the views of ecological modernism. In constructing the aqueducts in Thrace, existing technology was used but pushed to new limits. The development of the network of cisterns in the city also took a long-established technology and extended its function, not just the advances in building technology that allowed the enormous size of many of Constantinople's cisterns but also the sophisticated way that the cisterns were used together to create a reliable level of service in the city ([39], pp. 288–290). The potential for the city to support a large population was limited by local water resources but technological interventions created a city that could support a large population and meet the expectations of water use in a capital of the Roman Empire.

Although technology created a city where people not only survived but flourished, a long-term perspective of Constantinople suggests caution and possible lessons for ecological modernism. The technology built in Constantinople and its hinterland in the fourth to seventh centuries created a successful city in the short and medium term. The system which created that success was sustainable for the time and for the population that relied on it. However, that success turned Constantinople into such an important place to live that it created the conditions we see today: Istanbul the megacity, with a rapidly expanding population relying on continual technological innovation to solve the problem of reliable water for 15 million people. Analysis of Istanbul's water resources highlights that while the city is currently successful in providing sufficient water supply for its inhabitants, this is under threat from continued rapid urbanisation, increasing water consumption and changes in climate [19,38]. The water supply of the city now draws from large catchments on both the Asian and European sides of the city and uses a network of almost 19,000 km of water distribution pipes [19].

This idea that the use of technology to solve earlier problems has created the conditions that have created the need for more technology to solve more complex problems is perhaps not seen as something to be wary of by ecological modernism. However, from another perspective on sustainability this is as an example of "lock-in" – the success and stability of Constantinople created inertia, making it more likely that the location will continue as an important population centre despite the mounting challenges involved to make it liveable.

2.4.2 Socio-technical systems

The socio-technical systems approach emphasises the need to consider hard and soft systems together; that physical infrastructure and cultural behaviours and values shape each other. There is no doubt that development of Constantinople's water supply system was influenced by cultural expectations of what the capital of the Roman Empire should provide. Roman bathing culture placed high demands on water supply; 11 aqueducts supplied the abundance of water in Rome, allowing multiple Imperial baths across the city; some such as the Baths of Caracalla had a dedicated aqueduct branch and large cistern to balance the supply and demand ([13], p. 16). In Constantinople, there is evidence of similar importance being placed on water and bathing. Even before construction of the Aqueduct of Valens was complete, major works were under way on the Imperial baths of Constantianae, and at least eight major public baths were operational in the early fifth century when the *Notitia Urbis* was written [27].

There is some textual evidence of the institutional organisation involved in operating and maintaining the water supply. The law codes state that inspectors and guardians of the water – *hydrophylaces* – were to be branded so that others would know not to impede them in their duties (Codex Justinianus 11.42.10, trans. [16]). Ward [39] modelled the cisterns as an operational network, indicating what some of those duties might have been – to monitor water levels in the cisterns and to operate valves ensuring appropriate distribution of water across the network – there was not enough that all areas could get water all the time. These actions are familiar in modern contexts, closely mirroring the work of Mumbai's *Chaviwallas* (key turners) who operate the daily valve schedule to share water across the city ([2], pp. 101–107). In Constantinople, there were complex interaction between society and the water supply infrastructure key to socio-technical system framing. The water supply not only sustained the city by providing for its needs but also required and sustained a workforce that must have been sizeable, one that was given special status because their work was so vital.

The visibility of the water supply – both the physical infrastructure and the system's workforce – has implications for sustainability in the socio-technical system framing. Constantinople's grand aqueduct, ornate cisterns and its branded workers made sure that the water supply was always a visible part of society. We know that the water supply was admired by visitors and citizens – "that marvellous work the underground and overhead river" (Gregory of Nazianzus, Or. 33.6; trans. in Crow et al. [11], p. 226). Contrast that with the modern city, where water supply is automatic, reliable and invisible, only coming into the population's awareness when the taps run dry. Mattern [26] discusses the implications of infrastructure systems and their interfaces becoming increasingly distant from society as they grow in complexity and automation, concluding that there is a risk of losing the ability to shape cities to humanity's values.

2.4.3 Political ecology

Political ecology stresses the importance of understanding the influence of power relations when considering the sustainability of a system. It recognises that physical infrastructure and its sustainability is inextricably linked to the political system that created and governs it and seeks to identify the winners and losers in terms of profit and loss, whether financially or otherwise [36]. When considered at one scale, including only the physical infrastructure, a system might appear to be sustainable, but at a broader more comprehensive scale that incorporates financial and political aspects, the same system's sustainability credentials can appear undermined by the compromises and alliances required to achieve successful completion. Even something as functional as a water supply system can have multiple, nuanced drivers and narratives, and some of these can be counter to what is considered sustainable. For Constantinople and its water supply, the Imperial system was instrumental in the development of the water system; that it depended on the Empire and drew on its wealth and resources is clear and influences understanding of the system's sustainability. However, it is possible to argue that this initial Empire-wide investment provided considerable value, outlasting the wider Empire, sustaining it as it contracted until the eventual Ottoman conquest, a period of more than a millennium.

The scale of construction required for the water supply system was monumental. Its thin and linear nature belies its scale, yet the volume of material required was equivalent to that of the Great pyramid of Giza ([35], p. 211). The construction planning and resource management must have been tremendously difficult – many hundreds of kilometres through heavily forested and undulating terrain. The scale of investment required is indicative of what is necessary to create a city-scale infrastructure, the values that underlie the development of the water supply system are quite complex. On the one hand, there is the sentiment recorded in the Justianic law codes:

> "It would be abominable for the inhabitants of this Beautiful City to be compelled to purchase water" (11.42.7 Codex Justinianus trans. [34])

This reflects the importance placed on universal access that is set out in the sustainable development framing and embodied in the UN sustainable development goal 6. On the other hand, historical texts, including the law codes, attest to the Emperor granting and removing the right to access water – typically this meant a piped water supply, available to those of sufficient status – which indicates unequal access to water and even in one case, monasteries' right to water being restricted as a punishment for rebellious behaviour [10]. Although the importance of water for all was recognised at a basic level, those with political power and connection to the Emperor were able to obtain additional supplies.

The sustainability of Constantinople's water supply should also take into consideration the political message that was conveyed by constructing an "overhead river" ([11], p. 226). The water supply system is a clear statement of power and capability that reverberated around the world for centuries after its completion. Visitors from the West during the Crusades report that the aqueducts drew their water from the distant Danube (William of Malmesbury, [11], p. 238). Though inaccurate, this shows that the scale of the achievement had retained its hold on the imagination. To build up to 565 km of channel, over 100 bridges, and 10 km of tunnels to navigate through difficult terrain while maintaining gradients of less than 1 in 1000 ([33], pp. 97–109) would be an outstanding achievement even today.

In terms of sustainability we might question the need for such an enormous investment in aqueduct channels and bridges that brought water into the city. Indeed the Ottomans, during the second phase of Constantinople's water supply relied on (lower yield) water sources much closer to home ([39], p. 132). However, had the system been governed by practical considerations alone it is doubtful that it would have been built at all and this raises the troubling question that our most durable human achievements are those that go beyond necessity and are driven by ambition rather than rational considerations.

2.4.4 Sustainable development

This framing of sustainability includes concepts of fairness, both geographically and temporally, and was set out in the UNCED Our Common Future report [37]. In brief, it states that we should ensure our needs are met across the world, but also that we do not prosper in the present at the expense of future generations. This aim is embodied in the UNs 2030 Agenda for Sustainable Development with 17 Sustainable Development Goals (SDGs), including availability of clean water for all (SDG6). The longevity of Constantinople makes it suited to consider these ideals and examine how water supply infrastructure can have an impact across generations.

Starting at the upstream end of the water supply system, Constantinople extracted water from a number of sites throughout Thrace. Although there is no absolute evidence of how and where water was diverted into the aqueduct channels, remains of structures near Pınarca suggest that inlet structures directed outflow from springs into one of the aqueduct channels ([33], pp. 56–59). At Pazarlı, the furthest extent of the aqueduct, Ruggeri ([33], pp. 56–59) and Crow et al. [11] propose that water was diverted from a spring-fed stream. In both cases, water was taken only from what was available at the surface, so could be considered renewable and sustainable. Although the impact of this water abstraction has not been studied extensively, it is not, perhaps, as great as more recent examples. Consider the impact on the populations of valleys in north Wales, such as the village of Llanwddyn, which was flooded by Lake Vyrnwy reservoir in order to supply drinking water for Liverpool ([5], pp. 143–148); and the significant changes made to the landscape of the Highlands of Scotland to generate electricity. The water supply system of Constantinople met the needs of the time without adverse impacts for the future inhabitants of the area, so could be considered sustainable in accordance with this framing.

The nature of this method of water abstraction – a "renewable removal" – is such that the population was vulnerable to the natural and seasonal variability of water availability. Incidents of drought are recorded ([11], p. 17) but it is notable that the response to this was not to extract more water from the environment but instead to construct storage cisterns, an approach that makes the most of what is available and minimises waste – again, this reflects the approach of the sustainable development framing.

The water supply infrastructure of Constantinople was robust and long lasting, indeed some of it remains in and under Istanbul today (see Fig. 2.3). The early city provided for later inhabitants by constructing infrastructure that remained functional for over a millennium. This approach could, in some senses, be taken as sustainable, as future generations do not

Fig. 2.3: Byzantine era stone pipes excavated from the main street of Constantinople (Photo: Ward, 2015).

have to invest more resources to meet their own needs. However, it could also be considered an example of unsustainable "lock-in", with future generations tied to using a system that only partially meets their needs or that could be replaced by a more sustainable option but will not be because of the cost of creating the new system. The lock-in argument is not absolute – because of the relatively simple level of technology and the significant investment cost, adaptation and reuse of infrastructure was common across Constantinople. As suited the needs of the times, houses and storerooms became cisterns, cisterns became factories, gardens and column drums became pipes ([11], pp. 125–126).

The world population approximately doubled from the inauguration of Constantinople in the fourth century to the conquest of the city by the Ottomans in the mid-15th century, increasing from just over 200 million to just over 400 million [18]. In contrast, since Constantinople officially became Istanbul in 1930, the global population has almost quadrupled, from 2 billion to 7.7 billion in 2019 [32]. Seen against this background, the scale of Constantinople compared to the resources it had to draw on is vastly different to the scale of Istanbul and the resources it competes for with other megacities. The capacity for Constantinople to make a damaging impact on its environment was minimal in a world that was large and abundant, relative to its human population.

2.4.5 Radical ecology

Viewed solely in terms of its water sources, Constantinople, in a radical ecology framing, would be seen as fundamentally unsustainable. The sources of Constantinople's water were both quite distant from the city, comprising spring flows that were not part of the local hydrological system. Moving water outwith its local hydrological system is counter to radical ecology principles (here it is useful to remember that Rome did draw water from within its own hydrological system, and as such could not be criticised by radical ecologists on these grounds).

The source of the Aqueduct of Hadrian was relatively local, if it had not run into the aqueduct, it would instead have flowed into the Golden Horn, the estuary that creates the northern boundary to the city. If this water was at a level too low to be useful for the inhabitants of the city, it was diverted at the top of the catchment and channelled to the city. The fourth and fifth century aqueducts brought water from more distant sources and had to use multiple tunnels and bridges to cross the catchments and geographical barriers between the source and the city. As discussed in the sustainable development section, it is not known how much of the total flow was diverted into the aqueducts and although this abstraction of surface flow could be considered a renewable source of water under other sustainability framings, from the perspective of radical ecology, it is not.

With insufficient water available locally, Constantinople brought water in from elsewhere; radical ecology challenges this placing of human needs above the wider environment. The radical ecologist would argue against using technology to allow a site to support more people than the local resources can bear. However, it is incorrect to characterise radical ecology as anti-city: the primary objection to Constantinople is its location – not that it should never have been built, but rather that it should not have been built where it was.

2.5 Resilience vs sustainability

Another aspect of the sustainability debate that can be explored using Constantinople as an example is the relation between sustainability and resilience. Although the terms are often used interchangeably, they are recognized in the literature as distinct. Zhang and Li [41] use bibliometric analysis of studies of urban sustainability and urban resilience literature to identify how academic fields typically conceptualize the key features and differences between the two terms. Meerow, Newell and Stults [29] use a similar approach but focus solely on urban resilience. Both studies note the fragmented and inconsistent conceptualization of these terms within and across different academic disciplines, reflecting the complexity and continual refinement of the ideas, as discussed by Davison [12]. The proposed definitions that emerge from these studies are considered by Elmqvist et al. [15] and resolved with the prevailing understanding of terms used in policy and practice to create definitions and a conceptual model of how sustainability and resilience relate to one another.

Elmqvist et al. [15] propose a definition of urban sustainability close to that of the sustainable development framing discussed above, stating that it encapsulates the management and enhancement of resources and systems in consideration of equity, justice and current and future generations' needs. They suggest that urban resilience is an attribute of a system representing its capacity to continue functioning in the same way despite disturbance and disruption. As such resilience can act in favour of increasing sustainability by strengthening the capacity of a sustainable option, or counter to sustainability by enabling less sustainable systems to continue as the rational method – an example of lock-in as identified in the socio-technical framing of sustainability.

When Elmqvist et al. [15] framework is applied to Constantinople, the nature of the water supply system, the robust construction materials and straightforward gravity-fed supply, give it a high degree of resilience. This resilience in the water system kept the city on the same pathway for over 1000 years, supplying the city despite numerous changes to population, cultural practice and institutional power alongside sharper disturbances such as droughts and sabotage. It can even be argued that the resilience of the water supply system outlives the Byzantine era city and carries over to the Ottomans, who use the same approach, refurbishing and using much of the same infrastructure to supply water to the city. This shows the potential power of a resilient system to maintain a specific path whether or not it is a sustainable one.

2.6 Conclusions

The water supply infrastructure of Constantinople was a key foundation of the city; without it, the city would not have been able to prosper and support its large population, as it did for at least 1200 years. By viewing it through the lenses of Bell's [4] five sustainability framings and Elmqvist et al. [15] consideration of sustainability and resilience, it has been possible to compare the full range of current perspectives on sustainability and draw our own conclusions. As Davison [12] states "the value of debate about sustainable development [is] not to be found in fixed definitions and assertions … [but]…in the movement and interaction of ideas and interests".

Technology and innovation enabled the city to overcome the local deficiencies in water resources, allowing water to be transported through a challenging environment and then distributed and stored in cisterns to minimize waste and manage fluctuations in supply. This is a clear example of the principles of ecological modernism, demonstrating the impact engineering can have in creating a flourishing city.

The water supply system occupied an important position in Constantinople's society. Its physical infrastructure was visible and admired; the human organisation responsible for the daily operation of the water system was likewise prominent. This resulted in an infrastructure system that was visible, comprehensible and a source of pride, reflecting the positive interactions between society and technical systems that are highlighted as important aspects in socio-technical framings of sustainability.

One of the major drivers of the development of the water supply system was a stated belief in the importance of universal, free access to water for the city's inhabitants, which in some senses reflects the principle of equity that is key to the sustainable development framing of sustainability. The method of obtaining water was essentially renewable, as it relied on surface runoff from springs and streams. This, and the use of cisterns to maximise the use of the available water rather than pursue new sources also embody the sustainable development principle to protect future generations' ability to meet their needs.

On the other hand, some aspects of Constantinople can be considered unsustainable. This is most apparent through the lens of radical ecology, which is fundamentally opposed to the need to draw resources from outside the immediate locale, something on which Constantinople depended to survive.

Although universal access was a driver for the creation of the water supply system, there was considerable inequity in access to water, with those of high social status able to access a private piped supply. This access could be granted or taken away by the Emperor, which runs counter to some of the principles of sustainable development. Like many megaprojects today, the motivation for Constantinople's extraordinary water supply system was not straightforward, being influenced as much by the desire to make a statement of power as the need for reliable water. The political ecology framing of sustainability draws attention to these conflicts and contradictions demonstrating that what can appear at first to be obviously sustainable may be more complex and require a more nuanced analysis to determine its sustainability.

Paradoxically, the success and endurance of the water supply infrastructure demonstrate the unsustainable aspects of its supply system in the socio-technical framing of sustainability, which connects with the model of resilience and sustainability proposed by Elmqvist et al. [15]. The durability of the physical infrastructure is an example of "lock-in" where future directions are contingent on the existing infrastructure because of the relatively high monetary and social cost of change. This also reflects how resilience can negatively impact sustainability goals, with a high-resilience system effectively trapping a city in a particular developmental pathway. A more desirable, more sustainable pathway may not be pursued because of the difficulties of breaking reliance on existing systems.

Constantinople serves as an exemplar for ecological modernism, but also as a warning. Technology solved the problem of living the life of a Roman citizen in a place with insufficient water but that success created new problems and challenges. The success of the city cemented its importance, turning Constantinople, the archetypal capital of Empire, into Istanbul the megacity, faced with the challenge of managing continued population growth, rising water consumption and a changing climate.

Thus, Constantinople's water supply system provides an effective way of exploring the diverse and sometimes contradictory framings of sustainability. Water supply infrastructure serves a vital function making towns and cities inhabitable. There are limitations on using this historical city to investigate sustainability – Constantinople's water supply was not designed to be sustainable and the extent to which it was sustainable is more a by-product of available technology and resources.

What has emerged from this Chapter is that the sustainability of a system or city is dependent on its context – something that is initially sustainable can become unsustainable or hamper improvements towards more sustainable paths. Constant review and the capacity to transform are the basis for moving towards sustainability goals.

The purpose of this Chapter is not to suggest which framing of sustainability is to be preferred, but rather to show the value of a range of perspectives by applying them to the single example of Constantinople. In this way, we can appreciate both how they interact and conflict so that the overall debate is better informed.

References

[1] K. Altuğ. İstanbul'da Bizans Dönemi Sarnıçlarının Mimari Özellikleri ve Kentin Tarihsel Topografyasındaki Dağılımı; (PhD), İstanbul Teknik Üniversitesi, Istanbul, 2013.

[2] N. Anand, Hydraulic City: Water and the Infrastructures of Citizenship in Mumbai, Duke University Press, Durham, London, 2017.

[3] J. Asafu-Adjaye, L. Blomqvist, S. Brand, B. Brook, R. DeFries, E. Ellis, et al., An Ecomodernist Manifesto, 2015, www.ecomodernism.org.

[4] S. Bell. Framing urban water sustainability, in: Y. Zhuang, M. Altaweel (Eds.), Water Societies and Technologies from the Past and Present, 2018, pp. 200–220, doi:10.2307/j.ctv550c6p.16.

[5] G.M. Binnie, Early Victorian Water Engineers, T. Telford, London, 1981.

[6] S. Bradbury, Selected letters of Libanius: from the age of Constantius and Julian, Liverpool University Press, Liverpool, 2004.

[7] C.M. Brand, Deeds of John and Manuel Comnenus (Brand, trans.), Columbia University Press, New York, 1976.

[8] K. Çeçen, Halkalı Suları, İstanbul Büyük Şehir Belediyesi, İstanbul Su ve Kanalizasyon İdaresi Genel Müdürlüğü, Istanbul: T.C., 1991

[9] K. Çeçen, The Longest Roman Water Supply Line, Istanbul: Türkiye Sınai Kalkınma Bankası, 1996.

[10] J. Crow, Water and late antique constantintople, in: L. Grig, G. Kelly (Eds.), Two Romes: Rome and Constantinople in Late Antiquity, 2012, doi:10.1093/acprof:oso/9780199739400.001.0001.

[11] J. Crow, J. Bardill, R. Bayliss, The Water Supply of Byzantine Constantinople, Society for the Promotion of Roman Studies, London, 2008.

[12] A. Davison, Contesting sustainability in theory–practice: in praise of ambivalence, Continuum: J. Media Cultural. Stud. 22 (2) (2008) 191–199, doi:10.1080/10304310701861598.

[13] J. DeLaine, The baths of Caracalla: a study in the design, construction, and economics of large-scale building projects in imperial Rome, Portsmouth, R.I, J. Roman. Archaeol. (1997).

[14] H.B. Dewing, G. Downey, Procopius: In Seven Volumes. 7: Buildings. General Index, Harvard University Press, Cambridge, Mass, 1940.

[15] T. Elmqvist, E. Andersson, N. Frantzeskaki, T. McPhearson, P. Olsson, O. Gaffney, K. Takeuchi, C. Folke., Sustainability and resilience for transformation in the urban century, Nat. Sustain. 2 (4) (2019) 267–273, doi:10.1038/s41893-019-0250-1.

[16] B.W. Frier, S. Connolly, S. Corcoran, M. Crawford, J.N. Dillon, D.P. Kehoe, N.E. Lenski, T.A.J. McGinn, C.F. Pazdernik, B. Salway, eds. The Codex of Justinian: A New Annotated Translation, With Parallel Latin

and Greek Text Based on a Translation By Justice Fred H. Blume, Cambridge University Press, Cambridge, United Kingdom; New York, 2016.
[17] A.T. Hodge, Roman Aqueducts & Water Wupply, second edition, London: Duckworth, 2002.
[18] Hyde, Population. History Database of the Global Environment, 2010. Accessed 26 May 2019, https://themasites.pbl.nl/tridion/en/themasites/hyde/basicdrivingfactors/population/index-2.html.
[19] ISKI, Annual Report 2017, Istanbul, Turkey: Istanbul Water Works and Sewerage Administration, 2018.
[20] E. Jeffreys, M. Jeffreys, R. Scott, The Chronicle of John Malalas, Australian Association for Byzantine Studies: Dept of Modern Greek, University of Sydney, Melbourne, 1986.
[21] A. Jones, J. Martindale, J. Morris, The Prosopography of the Later Roman Empire, Cambridge University Press, London, 1971.
[22] G. Le Strange (Trans.), Clavijo: Embassy to Tamerlane, Routledge Curzon, London; New York, 2005, pp. 1403–1406.
[23] C.A. Mango, Le dévelopement urbain de Constantinople, IVe-VIIe siècles, Diffusion de Boccard, Paris, 1985.
[24] C. Mango, The shoreline of Constantinople in the fourth century, in: G. Necipoğlu (Ed.), Byzantine Constantinople: Monuments, Topography and Everyday Life, Brill, Leiden, 2001, pp. 1–28.
[25] C.A. Mango, R. Scott, G. Greatrex, The chronicle of Theophanes Confessor: Byzantine and Near Eastern history, AD 284-813, New York: Clarendon Press ; Oxford University Press, Oxford, 1997.
[26] S. Mattern, Interfacing urban intelligence, Places J. (April 2014). Accessed 01 Jun 2019, doi:10.22269/140428.
[27] J. Matthews, The Notitia Urbis Constantinopolitanae, in: L. Grig, G. Kelly (Eds.), Two Romes: Rome and Constantinople in Late Antiquity, 2012, pp. 81–115, doi:10.1093/acprof:oso/9780199739400.003.0004.
[28] L. Mays, G. Antoniou, A. Angelakis, History of water cisterns: legacies and lessons, Water 5 (4) (2013) 1916–1940, doi:10.3390/w5041916.
[29] S. Meerow, J.P. Newell, M. Stults, Defining urban resilience: a review, Landsc. Urban. Plan. 147 (2016) 38–49, doi:10.1016/j.landurbplan.2015.11.011.
[30] A.P.J. Mol, D.A. Sonnenfeld, Ecological modernisation around the world: an introduction, Environ. Polit. 9 (1) (2000) 14, doi:10.1080/09644010008414510.
[31] M. Mundell Mango, Thermae, balnea/loutra, hamams: the baths of constantinople, in: P. Magdalino, N. Ergin (Eds.), Istanbul and Water, Leuven: Peeters, 2015, pp. 129–160.
[32] M. Roser, H. Ritchie, E. Ortiz-Ospina, World Population Growth, Published online at OurWorldInData.org, 2019, https://ourworldindata.org/world-population-growth.
[33] F. Ruggeri, Engineering the Byzantine water supply of Constantinople: mapping, hydrology and hydraulics of the long aqueducts outside the city (PhD), University of Edinburgh, Edinburgh, 2018.
[34] S.P. Scott, Trans., The Civil Law, vol. 17, Cincinnati, 1932.
[35] J.R. Snyder, Exploiting the Landscape: quantifying the material eesources used in the construction of the long-distance water supply of constantinople, in: H. Baron, F. Daim (Eds.), A most pleasant scene and an inexhaustible resource: steps towards a Byzantine environmental history: interdisciplinary conference November 17th and 18th 2011 in Mainz, Verlag des Römisch-Germanischen Zentralmuseums, Mainz, 2017.
[36] E. Swyngedouw, M. Kaika, E. Castro, Urban Water: a political-ecology perspective, Built Environ. 28 (2) (2002) 124–137.
[37] United Nations World Commission on Environment and Development (WCED), Towards sustainable development, in: G.H. Brundtland (Ed.), Our Common Future, Oxford University Press, Oxford; New York, 1987, pp. 43–65.
[38] K. van Leeuwen, R. Sjerps, Istanbul: the challenges of integrated water resources management in Europa's megacity, Environ. Dev. Sustain. 18 (1) (2016) 1–17, doi:10.1007/s10668-015-9636-z.
[39] K. Ward, An engineering exploration of the water supply system of Constantinople (PhD thesis), University of Edinburgh, Edinburgh, 2018.
[40] K. Ward, J. Crow, M. Crapper, Water-supply infrastructure of Byzantine Constantinople, J. Roman Archaeol. 30 (2017) 175–195, doi:10.1017/S1047759400074079.
[41] X. Zhang, H. Li, Urban resilience and urban sustainability: what we know and what do not know? Cities 72 (2018) 141–148, doi:10.1016/j.cities.2017.08.009.

CHAPTER 3

Wholesome water, and natural water sources

John Griggs
FCIPHE, Chartered Institute of Plumbing and Heating Engineering, Hornchurch, UK

3.1 Introduction

The term "drinking water" or "potable water", meaning water that you might carry in a "pot" to drink, were commonly used for water suitable for drinking. As legislation and terminology evolved, "Wholesome water" is now the preferred term. Wholesome water has specific parameters and can be used for water that is not only for drinking, but also for other purposes where water of a drinkable quality is required. Water is almost anywhere, such as in the clouds in the sky, in rivers and the sea, as well as in ponds, puddles and ditches. However, very little is suitable for drinking. For water to be suitable for consumption, it needs to be of a quality that will quench thirst without causing unwanted side effects, such as: diarrhoea, vomiting, and poisoning that can lead to illness or death. One aspect of water is its appearance. Clear water may look clean, but contain soluble substances that could be dangerous to health. Therefore, a number of factors are considering for water quality. The visual quality i.e. turbidity of the water is important, but other many man-made substances could affect water quality such as the runoff from fields that may contain pesticides, fertilizers or genetically modified crops. In addition, pharmaceutical products that may have passed through humans and animals may accumulate to levels that could have impact on groundwater or water bodies used for water supplies. Such pollution impacts on drinking water, as well as eco-systems forming parts of the food chain, flora and fauna.

Whilst good quality drinking water is essential for human health, good quality water is not needed for water that just provides a motive force in sewerage systems. Thus, flushing WCs or urinals with drinkable water is wasteful of the resources that were used to bring that water to such a high quality. Hence, water provision need only be of an acceptable quality for specific tasks, such as vehicle washing, irrigation, WC flushing or laundry by either not treating the raw water to such a high standard, or by re-using water that was previously used for bathing, washing-up or rainwater that has been collected locally. There is no one water quality standard for all such applications. For example, in irrigation, some plants will accept washing-up water, that will contain detergents, foodstuff and grease, but other plants will not

Sustainable Water Engineering. DOI: 10.1016/C2017-0-04301-X

thrive with such water. The quality of the water can also affect irrigation systems: drip feed systems may become clogged and the nutrients in the wastewater may counteract or enhance existing fertilizers. Where rainwater is not used for irrigation, mineral deposits can build-up within the soils adversely affecting plants over time.

Many feasible solutions to problems are avoided due to the fear and consequences of failure, especially when the result can be death or severe illness or the associated costs. For instance, the provision of "non-drinking water" for in-building purposes brings risks such as cross-contamination i.e. where the drinking water becomes contaminated by the non-drinking water. However, this risk can be minimized by careful and clear marking of pipes and terminal fittings. The doubling, or tripling, of pipework in a system to provide different water qualities may be more expensive than simply providing all outlets with drinking water. Therefore, the economics and cost/benefit, as well as health impact need to be considered when supplying water of lower quality.

Water quality is not constant. The quality can vary in situations where regularly interrupted supplies are provided, such as by tanker or pipework that operate during certain times of the day. Water in a completely filled pipe will suffer some degradation in quality if a suitable residual disinfectant, such as chlorine, is not used. Further, if the pipe is frequently filled and drained the air/water interface will create films of water that can encourage corrosion, biofilm production and proliferation of bacteria which may lead to failure of the pipework and contamination of the supply. The materials that are used in conveying the water can have a significant impact on the quality of the water delivered to the outlets. Evidence suggest that materials such as lead may cause various illnesses as well as brain damage and death [39]. Although copper is popularly used to convey water, it too can be dissolved in some waters, mainly very soft water, leading to perforations of pipes and increased copper levels in the water. Plastics pipes or plastics coated hybrid pipes are immune from corrosion issues, but can encourage biofilm development that can lead to bacterial contamination. So, whatever system of conveyance of water that is used there are issues that need to be addressed through design, installation and maintenance. Hence, water quality is a dynamic state that will vary in time, location and environment. Water of poor quality can normally be treated to improve its quality, but good quality water can easily loose its quality through poor transport, storage and contamination.

3.2 Drinkable water

The quality of drinking water will vary depending on where in the world and the availability of a water supply. In remote deserts, people may use a solar still or drink at an oasis, on a mountain there may be a freshwater spring or melt some snow, and in many countries people simply collect and drink rainwater. However, there is an expectation when paying for drinking water; be it from a bottle or a tap that it will be of a suitable quality.

People who grow up drinking the local water tend to develop immunity to any issues with that water supply. Whereas, visitors may suffer upset stomachs if they drink the same water [14]. Even in countries and areas where the drinking water is "safe" there can be variations in taste, colour and odour. Although water is relatively tasteless, various taints can be attributed to the water encountering various metals, plastics and minerals [7]. The temperature of the water can also affect its perceived taste and its reaction with additives such as tea, coffee and various alcohols.

Therefore, drinking water comes in a wide range of parameters and the transport and delivery equipment that is used can affect its quality, both actual and perceived.

3.2.1 Do not drink the water...

More important than sustainability is the need for water supplies to be safe to drink. Even today, many countries do not have adequate separation between water supplies and wastewater disposal, leading to contamination of drinking water. Alternatives to water have often emerged where good quality drinking water is not available. For example alcohol based beverages that could be produced locally and were safer to drink than the local water in England and many other countries. Water was generally only used for washing and cooking. The negative effects of alcohol were either ignored or tolerated. Hence the proliferation of Ale Houses and Gin Houses in various countries along with wine and other local drinks. In the middle of the 18th Century, water source contamination by industries in the USA, led to the start of the bottled water industry. In New York, water had to be collected from out-of-town wells and transported in buckets and barrels. This type of drinking water was known as "Tea water" and as it cost more than the poor people could afford [31].

In the middle of the 19th century, John Snow proved that some drinking water sources were being contaminated by sewage, creating health hazards such cholera [33]. This discovery lead to the installation of, free of charge, water points with filtered clean water, often provided by entrepreneurs or UK councils.

Historic drinking water fountains and troughs were often provided to refresh travellers and their animals, such as horses and dogs. Drinking troughs became common outside of English public houses in the late 1860s, as they encouraged the travellers to buy beer [36]. Below is a sign that was found close to drinking troughs.

> "All that water their horses here, must pay a penny, or have some beer!"

Many other countries also erected drinking fountains and troughs in the following decades. Drinking water fountains were typically provided with cups on chains, but over the years these were replaced by "bubble jets" to improve the hygiene of the provision.

3.3 Types of drinking water

It is useful to consider the various types of drinking waters before exploring the various quality specifications that may apply. Water can exist in three states; solid, liquid and gas. As ice, water is often used to cool drinks or serves as the basis of a cocktail or slush drink. It is however worth noting that ice can be a significant route of contamination if it is not created with wholesome water in a clean environment and stored at the correct temperature [16].

Liquid water is the most common form of drinking water, but water vapour, steam or as moisture within the atmosphere is also available. Since the beginning of the 20th century, many people have strived to extract water from air by various routes [41]. Methods include variants of de-humidifiers, but recently there have been developments using desiccants (that can be used in most humidity) and Warka towers that only work well in areas of high humidity [3]. Small solar powered electronic units that can be attached to bicycles have also been prototyped, which promise drinking water from humid air as the bicycle passes through it. [29]

3.3.1 Hardness

Water from the air tends towards being pure and often needs to have minerals added to make it drinkable. Depending on the soils and atmosphere through which it has passed, naturally-occurring liquid water, is often loaded with various minerals. Water that has passed through chalk will tend to be classed as "hard" as it will contain calcium chloride in various forms. Various gasses such as carbon dioxide, sulphur could have dissolved into it if it has passed through atmospheres of various levels of pollution. These reactions will modify the pH, solubility and stability of the water. One of the main characterizes of water that impacts many applications of water is how easy it is to produce a lather from soap, i.e. "hardness". Water hardness can be expressed in a number of ways. Table 3.1 shows the classification system used in the UK.

Table 3.1: Classifications of water hardness.

Water hardness classification	Concentration of calcium carbonate (mg/l $CaCO_3$)
Soft	Up to 50
Moderately soft	50–100
Slightly hard	100–150
Moderately hard	150–200
Hard	200–300
Very hard	Over 300

[*Source*: P.J. Jackson, G.R. Dillon, T.E. Irving and G. Stanfield, Manual on treatment for small water supply systems Updated Report, Final Report to the Department of the Environment, Transport and the Regions Report No: DETR/DWI 4936/1, WRc-NSF Ltd, June 2015, Table 1 /http://www.dwi.gov.uk/private-water-supply/installations/updated-manual-on-treatment-for-small-supplies.pdf].

Other measures for water hardness include: American hardness, German degrees, French degrees, English degrees (or Clark), mmole/litre $CaCO_3$, mg/litre CaO, mg/litre Ca, mg/litre Mg, grains/gallon, and mEq/L. In addition, there are different types of hardness; temporary (i.e. a measure of the dissolved calcium hydrogen carbonate removable by boiling and deposited as scale on heating elements and hot surfaces), permanent (that which is a measure of the dissolved calcium sulphate) and total (that which is a measure of bicarbonates, chlorides and sulphates of calcium and magnesium ions). It is vital to know what measure of hardness is being used to determine water quality and the impact any treatment or conditioning may have on that measure.

3.3.2 Materials in contact with water

Many water supply utilities or companies use waters from various sources through the year. They will combine waters from different sources, and add or remove various substances in order to provide a consistent water quality to their customers. A consistent quality is important for many plumbing installations as most drinking waters can dissolve metals to varying degrees. Some of the metals that water is able to dissolve are copper, lead, iron and aluminium; all of which have been used extensively for water supply networks. Some waters are particularly known to be cupro-solvent, copper pipes should be avoided in such distribution areas [45].

In addition to "simple" metal solvency in water, the issue of leaching out of some metallic elements from various established alloys that are common in water supply systems is becoming an area of concern, especially in Europe [1]. Although concerns have been raised, the difficulty in testing alloys and determining the low concentrations that are often specified has resulted in a lack of agreed procedures to evaluate the leaching of metals. The established testing procedures for materials in contact with water are complicated and contain numerous caveats that need to be understood [44]. In addition to metals in contact with drinking water, there are lists of suitable non-metallic substances such as organic and cementitious materials e.g. [4MS, 4MS Initiative Positive Lists for Organic Materials used in Products in contact with Drinking Water, 9 May 2017 update (lists only): 13 April 2018/https://www.umweltbundesamt.de/sites/default/files/medien/374/dokumente/pl_om_13_april_2018_0.pdf] and [4MS, Assessment of Cementitious Products in Contact with Drinking Water, 4MS Common Approach, Draft September 2018/https://www.umweltbundesamt.de/sites/default/files/medien/374/dokumente/cementitious_products_-_4ms_common_approach_jmc_final_draft_sep_2018_2_0.pdf]. So, what starts out as wholesome water, may become less than wholesome after traveling though pipes, fittings, valves, filters, flexible connections, storage containers and outlet fittings; unless water flow rates are maintained to minimize stagnation, care to ensure that only appropriate materials are used, and that the fittings and tools needed to build the system are clean and used correctly are necessary.

3.3.3 Additives to water

The substances that are naturally absorbed into water sources are a key element of the Spa or Mineral water industry that claims various benefits for their waters. Some of these waters naturally contain carbon dioxide, but many such waters will have CO_2 added to enhance them with bubbles.

In some countries and regions, water supply companies are expected to add fluoride to the water to help children develop stronger teeth. The first reported scheme was in Michigan in the USA in 1945, the UK followed, 19 years later, with its first scheme in Birmingham. In the UK, most schemes are in the Midlands and the North of England, where action is generally taken to raise the level of Fluoride to 1 mg/L in areas where such a concentration is not naturally occurring [25]. In some countries the contribution of Fluoride to the population's dental and other aspects of health have been difficult to assess; as some diets include root crops that also provide a Fluoride intake and air pollution from some coals can also add to the total Fluoride intake. While too little Fluoride is a problem for teeth and bones, too much Fluoride (>10 mg/L) can lead to crippling diseases such as skeletal fluorosis [46].

Therefore, some countries and towns in the USA have now banned the addition of fluoride to water supplies [20]. This "mass medication without consent" has recently been questioned in various countries around the world [17].

3.4 Drinking water standards

Most countries have legally enforceable standards to ensure that water supplies are safe to drink [10]. Standards are different for piped municipal supplies (mains supply), and other private water supplies (e.g. from a well or spring). Private water supplies tend to be controlled by local governments or municipalities. Such supplies can be small and used by a single home, but they may also be considerable and used for food and beverage production or by some industries, hotels and hospitals.

Most standards today are based upon the World Health Organization guidelines [47]; which currently run to 631 pages. The WHO guidance was first published in 1958 and it is regularly updated and revised. So, drinking water specifications are not static.

In England, the list of parameters that need to be monitored for safe drinking water are set out in Annex 3 of the DWIs Drinking Water Safety Guidance to Health and Water Professionals. The number of elements, compounds, contaminants and characteristics that are included in such specifications are far larger than might be thought by most people. For a substance that is "just water", its constitution can be far more complicated than its simple chemical formula, H_2O, might suggest. In fact water itself also dissolves into ions; so it exists with hydrogen cations and hydroxide anions, along with: chlorine, silicates, magnesium, calcium, aluminium, sodium and other trace molecules and ions [19].

The trace elements in water can influence the taste and also produce scales of different types. The scales that form in pipework are not always inert, and can produce corrosion that leads to the contamination of the water supply [48].

3.4.1 Different quality water supplies

Beyond good practice, a designer's specification for *"wholesome water throughout a building"* can be just to simplify the plumbing. If waters of different qualities are to be provided in a building or development, a robust system of separation of the different water types will be required to avoid the contamination of the wholesome water. Cross connections between wholesome water and reused rain or greywater can be a major concern in developments where water reuse systems are installed. Guidelines and standards for domestic and non-domestic water reuse systems, pipe and fitting markings are provided by most nations', water companies, the ISO and other organizations. However, there is no established world-wide convention for the colour or marking of water or water reuse systems due to the different utility and communication networks in each country. For example, water reuse systems for irrigation in Australia and Israel are a pinkish-violet colour, but a similar colour pipe is used for telecommunication networks in the UK

3.4.2 Legionella and other plumbing related health problems

The first recorded instance of Legionella was at the July 1976 Bicentennial Legionnaires' Convention in the Stratford Hotel at Philadelphia, USA [5]. Since then the number of reported instances has risen steadily. Most piped water will contain some legionella bacteria. It will not be a problem and remain dormant in cold water – below 20 °C. However, the Legionella will start to multiply if the water temperature is between 20 °C and 45 °C. Maximum reproduction of the bacteria occurs at human blood temperature around 37 °C. Historically, engineers have aimed to maintain hot water distribution temperatures at about 55 °C and stored hot water in excess of 60 °C, but paid very little attention to keeping cold water below 20 °C. The advent of more energy efficient buildings with warmer non-habitable spaces, such as pipe ducts and cupboards, and the proliferation of bathrooms, has led to infrequently used branches of water supply systems which hold water volumes for long periods of time. This results in warmed cold water and cooled hot water along with stagnant zones that are likely to encourage Legionella growth.

3.4.3 Water treatment

Water suitable for drinking is rarely found in nature. There are ranges of available technologies that can be used to achieve drinkable water from a raw water source. These include: filtration, disinfection, organic and inorganics removal, adsorption, clarification, coagulation and flocculation, floatation, and membranes [24]. Many of these are natural processes that, if designed appropriately, will be inherently sustainable due to the use of natural bacteria and gravity [2].

The processes used to produce the various chemicals used in water treatment are a mix of well-established tried and tested innovative procedures. Some of the chemicals are toxic and/or can produce by-products that can be harmful to people and the environment. More efficient and cheaper alternative options are often being proposed for processes that use expensive materials or large quantities of energy. However, the main issue is human safety. Previous processes and materials that were once considered safe, such as the use of asbestos, have had to be replaced. Some processes such as fluoridation and the use of alum for coagulation, have now become questionable in some areas, are now under scrutiny as more data on their use becomes available. Hence, it can be quite complicated to determine the sustainability of any water treatment process. Chemicals that are available locally will have less transport-associated emissions that those from further away, but they need to be safe to extract, use and dispose. The development of membranes may reduce the dependence on chemical treatments, but there will always be a place for the use of some chemicals in water treatment systems.

3.4.4 Chemicals for water treatment

Chemical treatment processes are typically used to change, or preserve the properties of the water. The chemicals used in water treatment include:

Chlorine (CL_2) – this is used to disinfect the water and maintain its quality during distribution in pipe networks. At the treatment works, the drinking water is dosed at around 5 parts per million (ppm) so that the residual delivered to the points of use is around 0.1–0.2 ppm chlorine in the form of Cl_2, HOCl or OCl^- (chorine, hypochlorous acid, or hypochlorite (free chlorine)). In addition to the disinfection of treated water, chlorine can be used to remove ammonia. In such circumstances, high concentrations of chlorine are used in a process called superchlorination (or breakpoint chlorination), after which the excess chlorine has to be removed by dechlorination.

Chlorine can be provided as a liquid (under pressure), a solid as sodium or calcium hypochlorite salt, or generated electrochemically on-site. The electrical energy used to produce chlorine is considerable [11] and roughly equivalent, per unit weight, to that for iron and steel [28]

Ammonia (NH_3) – Ammonia is widely produced as it is also used in the production of fertilizers [27]. Although frequently removed from water, it can be added to produce an even more stable disinfectant residual in a process called chloramination which produces disinfecting compounds known as chloramines.

Ozone (O_3) – Ozone can be very toxic if inhaled but it is often used as a very powerful oxidant that produces few unwanted by-products. It is used in conjunction with UV irradiation for advanced oxidation.

Ozone is generally produced on-site [13], but this is very energy intensive; typically using 20–25 kWh/kg of O_3 generated.

Aluminium ($Al(OH)_3$) and ferric ($Fe(OH)_3$) hydroxides – Used as coagulants to bring particles together so that contaminants such as phosphate can be removed using filtration techniques. They work by changing the surface charge characteristics of the particles so that they attract one another. Other coagulants include Alum, or Aluminium sulphate ($Al_2(SO_4)_3$•$18H_2O$) – which used to be the most popular coagulant, but has been replaced by other chemicals such as polyaluminium chloride (PAC), polyferric sulphate (PFS) and flocculant aids such as clays and activated silica that do not leave a residual of aluminium in the water.

Lime ($Ca(OH)_2$) – Lime is used to soften water by raising its pH, then Soda (Sodium carbonate (Na_2CO_3) is added to lower the pH and complete the removal of the calcium carbonate ($Ca(HCO_3)_2$). Lime is inexpensive compared to soda, hence the lime-soda process is often used in preference to simply using soda to deal with all the softening. The lime/soda treatment is normally conducted at normal temperatures (cold), but if the water is needed for boiler supplies it can also be carried out at near boiling point (hot) where the reaction will be faster [38].

Sand filtration – Sand filters were a common water filtration strategy. They were used in equipment such as pressure filters, where backwashing would regularly take place or built into the ground. The latter had a large land take and were difficult to service and maintain. Today, membranes are the main technology for water treatment. Membrane technologies used to be prohibitively expensive, but the cost is now competitive due to the economies of scale and frequent innovations.

3.4.5 Membranes for treating small volumes

A handheld compact membrane filtration product, could be used if only small quantities of drinking water are required, such as for personal use. This may be as simple as a normal drinking straw. The personal water filter is about the size of a small torch and can be used to filter over 700 L of water [42]. These devices use hollow fibre membranes which filter out bacteria and parasites that are larger than its 0.2 μm pore size combined with a carbon filter to absorb pesticides and chemicals such as Chlorine. They become clogged and the flowrate reduces over time, so when it becomes hard to suck water through it, it is probably time to get a new one [43]. Such devices are also available in larger sizes with small reservoirs for use by families and small communities in emergency situations.

Other personal water filters are available that use similar technologies but in different ways. Some systems work like a coffee cafetière, in that you push a filter through the collected water in a bottle to clean it, or hand-pump the water through the filtration system. Personal UV purification systems are also available, but need a power source and they may not work well with high turbidity waters. The level of performance of all personal filtration systems vary in terms of: quantity of water that can be treated, speed of treatment, size and weight of equipment, the level of filtration provided and the substances that can be removed, as well as cost [32].

3.4.6 Membranes for treating large volumes

Similar techniques can be used for situations where greater quantities of water are required, such as at permanent installations, and are normally operated with electrical power. A comparison of various types of membrane water treatment, except electrodialysis, and the type and size of particle that can be removed is shown in Table 3.2. The particles further to the right of the table are larger and require only low membrane operating pressures. Hence, the removal of bacteria is relatively easy and requires only a low operating pressure. Whereas to remove the smaller particles, by reverse osmosis (RO), can require around 50 times the pressure used for simple microfiltration.

Table 3.2: Membrane pore size comparison.

Range	Ionic	Molecular	Macro molecular	Micro particle	Macro particle
Particle size of pollutants (μm)	0.001	0.01	0.1	1–10	100
Molecular weight cut-off (kDa)[a]	0.1	5	500	–	–
Pollutants	Aqueous salts		Colloids	Bacteria	
	Small sand				
	Metal ion		Latex emulsions		
	Sugar	Viruses & proteins		Cryptosporidium oocysts	
	Atomic radius			Giardia cysts	Pollens
Purification process	Reverse Osmosis				
		Nano-filtration			
			Ultrafiltration		
				Microfiltration	
Usual operating pressures (MPa)	>0.5				
			0.05–0.3		
				0.01–0.2	

[*Source*: N. Shamsuddin, D.B. Das[a], V.M. Starov, Membrane-based point-of-use water treatment (PoUWT) system in emergency situations: A review, Loughborough University Institutional Repository, 2014 – referenced as from: Mulder, M. (2003) Basic principles of membrane technology. Kluwer Academic Publisher. /https://dspace.lboro.ac.uk/dspace-jspui/bitstream/2134/15252/3/Shamsuddin_et_al_SeparationScienceReview_14Oct_submitted_Marked_latest.pdf].

[a]MWCO is the lowest molecular weight of solute (measured in Daltons) where 90 percent of the solute is retained by the membrane.

There are five main classes of membrane treatment processes that can be selected according to the size of the particles removed:

- **Reverse osmosis (RO)**: reverse osmosis is where a partially permeable membrane is used to treat the water by forcing it through the membrane by pressure. The cleansed water passes through but particles larger than the pore size of the membrane are retained and sent to drain.

- **Nanofiltration (NF), ultrafiltration (UF) and microfiltration (MF)**: nanofiltration, ultrafiltration and microfiltration are all similar to RO, but with increasingly greater pore sizes and lesser pressures.

Electrodialysis (ED): Electrodialysis is defined as 'dialysis in which the movement of ions is aided by an electric field applied across the semipermeable membrane'. The process uses multiple alternating membranes that selectively pass only positive or negatively charged ions. As all membranes tend to become fouled over a period of time, a way of cleaning them is needed. In the electrodialysis process, the electric field is simply reversed to clean the membranes. When used on a large scale, electrodialysis reversal plants can produce clean water from sea and other brackish waters. The system often competes with RO systems, but tends to work better with lower salt concentrations. Electrodialysis may also be combined with RO systems to deal with the effluent from RO units [34]. However this process, as with other membrane treatments, requires extensive pre-treatment to be reliable, such as the removal of all particles bigger than 10 μm, as well as colloidal matter, hardness, large organic anions, manganese and iron oxides [23,35]. Electrodialysis is not the only process that can use electric fields to aid the operation. In electrophoresis, charged particles are attracted to an anode in a uniform electric field, and in dielectrophoresis uncharged solids are attracted to an anode in a non-uniform electric field. Magnetic separation can also be used if the solids have paramagnetic or ferromagnetic properties.

The relatively compact nature of membranes has enabled portable equipment to be developed. However, as with sand filters and other filtration devices, membranes do suffer from fouling and need to be back washed to maintain their performance. The sustainability of the technique becomes questionable if the membrane is simply discarded after it has ceased to function at its full performance. In most applications, membranes need to be used along with other technologies so the sustainability of the whole process needs to considered, not just isolated or key elements.

Screens, such as micro-strainers, may be needed in large-scale treatment plants to remove algae. Sedimentation and floatation tanks (including dissolved air flotation units (DAFs)) along with deep bed filters will help clarify the water. Membranes can remove organic contaminants, but the removed contaminants will have to be collected, thickened and dewatered before disposal.

3.5 Water distribution

Current water supply systems rely on the pumping of water to some extent. Compared to local wells and springs, systems that pump water for vast distances from reservoir to taps are more dependent on pumps. The greater the distance of pumping, the greater the likelihood of leaks and loss of water. However, large water supply systems come provide economies of scale [49] that can reduce the energy consumption and carbon emissions per litre of water supplied.

The cost of installing and operating an extensive distribution system can negate all, or some, of the economies of scale achieved by centralization. In addition to simple pumping, water aerating, mixing and reactants may require prolonged use of various types of powered mixers - adding to the net energy use. The topography and gravity can be used to minimize the use of pumps for the transport of water between treatment processes in treatment works that occupy a large area of land. In compact plants or those housed within a building, a greater use of pumps may be required to transfer the various fluids such as: sludge, scum, different qualities of wastewater and air. So energy use may have to be balanced with water needs, land take and other planning issues. Although data on energy consumption for the treatment and distribution of water for nations are patchy, steps have been identified to improve the energy consumption in all water supply systems [6]. These include:

- System optimization – To ensure that pump speeds are adjusted to be as efficient as possible for the loads applied and that aeration equipment is well maintained and controlled with reliable sensors and control programs
- Upgrading and improving – To minimize leaks from existing systems, optimize the size of systems – reduce oversizing of plant and pipework, replace worn-out equipment, utilize new technologies and techniques
- Energy management – To compare energy use with similar facilities and learn from each other, reduce demand for water by water conservation; in particular hot water that requires additional heating energy
- Energy generation – To use waste material to generate biogas or biofuel, install solar panels to produce electricity and, if possible, use natural water power to generate HEP. [8]

3.5.1 Non-piped supplies

Other means of obtaining drinkable water are needed if no piped mains water are available. The most established methods are: obtaining water from a water body (such as a stream, river or lake), collecting water from a spring or a well, using bottled water, collecting rainwater, cleansing wastewater or extracting water from the air. Some of these are simple and need no additional technology (e.g. bottled water), but for the others, purification tablets [15] or other methods for making the water safe for consumption may be needed. Depending upon the volumes of water required and available equipment, a selection of the appropriate technology can be made.

3.5.2 Piped water supply systems

Water is typically supplied through a network of pipes. It is important that water within the pipework is not left stagnant and the pipework does not degrade the water quality. Therefore, the pipe sizing of water supply systems is crucial to good hygienic efficient water distribution. Traditionally, engineers specified oversized pipes, and other water structures, to provide the desired service volumes and for fear of failure [37]. However, with the advent of water efficiency measures around the world, this could cause problems of heat transfer and stagnation resulting in hygiene and health issues [40]. Water saving measures such as installing spray taps, aerated showerheads, reduced-flush WCs and water-saving washing machines and dishwashers may be good for the environment and reducing water bills, but these measures could have adverse effects on water supply networks if implemented on systems that were designed for far greater turn-overs of water.

The oversizing of water supply systems and networks is slowly being addressed by the LUNA (Loading Unit Normalization Assessment) project [4]. The project is led by CIBSE (Chartered Institution of Building Services Engineers) and CIPHE (Chartered Institute of Plumbing and Heating Engineering) in the UK with input from Heriot–Watt University. In the meantime, the latest design procedures, e.g. DIN 1988 [9] and SANS 10252 [30] provide recommendations for far smaller pipe sizes.

Pipe sizing is important so that the supply at the outlets is of sufficient flow rate and pressure to enable it to carry out its required function. There are frictional losses that reduce the delivered pressure of the water in any pipework system. In a system where the water does not flow, the loss of pressure is simply the difference in height between the outlet and the reservoir (this can be a small cistern, a water tower, or a lake etc.), or the height between the mains supply and the point of delivery. This "head" of water is the maximum pressure that can be available at an outlet [26]. Once a water system is flowing, the additional frictional losses, caused by shear stress along the pipe's internal surface, turbulence and momentum changes due to fittings, bends and valves can significantly reduce the water's dynamic pressure and in severe cases prevent any flow of water [22]. So the terminal fittings and pipework sizing can determine if the appliance or terminal fitting will work satisfactorily, or not. Most shower valves, taps, washing machines etc. will include a specification for an optimal range of supply static or dynamic pressures and flowrates. Unless the outlets of the system are matched to the supply pressure and flowrate, they may not work well or at all. Therefore, a sustainable water supply system includes terminal fittings and appliances that are appropriate to the available water supply parameters.

3.6 Questions of sustainability

A number of factors should be considered and balanced to determine the sustainability of a water supply [21]. These include:

- Volume of water required (total or per unit time): criteria for a few litres of water will be very different to those for a supply for a city
- Quality and quantity of available raw water supplies

- Location of water supplies (distance and topography)
- Availability of treatment processes, supplies and expertise
- Ambient conditions (temperature, humidity, wind, sun and dust)
- Life cycles of components, chemicals,
- Budgets (capitol and operational, maintenance)
- Security (resistance to terrorist action, theft of equipment and reagents)
- Record of performance of selected technologies
- Quality of materials used for pipework and long-term reliability of joints
- Options for expansion, or contraction, in the future
- Reliance upon local operators and skills, or manufacturer's trained staff
- Ease of maintenance and replacement of components
- Resilience of the system to variations in load and demand as well as to flood and drought
- Toxicity of any chemical used or waste produced
- Possibilities for utilizing water reuse within the plant and delivering non-potable supplies to users and uses that do not require wholesome water
- Options for generating energy, nutrients and other useful by-products from the system
- Possibilities to use low-tech or high-tech as appropriate for the location and situation

Clear options for a system should emerge by considering the above issues, and possibly applying a suitable weighting to different factors [12,18]. Although a system of water supply can be produced that is sustainable, it will form part of a water cycle that will cover an area, catchment, country, continent or planet. The whole-system sustainability comes into question if the water that is produced is then used for questionable purposes or wasted. Questionable purposes include the use of wholesome water to transport of human waste, WC flushing, vehicle washing and irrigation. An integrated water management approach is needed so that all the elements of the water cycle are made as sustainable as possible.

3.7 Summary/conclusion

A sustainable water system needs to be of a consistent and appropriate water quality. In many situations the simplest to specify, although not to achieve, will be wholesome water. This utilizes the simplest plumbing and distribution system, but results in wholesome water being used for applications where a lower quality of water would suffice. If different water qualities are to be provided, clearly identifiable pipelines, valves and fittings should be used to maintain safety of the system throughout its lifetime.

Wherever possible, gravity as a motive force should be used in preference to a pump. Where pumps or mixers are required, efficient drive motors should be specified and, if available, renewable energy sources such as solar or hydroelectric power should be used.

Water systems should not be designed in isolation. But considered within the context of the catchment area, available resources (both conventional and "waste" resources), level of competence of maintenance personnel, economic viability and the need and expectations of the users. Hence, any water supply system should be reviewed regularly to ensure that it is still meeting the needs of the users in a safe, efficient and environmentally beneficial way.

References

[1] 4 MS Joint Management Committee, Procedure for the Acceptance of Metalic Materials for PDW, 4MS Common Approach, 11th Revision, Retrieved from Acceptance of metalic materials used for products in contact with drinking water: (2019, March 5). https://www.umweltbundesamt.de/sites/default/files/medien/374/dokumente/11th_revision_4ms_scheme_for_metallic_materials_part_b.pdf.

[2] J. Birkeland, The Emergence of Design for Sustainability: And Onward and Upward. The Handbook, Bloomsbury, London, 89, 2017.

[3] A.A. Chavan, Sustainable water harvesting technique by condensation of water through atmosphere in an optimized approach for future cities in India, Conference on Technologies for Future Cities (CTFC), 2019.

[4] CIBSE, Assessment of Loading Units Method For Sizing Domestic Hot & Cold Water Services, Retrieved from Chartered Institution of Building Services Engineers - Knowledge, 2017, February. https://www.cibse.org/knowledge/knowledge-items/detail?id=a0q0O00000CBW9lQAH.

[5] CIPHE, Safe Water Guide, Insight, Scald Prevention and Legionella, Chartered Institute of Plumbing and Heating Engineering, Hornchurch, 2016.

[6] B. Coelho, A. Andrade-Campos, Efficiency achievement in water supply systems—a review, Renew. Sust. Energ. Rev. 30 (2014) 59–84.

[7] K.H. Cohen, Taste threshold concentrations of metals in drinking water, Am. Water Works Assoc. 52 (5) (1960) 660–670.

[8] Copeland, & Carter, Energy-Water Nexus: The Water Sector's Energy Use, Retrieved from Congressional Research Service, 2017, 7–5700, https://fas.org/sgp/crs/misc/R43200.pdf.

[9] DIN, DIN Drinking Water Supply Systems: Pipe Sizing Practice , Deutsches Institut für Normung, Berlin, 2012, 1988–300.

[10] Drinking Water Inspectorate, Drinking Water Safety: Guidance to Health and Water Professionals, 2009, September. Retrieved from http://dwi.defra.gov.uk/stakeholders/information-letters/2009/09_2009Annex.pdf.

[11] C. Egenhofer, L. Schrefler, V. Rizos, F. Infelise, G. Luchetta, F. Simonelli, L. Colantoni…, Final Report for a study on composition and drivers of energy prices and costs in energy intensive industries: the case of the chemical industry – chlorine, Retrieved from, Centre for European Policy Studies FRAMEWORK CONTRACT NO. ENTR/2008/006 LOT 4, for the Procurement of Studies and other Supporting Services on Commission Impact Assessments and Evaluation, 2014, January 13. https://ec.europa.eu/docsroom/documents/4164/attachments/1/.../en/.../pdf.

[12] A. del Borghi, C. Strazza, M. Gallo, et al., Water supply and sustainability: life cycle assessment of water collection, treatment and distribution service, Int. J. Life Cycle Assess. 18 (2013) 1158.

[13] D.V. Franco, W.F. Jardim, J.F. Boodts, L.M. DaSilva, Electrochemical Ozone Production as an Environmentally Friendly Technology for Water Treatment, Clean Air Soil Water36, January 2008, pp. 34–44. Retrieved from Wiley.com: https://onlinelibrary.wiley.com/doi/abs/10.1002/clen.200700080.

[14] R.K. Frost, How clean must our drinking water be: the importance of protective immunity, J. Infect. Dis. (2005) 809–814.

[15] C.P. Gerba, D.C. Johnson, M.N. Hasan, Efficacy of iodine water purification tablets against cryptosporidium oocysts and Giardia cysts, Wild. Environ. Med. 8 (2) (1997) 96–100.

[16] P. Graman, G. Quinlan, J. Rank, Nosocomial legionellosis traced to a contaminated ice machine, Infect. Control Hosp. Epidemiol. 18 (9) (1997) 637–640.

[17] T. Guardian, Water fluoridation: what does the rest of the world think? Retrieved from, The Guardian Newspaper, 2013, September 17 https://www.theguardian.com/world/2013/sep/17/water-fluoridation.

[18] J.S. Guest, S.J. Skerlos, J.L. Barnard, M.B. Beck, et al., A new planning and design paradigm to achieve sustainable resource recovery from wastewater, Environ. Sci. Technol. 43 (16) (2009) 6126–6130.
[19] A.M. Helmenstine, The Molecular Formula for Water, Retrieved from Thought Co: thoughtco.com/water-molecular-formula-608482, 2018, December 3.
[20] S. Jerome, Fluoride bans on the rise, (2018, October 19). Retrieved from Water online, News feature: https://www.wateronline.com/doc/fluoride-bans-on-the-rise-0001.
[21] C.A. Kennedy, A. Racoviceanu, B. Karney, A. Colombo, Life-cycle energy use and greenhouse gas emissions inventory for water treatment systems, J. Infrastruct. Syst. 13 (4) (2007).
[22] T.J. McGhee, Water Supply and Sewerage, sixth edition, McGraw-Hill, NewYork, 1991.
[23] S. Moran, An Applied Guide to Water and Effluent Treatment Plant Design, chapter 7 - Clean Water Unit Operation design: Physical Processes, (2018, June 8) pages 69–100. Retrieved from Science Direct: doi:10.1016/B978-0-12-811309-7.00007-2.
[24] S. Parsons, B. Jefferson, Introduction to Potable Water Treatment Processes, Blackwell Publishing Ltd, Oxford, 2006.
[25] Public Health England, Water Fluoridation Health Monitoring Report for England 2014, PHE publications gateway number: 2013547Retrieved from 2014, April 1. pp. 2, https://assets.publishing.service.gov.uk/government/uploads/system/uploads/attachment_data/file/300202/Water_fluoridation_health_monitoring_for_england__full_report_1Apr2014.pdf.
[26] R.H. Garret, Hot and Cold Water Supply, Blackwell Science, Oxford, 2000.
[27] I. Rafiqul, C. Weber, B. Lenmann, A. Voss, Energy efficiency improvements in ammonia production—perspectives and uncertainties - energy, Elsivier (2005).
[28] R. Remus, M.A. Aguado-Monsonet, S. Roudier, L.D. Sancho, Best Available Techniques (BAT) Reference Document for Iron and Steel Production, Industrial Emissions Directive 2010/75/EU (Integrated Pollution Prevention and Control), European Commission, 2013, 2012, March. JRC Reference Report, 2012, March. Retrieved from, https://eippcb.jrc.ec.europa.eu/reference/BREF/IS_Adopted_03_2012.pdf.
[29] K. Retezár, Bicycle bottle system condenses humidity from air into drinkable water, (2019). Retrieved from IFL Science: https://www.iflscience.com/technology/bicycle-bottle-system-condenses-humidity-air-drinkable-water/.
[30] SABS, SANS 10252-1: Water Supply and Drainage for Buildings. Part 1: Water Supply Installations for Buildings, Pretoria: South African Bureau of Standards (SABS) Standards Division, 2012.
[31] J. Salzman, Thirst: a short history of drinking water, research paper no. 92 December 2005, Duke Law School Legal Stud. (2005) 17.
[32] N. Shamsuddin, D.B. Das, V.M. Starov, Membrane-Based Point-of-use Water Treatment (PoUWT) System in Emergency Situations, Loughborough University Institutional Repository, 2014, October. Retrieved from https://dspace.lboro.ac.uk/dspace-jspui/bitstream/2134/15252/3/Shamsuddin_et_al_SeparationScienceReview_14Oct_submitted_Marked_latest.pdf.
[33] J. Snow, in: On the Mode of Communication of Cholera, J. Churchill (Ed.), London, 1855.
[34] Strathmann, Membrane separations| Electrodialysis. Retrieved from Reference Module in Chemistry, Molecular Sciences and Chemical Engineering Encyclopedia of Separation Science 2000, 2003, December 6, pages 1707–1717. https://www.sciencedirect.com/science/article/pii/B0122267702051310#!.
[35] H. Strathmann, Desalination, Electrodialysis, A Mature Technology with A Multitude of New Applications, Elsevier, 2010.
[36] The Metropolitan Drinking Fountain and Cattle Trough Association, History of Drinking Fountains, 2011. Retrieved from http://www.drinkingfountains.org/: http://www.drinkingfountains.org/Attachments(PDF)/DFA%20HIstory.pdf.
[37] J. Tindall, J. Pendle, Are we significantly oversizing domestic water systems? CIBSE Technical Symposium, 2015, April 16–17. Retrieved from http://nrl.northumbria.ac.uk/23010/1/JT_JP_Oversizing_DCWS_CIBSE_ref_no_63.pdf.
[38] T. Tompson, A Guide to Sanitary Engineering Services, Macdonald & Evans Ltd, London, 1972, pp. 55.
[39] W. Troesken, The Great Lead Water Pipe Disaster, MS: Massachusetts Institute of Technology, Cambridge, 2006.

[40] S. Volker, C. Schreiber, T. Kistemann, Drinking water quality in household supply infrastructure—a survey of the current situation in Germany, Int. J. Hyg. Environ. Health 213 (3) (2010) 204–209.

[41] R.V. Wahlgren, (2017, July), Water from air Quick Guide, second edition, Atmoswater Research: Retrieved from http://www.atmoswater.com.

[42] Walters, A Performance Evaluation of the Lifestraw: A Personal Point-of-use Water Purifier for the Developing World, UNC University Libraries Carolina Digital Repository, 2008. Retrieved from doi:10.17615/acmj-fd22.

[43] A. Walters, A Performance Evaluation of the Lifestraw: A Personal Point-of-use Water Purifier for the Developing World, Ann Arbor: Proquest, Uni Dissertation Publishing, UNC University Libraries Carolina Digital Repository, 2011, Retrieved from doi:10.17615/acmj-fd22.

[44] Water Regulations Advisory Scheme, A Guide for Manufacturers, Suppliers ans Test Laboratories on the Application Requirements for WRAS Material Approval, Version 3.0 WRAS Material Guidance, 2015, March 10. Retrieved from https://www.wras.co.uk/downloads/public_area/documents/materials_guidance_v3.0_issued_10th_march_2015.pdf.

[45] P. Wells, B. Bremer, B. Webster, Biocorrosion of Copper in Potable Water, American Water Works Association, 2001, August 1. Retrieved from https://awwa.onlinelibrary.wiley.com/doi/pdf/10.1002/j.1551-8833.2001.tb09269.x.

[46] World Health Organisation, Retrieved from Fluoride in Drinking Water, Background Document for Development of WHO Guidelines for Drinking Water Quality, 2004 https://www.who.int/water_sanitation_health/dwq/chemicals/fluoride.pdf.

[47] World Health Organisation, Fourth edition Incorporating the First Addendum. Retrieved from Guidelines for Drinking Water Quality, 2017 https://www.who.int/water_sanitation_health/publications/drinking-water-quality-guidelines-4-including-1st-addendum/en/.

[48] F. Yang, B. Shi, J. Gu, D. Wang, M. Yang, Morphological and physicochemical characteristics of iron corrosion scales formed under different water source histories in a drinking water distribution system, IWA Water Res. 46 (16) (2012) 5423–5433.

[49] H. Youn Kim, R.M. Clark, Economies of scale and scope in water supply, Reg. Sci. Urban Econ. 18 (4) (1988) 479–502.

CHAPTER 4

Sustainable greywater engineering

Katherine Hyde
School of the Built Environment, University of Reading, Reading, Berkshire, UK

4.1 Introduction

The rational incentives for reusing greywater will continue to strengthen since increasingly the finite nature of clean drinking water supplies begins to more deeply reach into national and worldwide consciousness. Greywater now attracts a description that is widely agreed in much of the academic literature, relating to water from hand basins, showers and baths; having a relatively low organic loading and providing an important water source for reuse. In the context of this chapter, any flowing waste stream of sewage and/or gross faecal matter will be excluded. The definition of the term "sustainable greywater engineering" will focus on its purpose and application whether for resource processing, resource consumption, or operational sustainability.

In this chapter, sustainable greywater engineering refers to both greywater treatment processes, as well as the means of separately conveying untreated greywater flows from treated greywater flows. This ensures that once a stream of raw greywater has been treated, it is prevented from mixing with untreated greywater, thus maintaining and ensuring the acceptable standard of the treated greywater. Without this critical engineering design and precept, a greywater system and network can be compromised. In the UK, for example, this is a statutory requirement under Health and safety regulations for the use of greywater in or close by buildings. The occasional failing of greywater treatment has sometimes been accompanied by misinterpretation and a reversal in technological implementation.

Partly due to the relatively limited amount of contamination in greywater, treatment systems are not unduly complicated. The term "greywater systems" includes all collection and distribution networks, installations, civil works and facilities, pumps, other electrical infrastructure, storage tanks, valves as required to complete safe design, construction and maintenance. The installation of sustainable greywater systems and networks has been proven to generate both good quality and quantities of treated greywater supply. Discussions about the degree of sustainability delivered by any particular greywater supply system should normally be assessed

Sustainable Water Engineering. DOI: 10.1016/C2017-0-04301-X

on a case-by-case basis; different system designs and engineering configurations will have different utility values and local significance.

The value of greywater supplies to users and recipients needs to be more widely recognized and preferably incorporated into water sensitive urban designs and sustainable urban strategies and policies. Sapkota et al. [20] noted that an improved understanding of the interaction between centralized and decentralized urban water services "is expected to provide a better integration of hybrid systems by improved sewerage and drainage design". If investors should choose to invest in a greywater treatment and recycling system, providing a well-proven infrastructure lifespan of say, 40 years rather than an infrastructure norm of 25 years, the planning judgement concerning positive social outcomes for recipients, residents, businesses and agencies/government may become more expedient. Local initiatives such as declarations of climate amelioration should incorporate greywater recycling in order to better equip communities in understanding the basic "rules" and requirements surrounding the use of greywater. This requires specific initiatives in knowledge-based capacity building, and practical upskilling for water recycling engineering; enabling suitable preparation in local governance systems and initiating relevant preparatory sub-urban and urban support infrastructures.

Significant change is required in thinking, expectations, routine designs for the domestic market, and wider market opportunities for non-domestic greywater applications. Unfortunately, market policies exist in certain parts of the worldwide water industry that reject opportunities for greywater recycling. One likely explanation is that volumetric sales and related water product market economics, particularly for potable water supply, would be reduced/adversely affected by widespread greywater recycling. McClanahan [19] provides theoretical perspectives and comments on countries that have overtly "oppressive regulations governing rainwater storage and residential water recycling". Such regulations restrict the development of the greywater recycling market, hampering the migration of technology to market-ready status and entry to the market, particularly the domestic recycling market.

4.2 Approach to greywater systems engineering: policy and practice

The concept of greywater engineering has explicit meaning when optimizing water reuse, water efficiency in buildings and with respect to inflows and outflows in the recycling process and thereby, system sustainability. Commonly, various options may be proposed for addressing each water recycling opportunity since utility, design and longevity for greywater treatment also depends on local conditions and requirements as well as on economics and finance. Retrofitting brings challenges to the domestic market, and ultimately a combination of technologies and essential social education enable these challenges to be met.

The basic principles of water treatment science and engineering are well established, e.g. Tebbutt [25]; Brandt, et al. [2]; American Waterworks Association, https://www.awwa.org/Publications.

Technologies are available and are in development to apply these long-established principles to greywater. Water recycling and reuse opportunities require that:

i. equipment must be designed and fitted to each individual system layout,
ii. equipment should properly manage the volume of incoming and outgoing supplies.

Early in the development of greywater technologies in the UK, designs and installations in domestic buildings were occasionally abandoned during the following years, sometimes due to a lack of ongoing maintenance. In 2012 the EU-DGEnv project [9] reported expectations such that "While initiatives have been taken to increase the water performance of buildings.... progress in the uptake of the measures is still low". The report continued to describe a poor awareness of water supply and quantity issues, since water is "popularly considered an abundant commodity, whose prices do not reflect its value. This leads to insufficient attention being given to reducing wastage and the opportunities for recycling". Whilst the current situation has changed slightly since 2012, the EU report continues to be of significance for countries in setting a benchmark for expectations. It stated that insufficient attention was being given to reducing wastage and the opportunities for recycling. The report noted that "systems are expected to be introduced voluntarily in buildings, with the help of voluntary or mandatory labelling, or minimum requirements for putting systems on the market and/or installing them". This approach was supported by the prior publication of the British Standard BS8525: part 1 2010 [3], and part 2 2011 [4]. In fact, well proven sources of greywater data- experimental, operational and user evidence- demonstrate a long and growing record of good results concerning greywater reuse and treatment.

Almost all UK sectors, apart from military, aviation and shipping continue to lack firm regulatory incentives for the recycling of greywater. The requirements in those environments are necessarily rigorous concerning what must and must not be undertaken and installed on health and safety grounds. Nolde (Personal Communication, September 2016) has pointed out that excellent health and safety records provide a notable and dynamic incentive to the greywater industry, since the record of successful greywater performance demonstrates robustness.

In circumstances where the rationale for installing and maintaining greywater systems is a high priority, financial, research, technical and industry support exists. For example, in Israel it has been reported that up to 90 percent of water supplied from mains water sources is recycled for a range of purposes [22]. Despite slow progress in many countries, greywater engineering globally has been developing with commitment since 2011 with a more significant level of sectoral achievement in areas of the world such as the Middle East.

In countries where there are plentiful supplies of water from all sources, there is less incentive for recycling, installing and maintaining greywater systems. Although partly driven by the free market, the average amount of drinking water supply per person is diminishing across the world and there is little incentive in certain countries and economic sectors to introduce change. Water supply companies may also face lower profitability in their revenue streams should the proportions of reused water increase. Recent evidence suggests (Personal communication, 2019) that even though the water industry in England and Wales has been deregulated, large water providers can marginalize the opportunities for greywater recycling. If up to 40 percent of domestic treated water was made available for recycling with appropriate engineering and/or retrofit, it could potentially lead to an effective loss in terms of production and turnover for water companies.

Several exemplar projects in sustainable greywater engineering were developed in the UK from around 1990. However, recycled greywater originating from hand basins and showers as at the Millennium dome (O_2 building) [23] needs to be distinguished from projects such as those blending recycled blackwater (urinal waste and sewage) with recycled greywater at the Olympic Stadium [13]. Comparing a greywater system with a blackwater system, the treatment of blackwater which contains raw sewage solids is significantly more capital intensive since additional stages of screening and sedimentation and more, are required. Furthermore, in some communities the uptake of greywater recycling may be slowly increasing and being assimilated, but culturally the idea of using recycled blackwater is less palatable and acceptable. There is evidence that a lack of understanding of the differences between greywater and blackwater adversely influences public perceptions about the benefits of recycling greywater [15].

This critical difference demonstrates the need for using the most appropriate terminology so that the public can understand and gain confidence about reuse, greywater quality and meeting greywater standards in small domestic units recycling hand basin and shower/bath water. The British Standard defines appropriate standard terminologies in BS8525-1:2010 [3], the code of practice, and in BS8525-2: 2011 [4], testing they conform to the standards. Other international standards are now available, although they function differently and have different purposes. Some greywater standards are prescriptive such as those in USA (e.g. turbidity < 2 NTU), UK (e.g. turbidity < 10 NTU), China (e.g. turbidity < 20NTU), depending on the reuse application. In contrast, the International/UK water reuse standard for urban areas BS ISO 20760 (2018) includes descriptive principles for safety evaluation for various water quality constituents, including some from highly polluted sources.

Zadeh et al. [27] assessed the potential for an innovative cross–connected water system for water resource savings and economic investment based on theoretical economic feasibility. They evaluated the collection of greywater from residential buildings for recycling in toilet/urinal flushing in both residential and office buildings. The scenarios evaluated are shown in Fig. 4.1.

The scenarios state that mixed-use developments such as accommodation buildings (e.g., residential, hotels, student halls of residence) produce more greywater than they need and the excess can be re-used in other types of buildings where greywater production is lower than

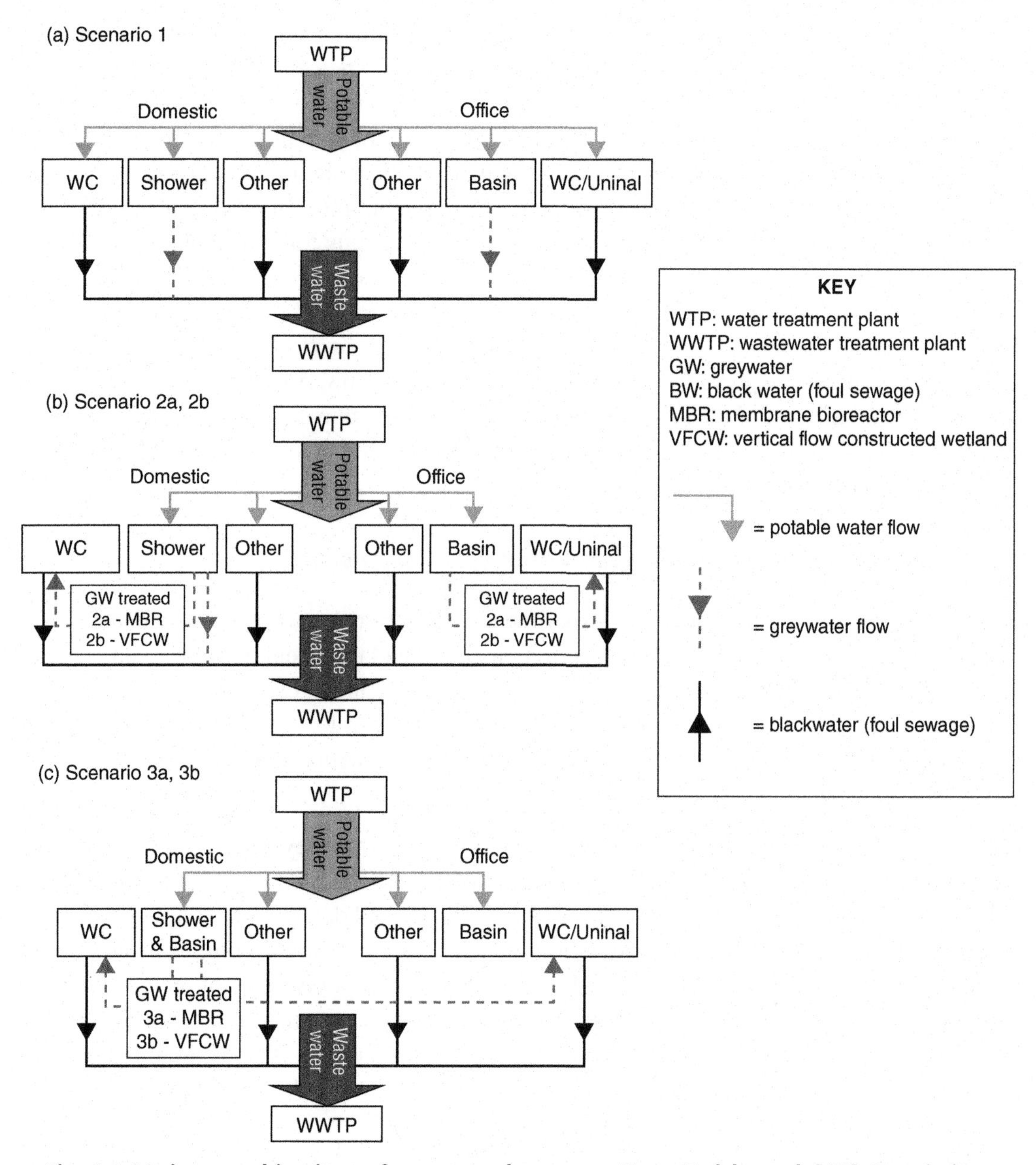

Fig. 4.1: Various combinations of water supply sources. From Zadeh, et al. [27]. Permission: Dexter Hunt, 2020.

demand, such as offices or retail buildings. Therefore, this represents the best prospects for greywater system use.

There are numerous case studies of installed greywater systems in individual family dwellings, multiple housing dwellings, multi-storey office buildings and individual (multi-room) hotel buildings. Toilet flushing is a frequently cited greywater application, not least because toilet flushing in a typical home accounts for approximately 30 percent of domestic water use and can reach over 60 percent in offices. The high volume of greywater generation in domestic properties, accounting for approximately 50 to 70 percent of total domestic daily water discharges, usually exceeds the capacity for using it. Toilet flushing utilises only approximately 20 to 36 percent of the total daily domestic requirement [16]. In other words, up to 50 percent of the daily greywater produced may remain after meeting toilet flushing demands. In contrast, the greywater produced in commercial, retail and other non-residential buildings, which accounts for approximately 21 percent of water discharges (21 percent being from hand basins alone), is substantially less than the requirement for greywater use. Toilet flushing in these properties requires from 43 to 65 percent water inflow. A deficit of greywater is experienced and the cost of the infrastructure and treatment equipment is unlikely to justify the long pay-back periods under current water pricing. However, this is very much influenced by the type of greywater treatment system adopted [12,27].

4.3 Principles of sustainable greywater systems design

Well-proven sources of greywater data–experimental, operational and user evidence–demonstrate a wide range of good results concerning its reuse and treatment. However, almost all UK sectors, apart from military, aviation and shipping continue to lack firm regulatory requirements (or even incentives) for the recycling of greywater. Key knowledge about greywater engineering from these applications could potentially benefit sectoral preparedness, graduate training and industry upskilling.

One basic principle of greywater treatment technologies is that they are very important for providing a flowing (often continuous) source of water. This is in contrast to rainwater supplies that are intermittent, with less easily anticipated and determined volumes, even in the context of accurate weather predictions. There is no guarantee about how much rain will fall at any specific location and in what quantities, under a given set of meteorological conditions. In contrast, greywater comprises a relatively constant stream of water from baths, showers and hand basins particularly at domestic properties. This is an important factor for supporting investment decisions concerning the uses and reliability of treated greywater supplies.

Sustainable greywater systems engineering cannot be separated from greywater quality characteristics and parameters as it flows through the system. The principle that the greywater resource is often obtained from a flow of previously treated piped mains water presents a key factor. Since the water quality of the treated mains water is reasonably well known, the proportion of contaminants acquired through use, whether showering, bathing or handwashing, can be assessed by taking analytical spot measurements of the constituent chemical concentrations in the greywater stream. The difference between these values and their concentration in average treated mains water can then be calculated. In this respect greywater also presents an important contrast with rainwater, often containing higher concentrations of the plant growth nutrients, phosphate, nitrate and potassium. This factor should be quantitatively valued for its considerable benefits for irrigation using treated greywater, although the phosphate concentration in the treated irrigation water should be periodically tested.

In Israel, Luckmann et al. [18] reported that reductions in freshwater resources have led to increased demand for reclaimed wastewater, in this case reclaimed and treated sewage rather than treated greywater. The greater willingness to pay, in a country in which the limited supply of freshwater together with increasing water demand presents such a "link scenario", is used to estimate the marginal value of unpurified sewage [18].

Zadeh et al. [27] provides useful system diagrams as shown in Fig. 4.1 where scenario 1 shows conventional domestic and non-domestic systems having greywater directly combined with more contaminated discharges of foul sewage, in an essentially non-reversible configuration. Scenario 2 shows a domestic situation where some of the treated water from showers is being used for flushing toilets; in non-domestic situations, all of the treated greywater from hand basins is delivered for flushing toilets and urinals. In both cases the greywater is treated using a membrane bioreactor (MBR), and vertical flow constructed wetland (VFCW) treatment units [24]. Greywater research at the University of Reading [14] has demonstrated that MBR treatment will improve greywater quality to meet the BS8525 standard [4], for flushing toilets and urinals without additional VFCW units. Scenario 3 shows all the greywater supply being collected from the shower and hand basin in the residential property, being treated together in a centralized plant and then supplied for flushing toilets and urinals in both domestic and non-domestic properties.

4.4 Significance for sustainability using chemical and biological standards for treated greywater quality enhancement and risk avoidance

Given clear design, and diligent installation and maintenance procedures, the following risks will be eliminated, or where appropriate, minimised;

1) risks to users of mains water arising from potential cross-contamination of mains systems through misconnection or other errors;
2) risks to users of greywater from contamination that could arise from the greywater itself, which might occur if the greywater was not treated appropriately prior to use;
3) other risks to users of greywater described in the literature, for example avoidance of pathogenic aerosols.

All the above risks impose routine demands on greywater manufacturers, installers and users, to make regular checks and tests for greywater quality and system integrity, from installation, through daily and weekly operation and final de-commissioning. Consideration must be given to the use of programmed, frequent measurements of turbidity, together with pH and other routine parameters including microbiological testing.

Many authors have described the concentration and profile of greywater constituents, for example Diaper et al. [5], Diaper et al. [6], Dixon et al. [7], Dixon et al. [8], Friedler [12], Lazarova et al. [16]. However, it is difficult to directly compare outcomes from various studies because different authors list different parameters, use different methodologies and measure constituents from a range of sources including separately from hand basins, laundry, kitchen etc. Also, studies now span approximately 20 years and during this time the constituents of personal care products, and the mix of pharmaceuticals have changed. However, an example of the kind of constituents in a variety of household sources and their concentrations are given in Table 4.1, taken from a review by Abed and Scholz of twelve studies. This shows the large ranges of values, and the differences in concentrations from various household sources.

Table 4.1: Constituent concentrations of typical household greywater, adapted from [1]. The total number of studies is given in parenthesis after the household location, the total number of data points is given in parenthesis after the concentration.

Parameter	Units	Bathroom (3)	Laundry (3)	Kitchen (2)	Hand basin (1)	Shower (2)
pH	na	6.4-8.1 (3)	8.1-10 (2)	5.9-7.4 (2)	nd	7.3-7.8 (2)
Turbidity	NTU	44-375 (2)	14-444 (3)	298 (1)	nd	23-42 (2)
Total suspended solids	mg/L	68-465 (3)	68-465 (3)	134-1300 (2)	nd	29.8 (1)
Electrical conductivity	μS/cm	82-250 (2)	190-1400 (2)	nd	nd	1317 (1)
Dissolved oxygen	mg/L	nd	nd	2.2-5.8 (1)	nd	nd
Biological oxygen demand	mg/L	50-300 (3)	48-472 (3)	536-1460 (2)	109	166-78 (2)
Chemical oxygen demand	mg/L	100-633 (2)	12.8-2950 (2)	3.8-2050 (2)	263	170-575 (2)
Ammoniacal nitrogen (NH_4-N)	mg/L	≤0.1-15 (2)	0.04-11.3 (2)	0.002-23 (1)	9.6	1.0-3 (2)
Nitrate as nitrogen (NO_3-N)	mg/L	0.28-6.3 (2)	0.3-2 (1)	0.3-5.8 (1)	nd	0.05-7.5 (2)
Orthophosphate as phosphorus (PO_4-P)	mg/L	0.94-48.8 (1)	4-171 (1)	12.7-32 (1)	nd	0.02-1.3 (2)

nd, no data.

Common standards define the water quality breakpoints at which quality is acceptable or not acceptable, and against which the test results are measured and compared. When greywater contamination events have occurred, the specific details may be highly localized in terms of the event pathways, and the outcomes. Details of a few such events have been published, for example Fernandes et al. [10] found after maintenance work in the Netherlands in 2001, that the public drinking water system had been unintentionally connected, during maintenance work, to a greywater system carrying untreated greywater. The connection was made in order to use mains water to flush out and clean the renovated pipework. During the operation, bacteria were allowed to contaminate the clean mains water system because fail-safe procedures, labelling and controls were missing.

In that example, the application of chemical and biological sampling and testing for compliance was used during normal operation of the mains system. When the contamination event occurred, sampling and analysis identified the problem, showing that standards had not been maintained. Then remedial action was implemented, quickly applying disinfection and re-testing. Applying standards for treated greywater systems, for quality and risk avoidance led to a change in practices to prevent any future such procedural errors. The significance of the rigorous design and operation of treated greywater systems, and the separation of these from untreated greywater flows, demands stringent controls. The capability to maintain such standards and controls is now routinely demonstrated by greywater systems.

The application of the two parts of the British Standard, BS 8525-1: 2010 [3] and BS 8525-2: 2011 [4] has provided industrial and public confidence in the selection and use of standard greywater system tests and their accurate application for imposing quality measurement and control.

The measurements for permitted concentrations in treated BS8525–compliant greywaters are shown in Tables 4.2 and 4.3 below.

Table 4.2: Chemical and physical quality requirements for treated greywater for type test.

Parameter	Spray applications	Non-spray applications			Testing
	Pressure washing, garden sprinkler use and car washing	WC flushing	Garden watering	Washing machine use	
Turbidity NTU	<10	<10	N/A	<10	BS EN ISO 7027
pH	5–9.5	5–9.5	5–9.5	5–9.5	BS 1427
Residual chlorine[a] mg/L	<2.0	<2.0	<0.5	<2.0	BS EN ISO 7393-2
Residual bromine[a] mg/L	0.0	<5.0	0.0	<5.0	Blue book 218, method E10 [N1]

[a] Where chlorine or bromine is used in the treatment process.

Table 4.3: Microbiological quality requirements for treated greywater for type tests.

Parameter	Spray applications	Non-spray applications			Testing	
	Pressure washing, garden sprinkler use and car washing	WC flushing	Garden watering	Washing machine use	Spray applications	Non-spray applications
E. coli number/100 mL	Not detected	25	25	Not detected	BS EN ISO 9308-1	BS EN ISO 9308-3
Intestinal enterococci number/100 mL	Not detected	10	10	Not detected	BS EN ISO 7899-2 or BS EN ISO 7899-1	BS EN ISO 7899-1

4.5 Degree of consumer acceptance of greywater quality and control

The quality assessment of treated greywater parameters is of critical importance for establishing not only whether greywater meets the relevant quality standards but whether it can satisfy the aspirations of users. Of course, satisfying users' aspirations is an in-principle evaluation; ideally, no-one should be obliged to reuse water if they would rather not, even when the water can meet all the very highest quality standards.

The reclamation of wastewater is an increasingly important water source in many parts of the world. However, the quantity of wastewater or greywater that can be reclaimed depends on the network connections available for diverting greywater away before it drains to foul sewers; or as Zadeh et al. [27] state, before it is connected as an economic agent to the sewage system.

Tortajada and Ong [26] described how public acceptance continues to be a critical barrier to the introduction of recycled water schemes. Survey results from Hyde, Smith and Adeyeye [15] showed that if a greywater supply was difficult to access, or if it was difficult to assimilate into conventional systems whilst accounting for behavioural norms, people could be dissuaded from using the treated greywater. Fielding et al. [11] collated evidence to show that despite,

i) the "clear need to address water security challenges by making alternative water supplies available",
ii) research evidence that "demonstrates the safety of potable recycled water", and
iii) the design and operation of water recycling schemes for example, in Singapore [17], the United States [21], and Windhoek, Namibia [28], a high public acceptance threshold continues to be a barrier to take-up of alternative water supplies and greywater reuse.

4.6 Conclusions

Opportunities for the implementation of greywater use and treatment have continued to strengthen in the first decades of the 21st century. Exemplar buildings implementing greywater systems are gaining BREEAM outstanding and excellent certification, for example in the business centres of major cities such as the City of London, the Netherlands and China. At the high end of building design, greywater recycling and rainwater capture have merit for demonstrating high aspirations in water resource stewardship. Such excellence in water stewardship rightly demands prominence.

There is much opportunity and need for design and implementation to be achieved in harnessing smaller, lower capacity, yet robust greywater treatment units in many applications in the built environment. Sustainable urban strategies and policies that include and thereby, value greywater supplies for individual users and communities need to be recognized. Water sensitive urban designs are being developed that use recycled greywater. Strong, directive and visional messages are required from Governments, together with appropriate policies that support user interfaces, ensuring common understandings that greywater is not for human consumption.

Certain commercial and market policies are rejecting opportunities for greywater recycling, whilst at the same time, local initiatives such as declarations of climate emergency require greywater recycling in order to equip communities, resources and infrastructure owners. Some small-scale projects were abandoned due to the lack of robust maintenance solutions during the early years of the 21st century, and the means to address these problems are now themselves more robust.

Sustainable greywater engineering together with appropriate treatment units, circulating plant/pumps and pipework are all available, though further design development will certainly enable more small scale implementation. Significant and expanding changes are required in expectations, habits and routine designs for domestic and non-domestic markets, and in wider market opportunities for water engineers.

References

[1] S.N. Abed, M. Scholz, Chemical simulation of greywater, Environ. Technol. 37 (13) (2016) 1631–1646.
[2] M.J. Brandt, K.M. Johnson, A.J. Elphinston, D.D. Ratnayaka, Twort's Water Supply, 7th ed, Butterworth-Heinemann, 2016.
[3] BSi Standard, Greywater Systems, Part 1. Code of Practice. BS 8525-1, BSi., London, 2010. ISBN 978 0 580 63475 8.
[4] BSi Standard, Greywater Systems, Part 2. Domestic Greywater Treatment Equipment: Requirements and Test Methods. BS 8525-2, BSi, London, 2011. ISBN 978 0 580 63476 5.
[5] C. Diaper, B. Jefferson, S.A. Parsons, S.J. Judd, Water recycling technologies in the UK, Water Environ. J. 15 (4) (2001) 282–286.
[6] C. Diaper, M. Toifl, M. Storey, Greywater technology testing protocol. CSIRO: Water for a Healthy Country Research Flagship Report Series, 2008. ISSN: 1835-095X.

[7] A. Dixon, D. Butler, A. Fewkes, M. Robinson, Measurement and modeling of quality changes in untreated greywater, Urban Water 1 (4) (1999) 293–306.

[8] A. Dixon, D. Butler, A. Fewkes, Guidelines for greywater re-use: health issues, Water Environ. J. 13 (5) (1999b) 322–326.

[9] EU-DGEnv, Water Performance of Buildings. Final Report Prepared for European Commission, DG Environment by BIO Intelligence Service, Paris, 2012.

[10] T.M.A. Fernandes, C. Schout, A.M.D.R. Husman, A. Eilander, H. Vennema, Y.T.H.P. van Duynhoven, Gastroenteritis associated with accidental contamination of drinking water with partially treated water, Epidemiol. Infect. 135 (5) (2007) 818–826.

[11] K.S. Fielding, S. Dolnicar, T. Schultz, Public acceptance of recycled water, Int. J. Water Resour. Dev. 35 (4) (2018) 551–586, doi:10.1080/07900627.2017.1419125.

[12] E. Friedler, Quality of individual domestic greywater streams and its implication for on-site treatment and re-use possibilities, Environ. Technol. 25 (9) (2004) 997–1008.

[13] S. Hills, C. James, The Queen Elizabeth Olympic Park Water Recycling System, London. Chapter 15 in Alternative Water Supply Systems, in: F.A. Memon, S. Ward (Eds.), IWA Publishing, 2014, pp. 309–349. https://iwaponline.com/ebooks/book-pdf/520888/wio9781780405513.pdf.

[14] K. Hyde, M.J. Smith, Greywater recycling and reuse, in: S. Charlesworth, C. Booth (Eds.), Urban Pollution-Science and Management, John Wiley & Sons Ltd, 2019, pp. 211–220 2019. LCCN 2018028469 (ebook). ISBN 9781119260462 (pdf). ISBN 9781119260509 (epub)9781119260486 (cloth).

[15] K. Hyde, M.J. Smith, K. Adeyeye, Developments in the quality of treated greywater supplies for buildings, and associated user perception and acceptance, Int. J. Low-Carbon Technol. 12 (2017) 136–140.

[16] V. Lazarova, T. Asano, A. Bahri, J. Anderson, Milestones in Water Reuse: The Best Success Stories, IWA Publishing, London, 2013. ISBN 9781780400075.

[17] H. Lee, T.P. Tan, Singapore's experience with reclaimed water: NEWater, Int. J. Water Resour. Dev. 32 (4) (2016) 611–621, doi:10.1080/07900627.2015.1120188.

[18] J. Luckmann, H. Grethe, S. McDonald, When water saving limits recycling: modelling economy-wide linkages of wastewater use, Water Res. 88 (2016) 972–980, doi:10.1016/j.watres.2015.11.004.

[19] B. McClanahan, Green and grey: water justice, criminalization, and resistance, Crit. Criminol. 22 (2014) 403–418, doi:10.1007/s10612-014-9241-8.

[20] M. Sapkota, M. Arora, H. Malano, M. Moglia, A. Sharma, B. George, F. Pamminger, An overview of hybrid water supply systems in the context of urban water management: challenges and opportunities, Water 2015 7 (1) (2015) 153–174, doi:10.3390/w7010153 .

[21] R. Sanchez-Flores, A. Connor, R.A. Kaiser, The regulatory framework of reclaimed wastewater for potable reuse in the United States, Int. J. Water Resour. Dev. 32 (4) (2016) 536–558, doi:10.1080/07900627.2015.1129318.

[22] S.M. Siegel, Let There be Water: Israel's Solution for a Water-Starved World, Thomas Dunne Books, St Martin's Press, 2015.

[23] A. Smith, J. Khow, S. Hills, A. Donn, Water reuse at the UK's millennium dome, Membrane Technol. 2000 (118) (2000) 5–8, doi:10.1016/S0958-2118(00)87565-5.

[24] A. Stefanakis, C. Akratos, V. Tsihrintzis, Vertical Flow Constructed Wetlands: Eco-Engineering Systems for Wastewater and Sludge Treatment, Elsevier Science, Amsterdam/OxfordMA, 2014 ISBN: 9780124046122 eBook ISBN: 9780124046870.

[25] T.H.Y. Tebbutt, Principles of Water Quality Control, 5th ed, Butterworth-Heinemann, Oxford, 1997.

[26] C. Tortajada, C.N. Ong, Reused water policies for potable use, Int. J. Water Resour. Dev. 32 (4) (2016) 500–502, doi:10.1080/07900627.2016.1179177.

[27] S.M. Zadeh, D.V.L. Hunt, D.R. Lombardi, C.D.F. Rogers, Shared urban greywater recycling systems: water resource savings and economic investment, Sustainability 5 (2013) 2887–2912, doi:10.3390/su5072887.

[28] P. van Rensburg, Overcoming global water reuse barriers: the Windhoek experience, Int. J. Water Resour. Dev. 32 (4) (2016) 536–558.

CHAPTER 5

Technical and non-technical strategies for water efficiency in buildings

Kemi Adeyeye[a,*], Ines Meireles[b], Colin A. Booth[c]

[a]Department of Architecture and Civil Engineering, University of Bath, UK [b]Risco, Department of Civil Engineering, University of Aveiro, Portugal [c]Centre for Architecture and Built Environment Research (CABER), University of the West of England (UWE), Bristol, UK

**Corresponding author.*

5.1 Introduction

Climate change, increasing urbanization, population growth, human development and industrialization, and the changing societal behaviours and resource demand patterns are increasing the strain on ecosystems and natural resources, including water [1,4]. The increase in human population and the rapid expansion of urban infrastructure have negatively affected the hydrologic cycle. As societies become more affluent and their lifestyles continue to evolve, humans around the globe are now striving for additional comfort, which demands an abundance of available resources, one of which is water [5,6]. Water supply challenges demand joined up solutions to minimize or even reverse their negative impact. As the need for greater sustainability in urban centres becomes increasingly apparent, new technologies and behaviours, which can contribute to a more sustainable consumption of important resources, are becoming of increasing interest. In particular, ways to reduce consumption of energy and water require the attention of researchers, industry professionals and consumers alike [10].

Integrated solutions for water include better public awareness and engagement, better policies regulations and standards, engineering and technological innovation and improved social, cultural and industrial practices. Water efficient practices by end-users inherently reduce the energy used in the abstraction, treatment, distribution of water. It also reduces water and energy demands associated with wastewater processes, including reducing pollutants and greenhouse gases. To this end, there are standards and regulations to address water practices across domestic and non-domestic sectors, as well as targeting water supply and demand processes. These standards, regulations and guidelines primarily address product engineering and system design and, to a limited extent, installation and use. Addressing the "downstream" aspects of water use can be challenging—multifaceted types of users, behaviours, actions and processes to manage and/or control. However, it has also been realized that water efficiency

Sustainable Water Engineering. DOI: 10.1016/C2017-0-04301-X

cannot be delivered through engineering solutions and the techniques of reduced waste-water flow, reduced piping and fluid machinery size, as well as their installation costs alone [17]. A holistic approach is, therefore, required.

This chapter, therefore, focuses on the design and engineering aspects of domestic water products. It also explores the socio-technical aspects of engineering innovations in water fittings and products to promote water use efficiency. It is widely recognized that technology or engineering solutions alone are insufficient for resource use efficiency, especially if users interact poorly or misuse products and systems. Hence, highlighting the collective importance of good design, installation, maintenance, and use to achieve positive water efficient results.

5.2 Water efficient fittings and products

Water for domestic use is typically consumed through fittings, such as kitchen, basin and bath taps, toilets, showerheads and urinals and products such as washing machines and dishwashers. For some buildings, water can be used in heating, ventilation and air conditioning (HVAC) and landscape irrigation systems. Many agricultural and industrial processes may also use significant volumes of water, but these are outside of the scope of the chapter.

Water efficiency can be achieved through technical and operational interventions. Technical measures include new or retrofitted water-saving fittings (such as, installing low-volume toilets and urinals, waterless urinals, low-flow showerheads and faucets; water-efficient clothes washers and dishwashers in residential domestic water usage…); efficient irrigation systems (such as, the ones having automatic shut-off hoses, rain sensors…). It also covers improved processes, such as replacing wasteful industrial processes with more water efficient ones (such as, cooling towers with recirculated water, systems, which use rainwater or recycled process water…), using drip irrigation systems, recovering tail-water in agriculture, using intelligent water supply systems, which detect leakages and the plumbing element that is to be replaced or repaired [17].

These technological interventions require changes in operational or human practices to succeed. Therefore, water efficiency interventions should include education programs and campaigns to raise awareness of insufficient water supply resources, educating customers about wasteful water use practices. This includes highlighting the importance of shutting off unnecessary flows from showerheads and faucets, using dual flush toilets properly, running cloth and dish washers only in full loads, using water-efficient equipment and of preventing water losses due to leakages and dripping from pipes and fittings, using rainwater, reusing water (greywater usage). Technological solutions like sensors, digital and auditory displays etc. also support education and behaviour change objectives. Lastly, policy and regulatory measures such as dis-incentivizing metering and pricing policies can be used to encourage or "nudge" consumers to save water [17].

The focus of this chapter is on water efficient fittings and products only; regulatory and market forces mean that inefficient devices will likely cease to be available in due course.

5.2.1 Concepts and terminology

Water consumption in fittings is determined by the frequency of use, duration of use (event-duration) and the flow rate of the tap. The frequency of use and the event duration are dependent on user behaviour, while the flow rate is determined by the technology, which in turn is governed by several physical factors, including water pressure and specific tap design [7]. Therefore, in addition to design factors, such as form and aesthetics, the following key hydraulic concepts and principles are necessary to understand the functionality and performance of water fittings and devices.

- Volume: The volume of water (SI unit is m^3. m = metres) is the area of the water multiplied by its height.
- Density: The density of water (represented by the Greek letter "rho" ρ or D; SI unit is kg/m3. kg = kilograms) or volumetric mass density of water is its mass divided by its volume. Therefore, increasing the volume of water only increases its mass, not its density. However, the density of water will change based on its temperature and pressure.
- Pressure: Water pressure (SI unit is Pascal (Pa). 1 Pa = 1 N/m^2. N = Newton) is the perpendicular force exerted by water on a surface divided by the area of that surface. Pressure can be absolute pressure or gauge pressure; the latter is relative to atmospheric pressure. Pressure is equal at any given depth in all directions irrespective of the volume of shape and always acts at right angles. Water pressure is also dependent on flow – whether open channel (e.g. in a river or closed conduit (e.g. in a pipe)). The energy transfer will be different in these situations.
- Flow, Flow rate and Velocity: There are three major types of flow in a water conduit or pipe: laminar flow (i.e. smooth and steady flow in the same direction); turbulent (i.e. fluctuating and agitated flow, which occurs when there is an increase in velocity and/or pipe size, or decrease in viscosity). As well as transition flow, which is an interstate between laminar and turbulent flow. The flow rate is the measure of the flow of water volume in a given time (flow rate $Q = V/t$). Thus, flow rates can be measured in m^3/s or in litres per second (1000 L = 1 m^3). Flow velocity quantifies the rate at which an element of water changes its position; and can be measured in meters per second. The mean velocity of water in a pipe or channel is measured as the flow rate divided by the cross-sectional area of the flow (mean velocity $U = Q/A$).
- Energy: It is common for hydraulic engineers to present the total flow energy in terms of head (SI unit is m). The total head is the total mechanical energy per unit weight and is the sum of the velocity head, the elevation head, and the pressure head. Each of these head terms consists of an energy or work per weight of fluid, respectively, kinetic energy, potential energy and flow work per weight of fluid. The Bernoulli Theorem simply states

that the total head is constant along a streamline for steady, constant density, frictionless flow. When dividing the total head by the acceleration of gravity, the equation is presented in the pressure form: total pressure = dynamic pressure + hydrostatic pressure + static pressure.

A more thorough explanation of these principles and their applications can be found in [13,9,8].

5.2.2 Design and performance specifications

The first step towards successful water management is to establish conservation goals to reduce daily consumption. Engineering products in general work within the constraint of materials, technological, economic, legal, environmental and human-related factors in order to derived optimized solutions [16]. These combined factors are therefore, typically embedded as design and performance requirements in national and international standards, such as:

- BS EN 200:2008 – sanitary tapware – single taps and combination taps for water supply systems of type 1 and type 2 – general technical specification.
- BS 6340-4:1984 – shower units. Specification for shower heads and related equipment.
- BS EN 14055:2018 – WC and urinal flushing cisterns.

Technical guidance for taps, showerheads and WC cisterns are summarized in this section. It is worth noting that the plumbing and water supply system and type (e.g. System 1 or 2), nature of the installation of fittings (e.g. kitchen sink, bath or basin, pipe material, sizing and connectors (connections)), in addition to use, impacts on the performance of water products and fittings.

Taps: Taps (or faucets) are installed and used with sanitary appliances in kitchens, cloak/washrooms, shower-rooms and bathrooms to dispense water for cooking, washing (food, cutlery and crockery), aid cleaning e.g. of hands, teeth etc. They supply water to sinks, wash-basins, baths, bidets etc. Water efficiency in taps is achieved in various ways: Stop taps where water flow stops after a pre-set time (e.g. 3 s). Low flow taps – with reduced control flow. Push action "non-concussive" taps for controlled timed flow water use – water flow reduces over a set time until it stops. Push level flow control for use by hand, knee or foot. Adjustable temperature and flow taps or mixer taps. Sensor or capacitive (touch) taps (see – *Operation*). Aerators can also be connected to the sprout of existing taps to aid water conservation during their use.

- **Form:** The forms of taps are broadly classified as pillar, bib, single (mono-bloc) and multi-hole combination taps. The pillar and bib taps typically provide single supply – hot or cold water. They are less popular with the prevalent use of combination or mixer tabs. In the UK, bib taps are typically used outside the building (e.g. garden taps). When viewed from the front, the global standard in all taps is for cold water to be dispensed via the right tap (or operation to the right in combination taps) and hot water by the left tap (or operation to the left in combination taps).

- **Materials:** Taps are made of a wide range of materials including brass, zinc and steel. Most tap surfaces are typically chromium-plated. All materials used in a tap assembly must comply with relevant standards and should not affect human health or the quality, taste, appearance or smell of the water.
- **Outlet:** Tap outlets can be fixed, moveable, split and can include flow diffusers, aerators, restrictors or regulators.
- **Performance:** In addition to water efficiency, taps are required to function effectively within pressure, temperature and flow parameters. They are also required to endure extensive use/abuse by users. Most taps now incorporate backflow protection devices to minimize water contamination. The type of plumbing system or setup will determine the dynamic and static pressure range and flow rates of the different types of taps. The manufacturer determines pressure ranges by testing the mechanical performance of the tap under extreme pressure. The flowrates for a partially or fully open tap are specified according to system type and pressure. The typical flow rates for combination taps are summarized in Table 5.1.

Table 5.1: Typical flow rates for combination taps (excerpt for combination taps only from BS EN 200:2000 Table 10).

Combination taps (each side tested separately)	Taps for supply system of Type 1 Flow rate	Taps for supply system of Type 2
Basin, Bidet, Sink (water saving)	4–9 L/min (0.066–0.15 L/s)	3-6 L/min
Basin, Bidet, Shower, Sink	For taps with a pull out spray attachments or flexible hoses, a maximum flow rate of 0.15 L/s applies	7.5 L/min
Bath, hot or cold mixed	19 L/min (0.32 L/s)	15 L/min
position, valves fully open	20 L/min (0.33 L/s)	22.5 L/min

Showerheads: Due to the diversity in human preferences and socio-demographical needs, showers are now comparatively more popular than having a bath. Studies, like [2], have shown that the design (i.e. colour, shape, and size), as well as the performance criteria (i.e. flow and pressure of a showerhead product) can influence user perception, acceptability and experience of water efficient products and this can in turn affect the extent to which these products promote in-use water efficiency.

Showerheads are typically mounted perpendicularly to the wall. A shower assembly typically consists of the showerhead spray plate and arm, a fixed height riser/pipe, or a flexible hose with fixed or variable riser or wall holder. Risers are typically fixed at 1.2 or 1.5 m above the finished floor level. A mixer or thermostatic tap or controls (valve) to which the hose or riser is connected may be included as part of the assembly.

- **Form:** Showerheads are typically circular in shape but they now come in different shapes and sizes including donut, oval, rectangular. The main types of showerheads are; removable, swivel or fixed. The various types now available in the market as listed below [3]:
  - **Single head:** This type of showerhead may have a single setting or more than one setting. Settings often include more and less focused sprays and a pulsating spray.
  - **Multiple-head shower:** These fixtures may have two or more spray nozzles connected to one pipe. They can easily replace a single head fixture.
  - **Cascading showerhead:** They often are mounted overhead, such that the water drops straight down. They typically give a softer spray and have diameters of 15–20 cm.
  - **Shower panel or shower tower:** These are designed to spray water from more than one location having more than one showerhead. They may operate sequentially.
  - **Rain systems:** Rain systems simulate rain by allowing water to fall from an overhead fixture.
  - **Body spas:** Body spas consist of multiple showerheads and are described by some as the vertical equivalent of a spa. The showerheads may be activated sequentially or intermittently.
- Although the removable shower handset is popular in houses, some shower assemblies offer combined options of removable or fixed. Showerheads in particular have become increasingly complex over time and this has influenced the volume of water uses in showering. As shown below, even eco- or water efficient showerheads come in varying types and offer different flow rates/patterns, pressure, temperature ranges etc.
- **Materials:** Showerheads can be made of metals including copper, steel and ABS plastics. They are typically chrome or nickel-plated. Any materials used in showerheads should ensure safe use including when handled.
- **Outlet:** The type of spray outlet is an important technical criterion for water saving. Most showerheads offer a wide range and arrangement of sprout types to promote water efficiency without compromising water use efficiency. The spray outlets also support the single, dual or multi-mode functions of the showerheads. The form, mode and trajectory showerheads include atomizers, aerators i.e. mixing water and air, or form-induced or mechanical means to reduce the amount of water delivered without compromising the user perceived distribution, temperature, pressure and volume of water delivered to the body.
- **Performance:** The spray type, form (distribution) and trajectory have been found to be important performance criteria, which could affect user satisfaction [2]. Therefore, the *BS EN 6340-1:1984 Shower units. Guide on choice of shower units and their components for use in private dwellings* specify the testing criteria for these parameters. For instance, guidance is different for each assembly type, and includes requirements like avoiding

leakage where bath tap are part of the assembly. The BS EN 6340-4:1984 also states that for the two types of standard showers (type 40 and 70) and under the minimum pressure head conditions specified by the manufacturer, the combined flow rate performance should be:

- Type 70. Flow rate should be 12 L/min (0.2 L/s); with the controls set to give a temperature of 40 °C at the shower head the water flow rate should be not less than 0.07 L/s at 0.1 bar pressure.
- Type 40. Flow rate should be 4 L/min (0.1 L/s); with the controls set to give a temperature of 40 °C at the shower head the water flow rate should be not less than 0.04 L/s at 0.1 bar pressure.
  NOTE: Where a type 70 or type 40 incorporates an instantaneous water heater, the flow rate referred to above relates to incoming water supply to the heater at 5 °C. With inlet water temperatures above 5 °C and with higher energy rating heaters, the water flow rate at a satisfactory outlet temperature will be higher.

WC flushing cisterns: Water for flushing toilets can be 30–35 percent of the total water consumption of a household. This is not only significant in terms of using highly treated water to flush toilets but also its environmental and economic implications. A toilet assembly consists of the cistern, the toilet bowl, flushing apparatus and water supply/waste connections.

- **Form:** The form of the cistern depends on its material. Traditionally, metal and ceramic cisterns were used. However, plastic cisterns are increasingly more common. This is due to the increased versatility in how WCs are deployed in buildings (e.g. cisterns hidden in walls or cabinets).
- **Materials:** WC systems can be made out of ceramics or plastics. The thickness for vitreous and non-vitreous ceramic cisterns should not be less than 10 mm and 12 mm respectively. Plastic systems should be non-corroding but also protected against the corrosion action of the water. The possibility of electrolytic action should also be avoided by design or manner of installation.
- **Performance:** Most modern cistern offers dual-flush capabilities to minimize waste during flushing. Therefore, the first category of performance of a cistern is defined by its capacity 3 L/6 L or 4.5 L/9 L. Most WCs are now designed in line with national standards or to achieve building performance ratings. Although float operated valves were common, valve-less siphonic flush mechanisms are now prevalent to maximise water savings. In addition, dual flush cisterns are provided with air inlet devices, which interrupt the siphonic action to deliver the shorter/lower volume, flush. Other performance metrics are rate of discharge, appearance, colour, opacity, dead load and impact capacities. For instance, the BS 1125:1987 Specification for WC flushing cisterns (including dual flush cisterns and flush pipes) states a cistern should not distort if subjected to a dead load of 23 kg for 30 s. All cisterns should be watertight and prevent drips or leakage into the toilet pan.

5.2.3 Installation and maintenance

All fittings should be mounted or installed such that they are mechanically secure, leak and drip-proof. This means the correct sizes of holes, bases, flanges, and outlets should be used. These sizes are specified in national and international standards. It is also good practice that a hydraulic test is conducted when the system is commissioned. Adequate and safe means of access should be provided for maintenance and repair.

5.2.4 Operation

Fittings (such as taps (or faucets), showers, toilets) can be operated manually or using sensors, automatically. Manual operation is typical in domestic buildings, whilst automatic operation of fittings is increasingly popular in non-domestic buildings. If a water device is to be manually operated, the ergonomics and tactility through form and materials are important engineering design and performance principles. The ease of reach and handling should be considered for a wide range of users, including people with disabilities, especially those with mobility and dexterity challenges. Fittings and systems should also be intuitive to use as people rarely have the time, inclination or patience to read complex or multi-page manuals.

Sensor-activated systems help to promote water savings, deliver operational standards, improve durability, reduce vandalism, minimize repairs, and deliver better hygiene. The latter due to reduced contamination or re-contamination and the spread of germs and other infections from users touching fittings in high-traffic kitchens or restrooms.

Two main types of sensing technologies are used: Active Infrared (IR) and Capacitance (C). Active Infrared sensing operates when a part of a user's body (e.g. hand) reflect an invisible light beam, activating the flow of water. Infrared models are designed to provide easy, above deck access to key components, and can offer additional user enhancements.

Capacitance sensing utilizes the human body's own natural conductivity whereby the fitting is activated when it senses the body. There is no sensor window and critical components are protected in a watertight, below-deck box. In some instances, metal fixtures, or the proximity of large metal objects, can inhibit proper functioning of capacitance technologies [12].

Water devices with sensors also deliver other associated functions. These include timing – e.g. a timer to control the dispensing duration, or to stop the flow of water when the user is no longer present. Volume and flow control – i.e. how much water is dispensed at what rate per each activated period, mixing – the percentage of hot and cold water dispensed, data and user-experience logging etc. The choice of sensors should consider location, user group, aesthetics, ergonomics, and cost. Power supply is also an important consideration and opportunities to capture/recapture energy from renewable sources should be considered without compromising the reliability of systems.

It is worth noting that for both manual and automatically operated fittings, the manner and degree of affordances offered to the user is important to ensure correct use and to achieve the desired results. In engineering design, affordances are the actions or functions, which are offered or presented to users (or which they perceive are available to them) and the constraints or limits on their behaviour provided by the system [11]. As the following case studies will show, poorly perceived affordances will result in user dissatisfaction and negative environmental behaviours. For instance, poorly timed showers in sporting centres, ineffective low-temperature quick wash cycles on washing machines etc. Other features of automatic appliance include visual displays to provide performance information, persuasion or use feedback towards efficient behaviours. Although the rebound effect – where users respond negatively by consuming more – has been reported, in many cases, extremely simple informational descriptive feedback – even more useful or easily visible markings or calibrations on a product – have the potential to make it easier for users to be more efficient [11].

5.3 Case study

The case study building is the Civil Engineering Department (DECivil) building in the University of Aveiro (UA) Santiago Campus (Fig. 5.1). Approximately 300 students and staff use the building, with around 180 people occupying the building at one time during office hours. The building is rectangular, as most of the buildings on the campus. The ground floor is largely occupied by a laboratory, the second floor by classrooms and the third floor by classrooms, offices and a smaller laboratory.

This case study example is used to present the interaction between the technical and use factors in achieving water efficiency. The methods include water efficiency audits, water efficient tap intervention, followed by user feedback through surveys.

Water efficiency audits are used to calculate or determine the water savings potential in any given building. Audits include determining the existing flow rates of appliances followed by quantifying the volumetric water savings in each efficient fixture. Volumetric water savings are typically estimated through the mass balance between the original water consumption and the corresponding savings [14] as shown in Eq. (5.1):

$$V_x = \sum_{i=1}^{n}\left[\sum_{j=1}^{12}\left(MC_j \times PC_i \times PR_i^X\right)\right] \tag{5.1}$$

where:

V_x is the potential for annual water savings due to water efficiency measure X (m^3)

MC_j is the water consumption in the building in the jth month (m^3)

PC_i is the water consumed in the ith fixture in percentage of the water consumed in the building (percent)

PR_i^X is the effective reduction of water consumption in the ith fixture due to the water efficiency measure X (percent).

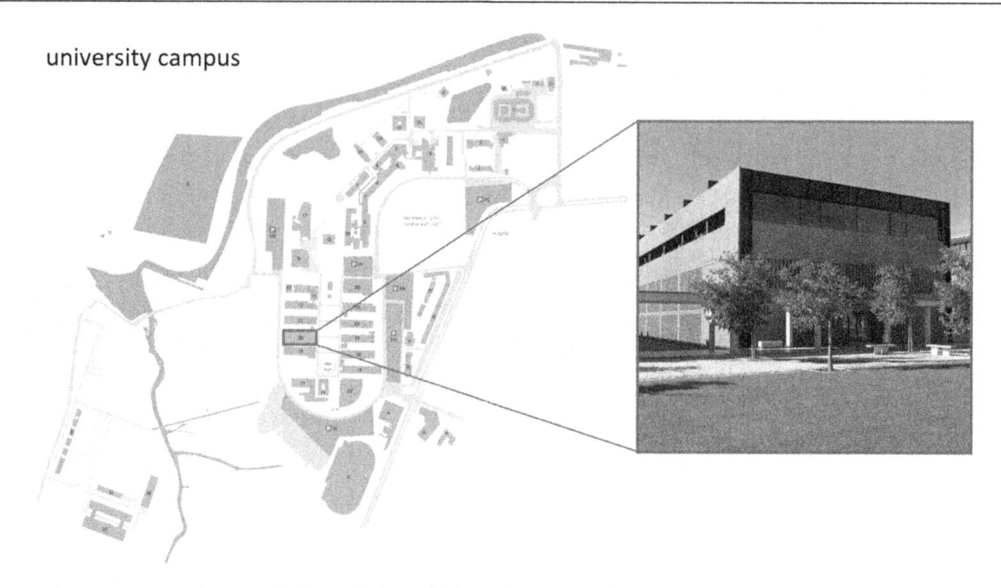

Fig. 5.1: Location and view of the Civil Engineering department building at the University of Aveiro.

This approach generically assumes a proportionality between the volumetric flow rate of the device (or volume, for the case of cisterns) and that of the efficient device. This could result however, in a significant overestimation of the potential savings. Therefore, the following statements apply:

1. A change to an efficient fitting will not directly contribute to water savings if the water use is volume driven (e.g. floor cleaning, cooking, laboratory use).
2. If the use is linked to human satisfaction (e.g. showering, washing hands), the importance of the satisfaction with the device is much more important than the volume flow rate/volume delivered.
3. A significant reduction in the design volume flow rate/volume of the device has to be linked to an improvement on all the connected devices. Otherwise, malfunction or poorer performance may inadvertently result in reduced savings, or an increase in water consumption. For instance, a low flow showerhead can result in longer showers duration. Similarly, a low volume cistern has to be designed in connection with the toilet bowl, or it will not function correctly, and force the user to flush more times to clear the bowl.

This chapter considers the technical design and performance of water efficient fittings in connection with user interaction with the product. This case study utilizes tap/washbasin audits, interventions and user feedback in the case study building to demonstrate the interaction between these factors. Apart from the changing and shower facilities from the main laboratory, the building has a set of toilets at the main entrance of the building, in the first floor and in the second floor. Water is used in the changing and shower facilities, toilets, and laboratories (Table 5.2).

Table 5.2: Characterization of the water devices in the DECivil building.

Changing and shower facilities	Floor	Washbasin	Toilet	Urinal	Shower
Laboratory 1					
LB1	0	2	1	2	1
LB2	0	2	2	–	1
Toilets	Floor	Washbasin	Toilet	Urinal	
Gentlemen					
GT1	1	2	2	4	
GT2	2	3	2	3	
GT3	2	2	1	3	
Ladies					
LT1	1	3	2	–	
LT2	2	2	2	–	
LT3	2	2	2	–	
Disabled					
DT1	1	1	1	–	
DT2	2	1	1		
Laboratory 1					
LB1	0	2	1	2	
LB2	0	2	2	–	
Secretary					
ST	0	2	1	2	
Laboratories	Floor	Laboratory taps	Sink tap	Hose tap	
Laboratory 1	0	2	2	1	
Laboratory 2	2	4	–	–	

The discussion centres only on the consumption of water in the washbasins, since the normal uses in the laboratories and cleaning are volume based and the volume spent in toilets and urinals are more directly connected to the characteristics of the products than to human behaviour. In particular, the study was focused on the 14 washbasins of the six main water closets (GTs and LTs in Table 5.1), since the consumption in the other water closets was negligible. The baseline situation was determined through audit and the four different water efficient fittings were then studied during two subsequent academic years.

The existing laminar flow push taps were found to have a discharge rate of 6.7 L/min and a shut off time of 6.1 s. Easily assembled aerators were then added to the push taps to reduce the discharge flow (Fig. 5.2):

1. Aerator A – aerated flow with $Q = 4.7$ L/min
2. Aerator B – spray flow with $Q = 3.9$ L/min
3. Aerator C – aerated flow with $Q = 3.4$ L/min; and
4. Aerator D – spray flow with $Q = 2.0$ L/min.

Note that these values are averages: Each tap was tested four times in different hours of the day, to take into account differences in pressure. This procedure was performed for the baseline case and for the four different aerators assembled to the existing taps. Differences of

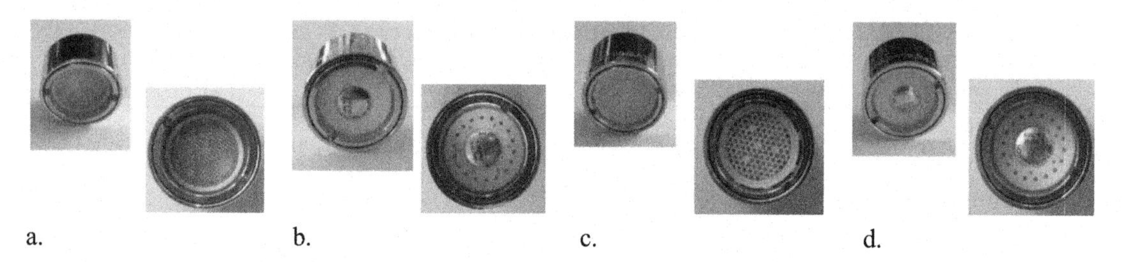

Fig. 5.2: Characteristics of the different aerators: (a) aerator A (aerated flow; Q = 4.7 L/min); (b) aerator B (spray flow; Q = 3.9 L/min); (c) aerator C (aerated flow; Q = 3.4 L/min); (d) aerator D (spray flow; Q = 2.0 L/min).

up to 7 percent in the volume flow rate and of up to 23 percent in the total volume of water consumed per push were observed.

The study was performed in two stages during two academic years: the first focused on analysing the effective reduction in the consumption of water when the different water efficient fixtures were installed, while the second focused on the satisfaction and preferences of the users. Metered data were combined with diary monitoring performed 8:30 am to 6:30 pm every Tuesday and Friday for 9 weeks during the summer semester. In addition, user feedback were captured using short surveys. Questions included perceptions of use and the number of pushes to activate water flow. After the study, the taps were returned to the original baseline situation.

In the subsequent academic year, a similar study was undertaken except that the user feedback captured water consumption, the users' sensitivity to fixture changes and in the user's satisfaction when using the different aerated taps. Feedback was obtained using three online questionnaires, made available during the Summer exams period, to accompany the changes in the taps of all the main water closets: i) the first for aerator D, ii) the second for aerator C; and iii) the third for the baseline tap. Each questionnaire was made available after a usage period of at least one week, and the questions focused on the users' sensitivity to the change and their levels of satisfaction. In each monitoring campaign, the potential population of users was the same, but there was no way to ensure that the samples were statistically equivalent at the onset. By using the Chi-Squared test to compare the sample of users in each monitoring campaign, [15] found there was no statistically significant differences in terms of age ($\chi^2(15) = 11.572, p = 0.711$) or gender ($\chi^2(4) = 2.306, p = 0.680$).

The findings are summarized as follows:

1. **Consumption reduction:** Irrespective of aerator type, a reduction in water consumption was observed when compared with the baseline. As expected, Aerator D performed better (i.e. dispensed less water compared to the others and the baseline scenario) [15]. Further, water consumption at the washbasins were on average 46 percent less than the flow rate reduction achieved with each aerator. In fact, while the aerators contributed to flow rate

Table 5.3: Relation between flow rate and consumption reduction.

Aerator	Discharge reduction	Consumption reduction (PR_i^X)	Rel. diff. discharge and consumption reduction
A	30 percent	15 percent	51 percent
B	42 percent	17 percent	60 percent
C	49 percent	27 percent	44 percent
D	70 percent	49 percent	30 percent

reductions between 30 percent and 70 percent, the reduction on water consumption was only between 15 percent and 49 percent (Table 5.3). The percentage of reduction in water consumption was not equal to the percentage of reduction of the flow rate when using a water efficient fixture, in comparison with the baseline. Therefore, it was confirmed that user factors influence water savings. It also confirms the risk of overestimating the savings potential of technological interventions alone.

2. **Sensitivity to changes.** Surveys were used to determine whether users were sensitive to the changes made to the taps (Fig. 5.3). The results showed the change between the baseline and aerator A was barely noticed, almost half of the respondents noticed a change between aerators A and B, and majority noticed the change to aerator C and raised questions about a possible problem with water pressure in the building soon after the change. Almost all the respondents noticed the change to aerator D. It should be noted that no survey was performed after aerator D was removed and the tap returned to the baseline. The three online questionnaires were completed by 100, 86 and 81 users, respectively. Again, since there was no control over the users replying to each online questionnaire, the Chi-Squared test was performed to compare the sample of users in each case. There were no statistically significant differences in terms of age ($\chi^2(10) = 6.603$, $p = 0.762$),

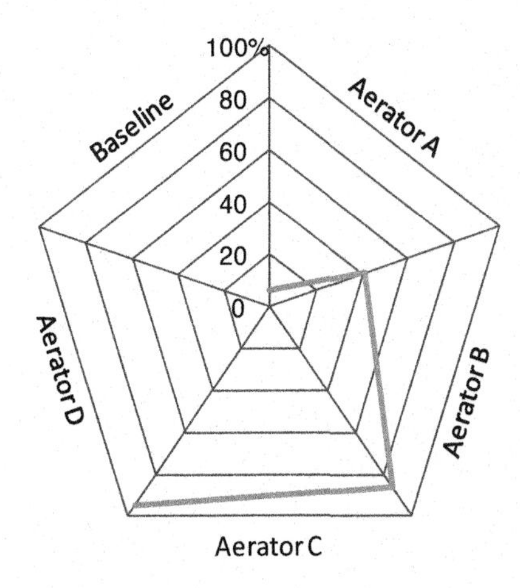

Fig. 5.3: Direct survey: sensitivity to changes in fixtures (in percentage).

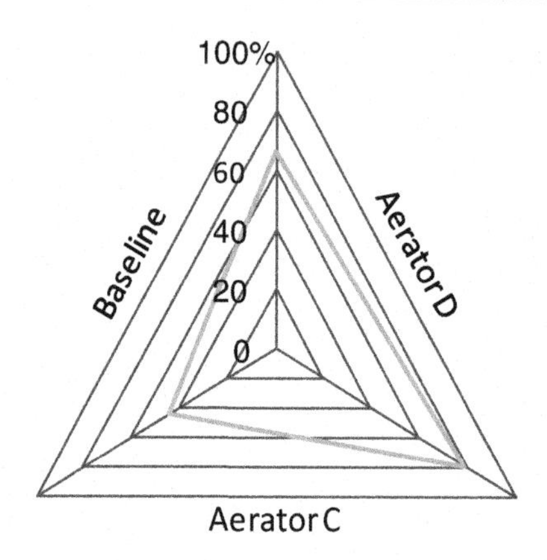

Fig. 5.4: Online survey: sensitivity to changes in fixtures (in percentage).

professional position ($\chi^2(8) = 5.270, p = 0.728$) and gender ($\chi^2(2) = 0.689, p = 0.709$) between the samples of respondents. The respondents' perception of fixture change and preference results were found to be statistically significant between the baseline and the aerators (perception: $\chi^2(2) = 18.138, p < 0.000$; preference: $\chi^2(8) = 15.852, p = 0.045$). Between aerators, the respondents had a statistically weak perception of flow rate change ($\chi^2(1) = 3.217, p = 0.073$) and there was no statistically significant difference on the preference results ($\chi^2(4) = 0.601, p = 0.963$). The respondents were mainly sensitive to flow change (Fig. 5.4), especially between aerators C and D (79 percent), followed by the change between the base situation and aerator D (66 percent) and the change between the base situation and aerator C (44 percent). This finding suggests users might be more sensitive to flow type than flow rate.

3. **Satisfaction.** The users were asked to rank their satisfaction with the aerators in a scale of 1 to 5 (1 – very dissatisfied to 5 – very satisfied). None of the users were very dissatisfied with any of the aerators but the satisfaction rate progressively reduced from aerator A to D (Fig. 5.5). It should be noted that the satisfaction with the baseline tap was not directly surveyed. Using an online survey, the satisfaction with the use of aerators C and D were investigated, then compared to the satisfaction with the baseline tap (Fig. 5.6). Less than 15 percent of the online questionnaire respondents stated they were very satisfied with the use of aerators C and D, as opposed to the 46 and 45 percent of the users in the direct questionnaire. A possible explanation is the fact that for the direct survey, the aerators were installed by decreasing discharge, so the users had some time to progressively adapt to lower discharges. For those completing the online questionnaires, the aerators were installed by increasing discharge, and the users were faced with a drastic change from the baseline to aerator D, the one with the lowest flow rate. Nonetheless, only about 12 percent of the users negatively classified aerators C and D in the online questionnaires. In addition, less than 20 percent considered the base situation

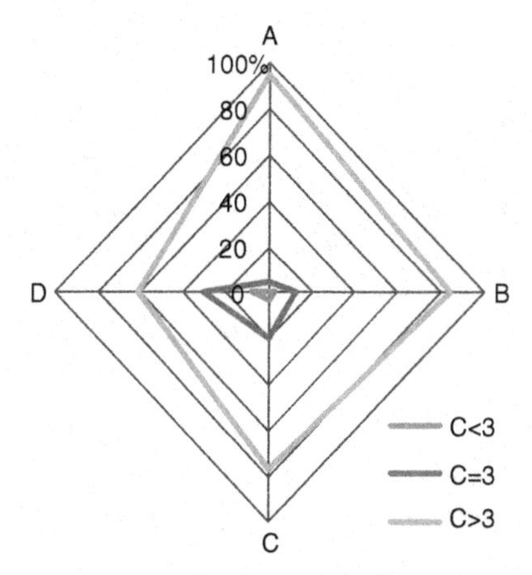

Fig. 5.5: Direct survey: User satisfaction with the use of the different aerators (blue – satisfaction's level below 3; red – satisfaction's level equal to 3; green – satisfaction's level above 3).

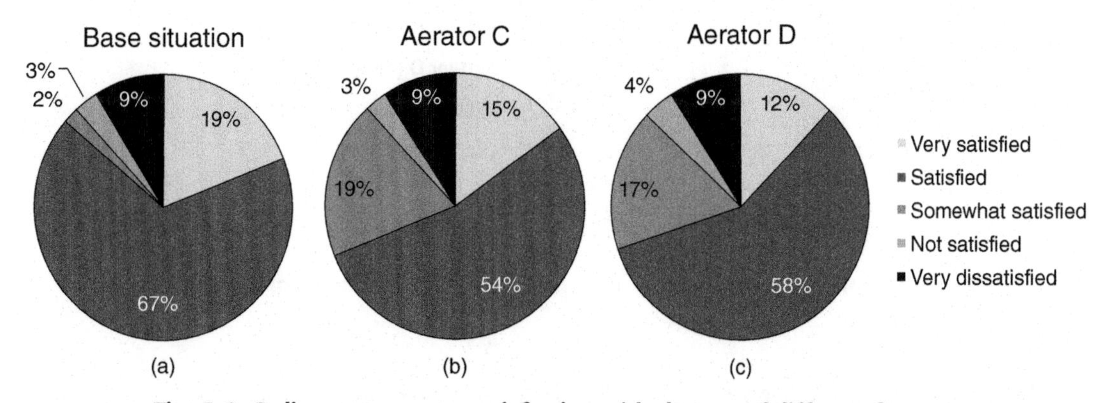

Fig. 5.6: Online survey: user satisfaction with the use of different fixtures.

very satisfying and 5 percent classified it negatively, even without the reports of water splashing occurrences.

4. **Preferences.** As presented above, no statistically significant difference was found on the preference results when aerators C and D were compared in the online surveys. For resolution and clarity, photographs of the taps with these aerators and corresponding flow was added to the questionnaire (Fig. 5.7). The results (Fig. 5.8) show that 50 percent of the respondents preferred aerator D to aerator C, compared to 28 percent that made aerator C their choice. This suggests some users would be happy with a lower flow tap perhaps in order to save water. This point is further explored in the next section.

Fig. 5.7: Type of flow: spray (left); aerated (right).

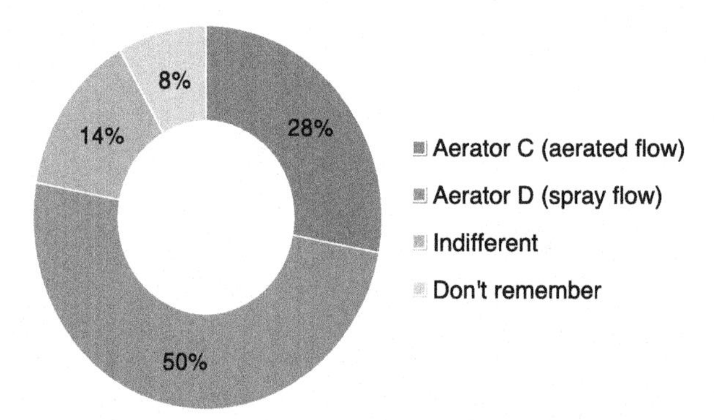

Fig. 5.8: Updated online survey: user preference between aerators C and D.

5.4 Further insights into the uptake of water efficient devices

Technological invention of new water efficient devices provides an opportunity for users to make sizeable water savings. However, to achieve this, they need to adapt or replace existing inefficient devices with new ones, and then use them as the designers intended. To reach that point, consumers need to know that new devices are available to buy, understand the advantages new devices provide and, most importantly, want to purchase and install them. To explore these issues, it is necessary to be conversant with market demands and user intensions. Therefore, a postal questionnaire, sent to 1000 households in a large UK city (with a 12.5 percent response rate) was used to gather the views of a wide cross–section of household residents about their opinions and intentions towards domestic water efficient devices and to also gauge their general attitudes towards domestic water shortage.

Analysis of survey results reveal most participants claim they know how to save water (78.4 percent) and recognize it is their own responsibility to save water in their homes (85.6 percent).

However, disappointingly less than half of them say they are interested in installing new domestic water saving products. Of those households who have already installed water saving products, many (44.0 percent) claim to save water because they are on a water meter. Most worryingly, almost a third of all survey participants share the belief they should simply be allowed to use as much water as they want because they pay their water bill.

A wealth of water saving devices are now widely available. Therefore, it is noteworthy to report purchasing intentions towards water efficient products for personal or household use. For instance, very few survey participants indicated they plan to buy water efficient washing machines (0.8 percent) or dishwashers (1.6 percent), water saving spray taps (0.8 percent) or self-closing taps (0.0 percent), water saving dual-flush toilets (2.4 percent) or low-volume cisterns (2.4 percent), and only 1.6 percent want to buy aerated shower heads. Moreover, despite the earlier claim that most survey participants know how to save water, when further questioned, a much greater proportion of participants claim not to know about the various water saving products available to them on the market. Products such as washing machines (21.6 percent) and dishwashers (25.6 percent), spray taps (38.4 percent) self-closing taps (40.0 percent), dual-flush toilets (11.2 percent), low-volume cisterns (15.2 percent), and aerated shower heads (30.4 percent). This illustrates a worrying situation where only a very low proportion of householders want to buy new domestic water saving devices and a much greater proportion of householders simply do not know about the many new domestic water saving devices available.

The survey attempted to understand further the decision-making process behind consumer choices towards new domestic water saving products. The survey participants indicated sale price (70.4 percent) and water saving potential (66.4 percent) are the most important factors when choosing their next domestic water saving products. It is reassuring to know that both factors are considered in almost equal importance. However, the participants also claim they will not refurbish or replace their existing fixtures and fittings until they are needed (60.8 percent) or unless the new devices are subsidized (67.2 percent). Therefore, it is unlikely there will be continuous improved reductions in domestic water consumption if the water industry is entirely reliant on the uptake of new domestic water saving products.

Household behaviours' and attitudes towards domestic water usage are also important aspects of understanding consumer decision-making. For instance, 32.0 percent of survey participants claim that they are not worried about water shortage, despite 46.4 percent of them believing that the frequency of national droughts are increasing. Most survey participants acknowledged public education (74.4 percent) and changing personal behaviours (70.4 percent) as essential approaches to managing future water usage and many (60.8 percent) recognized the adaptation of fixtures and fittings especially in homes as a necessity. However, the shared household opinion (63.2 percent) that poor water management by the water companies is the primary cause for water shortages probably needs attention for domestic water use changes to occur. This is further reinforced by many participants calling for greater water infrastructure investment (60.0 percent), including the construction of new reservoirs (66.4 percent), as the best means of solving drought and water shortage issues. This presumably underpins why very few survey participants claimed to make domestic water usage changes (behaviour or new products) even during (8.8 percent) or after (13.6 percent) drought and water shortage events.

5.5 Conclusion

Innovative water efficient devices offer a viable solution towards water savings in buildings. There is now a wealth of new water saving products and devices being designed and manufactured for market uptake. This chapter has provided an overview of the design and technical specifications of water saving taps, showerheads and toilets, and has given an indication of the benefits and potential water savings that can be achieved by improving the water performance of the fittings. Further, the chapter has highlighted that users are an important consideration in water efficiency solutions. They need to be considered during the design of the products with user-centred decisions made about the appearance, material, demographics, ergonomics and ease of installation. In addition, in-use preferences and experiential factors should be considered, as this will affect the extent to which users adopt and use the product to achieve its intended water savings.

The uptake of new domestic water saving products and devices is reliant on consumers purchasing, installing and using them. The drivers for the transition from existing non-efficient products to new efficient products also needs further study. This study has revealed that whilst households continue to believe poor water management is the primary cause for domestic water shortage, attempts to encourage changes in personal behaviours and attitudes towards domestic water saving is likely to remain a considerable challenge and substantial burden for the water industry—product designers, engineers and water providers alike. Unless needed or subsidized, some users would not buy or install water saving products, either because they do not know about them or because they do not intend to refurbish or replace their existing fixtures and fittings.

Transforming consumer thinking and behaviours from being wasteful to being efficient goes beyond advancement in engineering and technologically innovative solutions. Strategies should include user engagement, raising public awareness and implementing effective policy and market/marketing strategies.

References

[1] K. Adeyeye, J. Gibberd, J. Chakwizira, Water infrastructure for human settlements: urban – non-urban linkages and water access, Abstract Proceedings of 2019 International Conference on Resource Sustainability - Cities (icRS Cities), 2019. Available at SSRN https://ssrn.com/abstract=3403494 13 June 2019.

[2] K. Adeyeye, K. She, A. Baïri, Design factors and functionality matching in sustainability products: a study of eco-showerheads, J. Clean. Prod. 142 (2017) 4214–4229.

[3] P.J. Biermayer, Potential Water and Energy Savings from Showerheads, Lawrence Berkeley National Laboratory, 2005. Online at: https://escholarship.org/uc/item/79f5g8ds Accessed: 05 April 2018.
[4] C.A. Booth, S.M. Charlesworth, Water Resources for the Built Environment: Management Issues and Solutions, Wiley-Blackwells, Oxford, 2014.
[5] S.M. Charlesworth, C.A. Booth, Water resources issues and solutions for the built environment: too little versus too much, in: C.A. Booth, F. Hammond, J.E. Lamond, D.G. Proverbs (Eds.), Solutions to Climate Change Challenges in the Built Environment, Wiley–Blackwells, Oxford, 2012, pp. 237–250.
[6] A. Fernández-González, Water use and conservation. In: Low Energy Low Carbon Architecture, CRC Press;, 2016, pp. 215–236.
[7] A.M. Fidar, F.A. Memon, D. Butler, Performance evaluation of conventional and water saving taps, Sci. Total. Environ. 541 (2016) 815–824.
[8] D.D. Gray, A First Course in Fluid Mechanics for Civil Engineers, Water Resources Publications, LLC, 2000.
[9] R. Greeno, F. Hall, Building Services Handbook, Routledge, 2013.
[10] A. Liu, D. Giurco, P. Mukheibir, Urban water conservation through customised water and end-use information, J. Clean. Prod. 112 (2016) 3164–3175.
[11] D. Lockton, D. Harrison, N. Stanton, Making the user more efficient: design for sustainable behavior, Int. J. Sustain. Eng. 1 (1) (2008) 3–8.
[12] P. Lozier, Electronic sensor faucets improve hygiene and conserve water in commercial restrooms, Architect. 105 (1) (2016) 74–77.
[13] R. McMullan, Environmental Science in Building, Palgrave Macmillan Education, 2017.
[14] I. Meireles, V. Sousa, Assessing the environmental performance of water efficient measures in buildings: a methodological approach, Environ. Sci. Pollut. Res. (2020), E-Publication ahead of print. doi:10.1007/s11356-019-06377-3.
[15] I. Meireles, V. Sousa, K. Adeyeye, A. Silva-Afonso, User preferences and water use savings owing to washbasin taps retrofit: a case study of the DECivil building of the University of Aveiro, Environ. Sci. Pollut. Res. 25 (20) (2018) 19217–19227, doi:10.1007/s11356-017-8897-5.
[16] G. Pahl, W. Beitz, Engineering Design: A Systematic Approach, Springer Science and Business Media, 2013.
[17] M. Sönmez, Two offers to prevent excessive water consumption: a proposal for industrial design departments of universities, Int. J. Water Resour. Environ. Eng. 8 (2) (2016) 24–31.

CHAPTER 6

Cities running out of water

Jeremy Gibberd
Built Environment, CSIR, Pretoria, South Africa

6.1 Introduction

The United Nations indicates that water scarcity now affects every continent. Many areas have reached the limit of local water services because water use has increased at twice the rate of population growth [43]. It is estimated that a quarter of the population of large cities, or over 38,155 million people, have water supplies that are under stress [24]. Major global cities (figure in bracket indicates population) such as Tokyo, Japan (36,933,000), Delhi, India (21,935,000), Mexico City Mexico (20,142,000), Shanghai China (19,554,000), Beijing China (15,000,000), Kolkata, India (14,283,000), Karachi, Pakistan (13,500,000), Los Angeles, United States (13,223,000) and Rio de Janeiro, Brazil (11,867,000) are now considered water-stressed [24]. These cities accommodate economic activity to the value of $4.8 trillion or 22 percent of the global economic activity of large cities. However, despite its strategic importance, there are few studies on water stress in cities [24].

This chapter, therefore, explores water stress in cities. It aims to understand what happens when water becomes scarce, or runs out, in cities and describes the impact of this on government, business and communities. Reasons for temporary and long term water scarcity in cities are determined and described. Mechanisms to address water scarcity are identified and discussed to propose a simple framework that can be used to understand and address water scarcity in cities.

6.2 Cape town – a city that almost ran out of water

To understand what happens when cities face severe water shortages, it useful to look at a recent example. In 2018, Cape Town, a major city in South Africa with a population of over 4 million people was within 3 months of running out water. There was little sign of this in 2014 when the city's dams were full and water scarcity was not a concern. In 2015 the city was recognized as a leader in water management and conservation and given an award by C40 [21]. C40 is a network that aims to drive sustainable action on climate change in cities [7]. The city's award mentions initiatives related to raising public awareness, introducing new water tariffs, carrying out free plumbing repairs for low-income families and promoting alternative

Sustainable Water Engineering. DOI: 10.1016/C2017-0-04301-X

water sources such as borehole water and recycled water within the city. It also commends the city's efforts to improve water infrastructure including better asset management, water pressure management, pipe replacement, leak detection and improved meter management.

However, in January 2018, the city realized it had only three months' supply of water left and officials announced that 'Day Zero', the day the city would run out of water, would be in April 2018. Day Zero, the council indicated, would mean shutting off water to most businesses and homes and relying on an emergency water ration of 25 L per person that would be available from 200 standpipes in the city. Police and the army were told to be on standby to deal with unrest [21]. So how could a city that appeared to be managing water well in 2014 and 2015, run out of water only three years later?

This rapid change can be attributed to a range of factors. A drought had been developing in the Western Cape and rainfall between June 2015 and June 2018 was 50 to 70 percent of the long-time average [46]. Rainfall in 2017 was particularly limited with records indicating this was the lowest since written records started in the 1880s [46]. The city's water systems were also inappropriate for a modern city with a limited water supply [58]. For example, in 1998 households used 37 percent of the total water supply and 21 percent of this was used to irrigate gardens and fill swimming pools. The city had also experienced a steadily increasing population which grew from 2.4 million people in 1995 to 4.3 million people in 2018. Demand for water from agriculture and wine production, which attracted 1.5 million visitors annually, had also grown rapidly and used about a third of the available water in 2018 [37]. Despite warnings to increase the capacity and diversify water sources, limited additional water resources were developed and Gasson [19] indicated that the City would breach its supply limits as a result of growth by 2025.

Governance related to the allocation of water also contributed to water shortages. The responsibility for water supply in South Africa is divided between national and local government. In the case of Cape Town, these arms of government were run by different political parties. This not only led to a fragmented capability for long-term planning but also to a poorly coordinated response to the drought. As a result, the water supply to farmers was not restricted in the first two years of drought, leading to a severely diminished water resource [40,37].

Therefore, preserving the remaining water for as long as possible became an intense focus of the City who announced a water ration of 87 L per person per day in January 2018. By February, this was reduced to 50 L per person per day. The City developed projections that charted when water would run out and Day Zero would occur, as shown in Fig. 6.1 [15]. The Mayor had daily water meetings and a crises communications consultant was appointed to draw attention to the consequences of Day Zero [51].

In February 2018, the allocation of water to agriculture was reduced and farmers agreed to divert some of the additional water they had stored to the city [1]. The city also implemented a pressure reduction system, which led to water savings of 10 percent. As a result, a

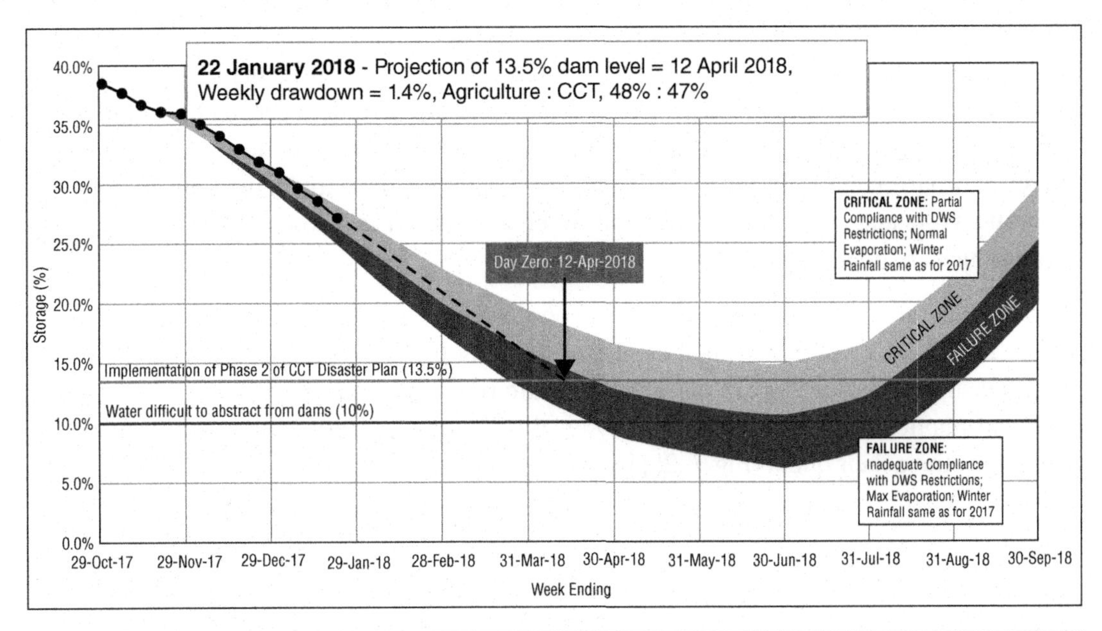

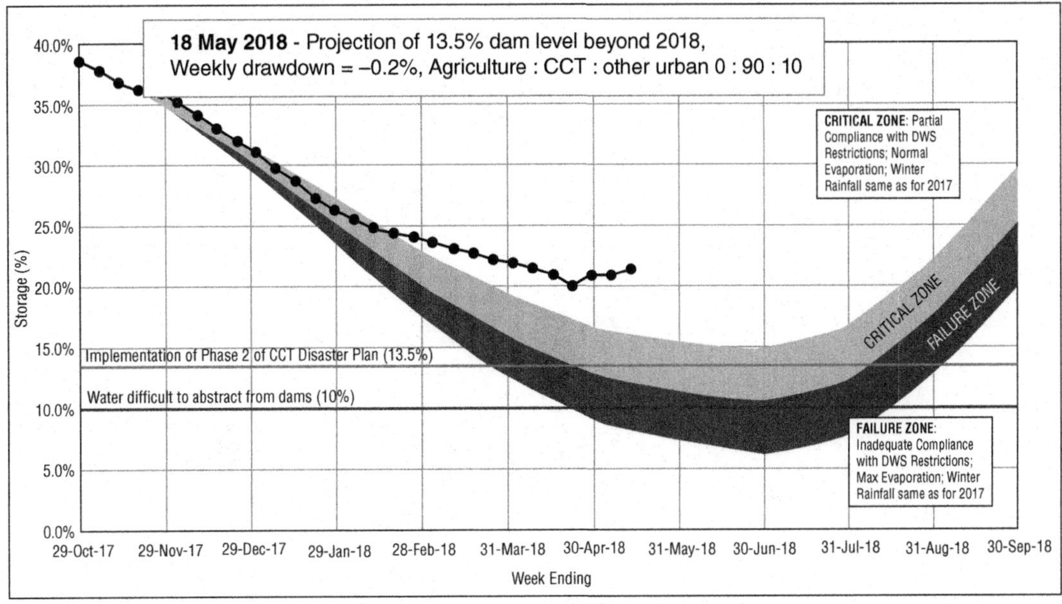

Fig. 6.1: City of Cape Town's Projection of day zero in January 2018 and the situation in May 2018 [15].

significant reduction in water consumption was achieved and by May 2018 the City was no longer in the critical zone and could last until rains fell in September 2019 [17].

While these macro factors help to explain how water shortages developed in the City, they do not explain the remarkable reductions achieved in domestic household consumption.

[44], show how water consumption from free-standing households in Cape Town dropped by almost 50 percent in three years from 540 L per household per day in January 2015, to 280 L per day in January 2018. They also show how wealthy households reduced consumption to the level of the poorest households. While investment in boreholes and rainwater harvesting systems may have reduced water use in some areas, this does not explain the radical reduction across all sectors and income levels. This has to be understood as a complex set of interacting factors, which are explored next.

A key contributing factor to reduced domestic consumption was the water restrictions instituted by the City. Cape Town banned the use of drinking water for irrigation and washing cars and instituted personal water consumption rations. Online maps were developed to show water consumption at an individual property level-scale which identified households who were going above their quota (Fig. 6.2; [15])

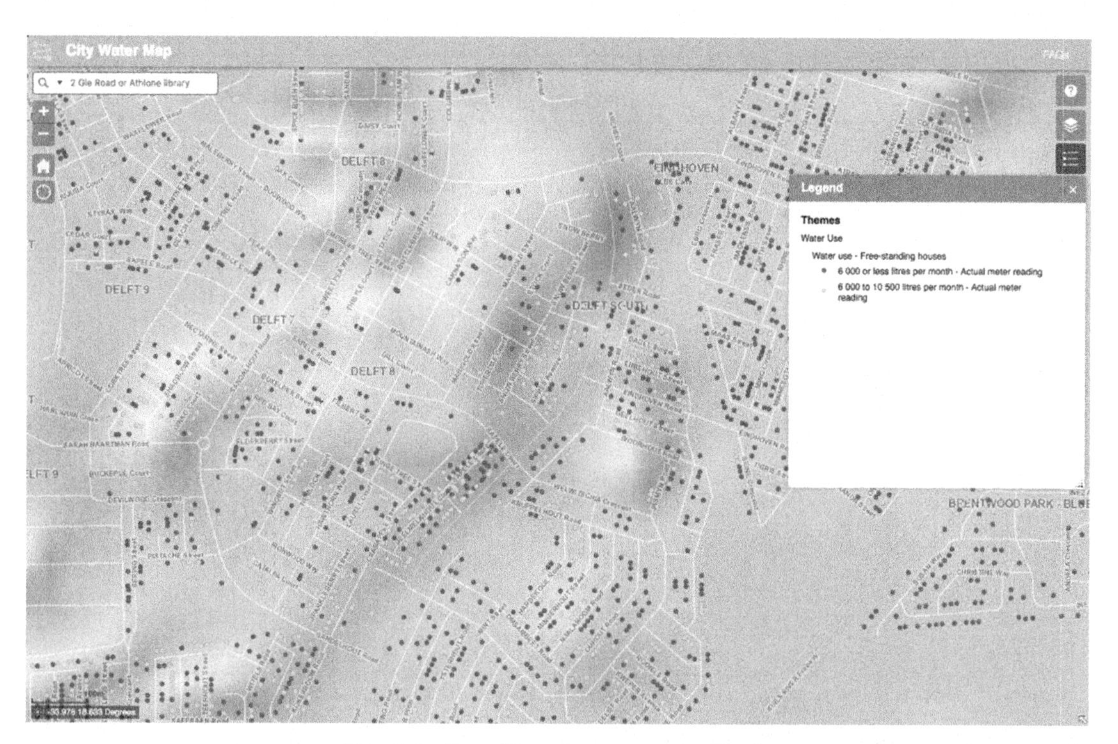

Fig. 6.2: Cape Town's water map [15].

Not only were restrictions widely publicized; they were also enforced through patrols and fines for infringement. Where there was an ongoing disregard for restrictions, the City installed water management devices that automatically cut off the water supply after 350 L had been provided per day and 18,000 of these devices were installed. Leading up to 2018, the City also radically raised water tariffs with a 400 percent increase for most users and an over 2,500 percent increase for large consumers [52]. Water pressure was also substantially reduced at specific times of the day by the municipality. This led to some higher-lying areas being without water at particular times of the day. Municipal swimming pools were closed and the topping up of private swimming pools was only allowed if this was done using water brought in by tankers from other areas. Hotels were asked to encourage tourists to reduce water use; in many hotels, baths were plugged to stop them being used and showers were encouraged instead. Restaurants changed the way they prepared and served food, with some developing 'Drought Kitchen' menus of low water food options served from cardboard plates [6].

There was a well-publicized 'Day-Zero' campaign that indicated what would happen if water supplies ran out. This described every-day experiences the people would face, including having to walk to fetch water in buckets from standpipes for daily use. Political leaders made a point of showing a commitment to saving water and went on popular TV shows to talk about what they were doing to save water [29]. Water ambassadors were also appointed that publically supported reductions in water and shared new ways that water could be saved [10]. The City and other organisations ran media campaigns on how to save water. Water-saving tips such as "If it is yellow, let it mellow" were widely shared to encourage people to reduce water consumption by not flushing toilets after every use [13]. NGOs like the World Wildlife Fund developed guides on how households could reduce water consumption [48]. Residents also started groups to share water-saving tips and how to access technologies such as composting toilets and greywater systems. Many of these groups grew rapidly through Facebook and other social media platforms.

The example of Cape Town shows that water shortages are possible even in cities that appear to have effective systems in place. It indicates how urban growth, drought and lack of coordination between bodies responsible for water supply can combine to lead to major water shortages. It also demonstrates how a city can implement very effective measures to rapidly reduce water consumption. Finally, it confirms the importance of having a population that is prepared to radically change their lifestyles to make very substantial water savings.

6.3 Why do cities run out of water?

As the example of Cape Town shows, water shortages in cities cannot be attributed to only one or two causes but are usually a result of a complex combination of factors. In the next section, these factors are identified and discussed.

About 78 percent of large cities obtain their water from surface sources [14]. In these cases, the city's water supply is directly affected by what happens in their supply watersheds, which can be many kilometers away [24]. Understanding of this watershed and exerting control over how this is managed is therefore important. For instance, the spread of high-water consumption, non-indigenous species within catchment areas can result in significant reductions in water flow. In South Africa, the spread of invasive species, such as Wattle, Eucalyptus and Pine, has led to concerns that this could result in reduced stream flows by 21–50 percent [39]. Erosion within catchment areas can also be a factor, as runoff containing silt reduces useable volumes of water in dams. For example, during the 2018 drought, the volumes of silt in the dams supplying Cape Town meant that the last 10 percent of water in dams was not useable [35].

Understanding variable weather requires capacity to comprehend water systems and forecast how they will behave under different conditions and scenarios. For instance, evaporative water loss in dams can be substantial but varies widely depending on local temperatures and wind conditions. In Cape Town, higher temperatures projected under future climate change conditions are predicted to lead to water losses through evaporation of up to 25 percent of the total volume of dams per annum [32].

Variable weather also makes it difficult for cities to manage water supplies and can lead to an ongoing cycle of shortages. For instance, Bulawayo, in Zimbabwe has experienced a cycle of scarcity for the last 80 years and the city has imposed water restrictions in 1938–43, 1947, 1951, 1953, 1968, 1971–73, 1983 and 1990 and rationing in 1949, 1984, 1987 and 1991 [11,26]

Climate change has made projections of local variability more complex. Traditionally, water supply forecasts could be made based on historical records but changing conditions under climate change has meant that this approach is inadequate and more accurate future projections are required [50]. However, global climate change models do not detect local climate variability and are of limited use in city water supply modeling [12]. Local climate change projections are, therefore, required and are being developed for some major cities, but are generally not available for many cities.

Rapid urban growth has often meant that rising urban populations often outpace the ability of cities to provide sustainable water supplies. Gasson [19], for instance, shows how growth would have naturally resulted in Cape Town running out of water by 2025. Urban growth in developing countries often occurs in the form of informal settlements and the rapid development and unregulated nature of these areas make it difficult to predict water demand with any accuracy in many cities in Africa, Asia and South and Central America.

With economic development, the consumption of water per capita increases rapidly as more people acquire technologies such as washing machines and dishwashers that use water [24]. This growth has resulted in China identifying water supply as being one of the key challenges facing many cities [41].

Water demand projections are also increasingly complex. Water consumption patterns can vary considerably with urban and climate conditions. Also, the extent to which water savings are achieved is closely related to behavior and the extent to which residents cooperate with local government and water utilities. Projected reductions can be highly inaccurate where there is little or no cooperation from users and this is not been taken into account in modeling [27].

Poorly constructed and maintained infrastructure can result in significant water leakage, which increases the frequency and severity of water shortages in cities. For instance, in South Africa, a water-scarce country, estimates indicate that 35 percent of the country's drinking water is lost through leakage from ageing water infrastructure [16].

Often there are competing demands for water. In the case of Cape Town, farmers were in direct competition with the city for water and late restrictions of their supply were identified as a key contributing factor to the major water shortages in the city in 2018. A policy to support urban agriculture in Bulawayo, Zimbabwe is also thought to have exacerbated water shortages [11]. Showers [33] shows how the need to maintain electricity supplies from hydroelectric plants may affect the availability of water supplies for local towns in several African countries.

Water policy and governance can have a major influence on water resource planning and management. A lack of clarity about the responsibility for planning and managing water resources between different spheres of government can lead to inadequate and uncoordinated action [27]. This was identified as one of the key causes of severe water shortage in Cape Town in 2018 [37]. Elected local government officials, such as mayors and councilors, may complicate this further by being reluctant to impose water restrictions and punitive demands on residents to curb water demand as this may affect their popularity and chances of re-election [27].

Many cities do not undertake detailed water resource planning and are therefore unprepared when conditions, such as drought occur and result in water shortages. There is a tendency only to address water resource planning when water supplies are already strained [27]. Water resource planners may also be unaware of water forecasts and how these can be used for planning. They may also distrust the accuracy of these reports [8]. Further, Ziervogel et al. [50] highlight that while water resource planners may be aware of climate change, they do not always have the tools to integrate this systemically into planning and management processes.

Water infrastructure in many areas is severely constrained by a lack of financial resources. Finances for increasing the capacity of water resources such as new dams and for reducing consumption by addressing leaks and for installing more efficient water equipment may be limited [42]. A lack of investment in water infrastructure over time results in the deterioration of equipment and significant maintenance backlog and investment arrears. For instance, in 2018, the National Water & Sanitation Master Plan of South Africa estimated investment arrears in water and sanitation infrastructure was R898bn as a result of a lack of investment. This is against an annual water infrastructure development budget for 2020/2021 of R13bn [16,38].

In many cities, there is a severe lack of capacity within water utilities and municipalities to plan for and manage water resources. For instance, a study of South African municipalities indicates that only 55 of 278 municipalities had engineers leading their technical divisions

[16]. A lack of capacity can also result in an increased reliance on consultants, resulting in reduced accountability and a loss of institutional memory [27].

A lack of foresight and structured contingency plans for water shortages make it difficult for cities to cope with future water shortages. Mukheiber [27], for instance, shows how towns in the Northern Cape failed to plan for water shortages and as a result overused their groundwater supply in 2000. This depleted an important resource that was needed for future droughts, resulting in many municipalities having to provide water by road tanker to communities in 2004.

6.4 Plans and practices for water-scarce cities

An understanding of why cities run out of water can be used to develop avoidance and mitigation plans. A review of the how cities plan and manage for water scarcity show that a wide range of approaches may be applied. These mirror the complexity of the factors that cause water scarcity in the first place. Approaches proposed to avoid water scarcity in cities are discussed next.

An independent water regulator has been proposed as an effective means to plan and regulate water resources. A regulator would act as a custodian of bulk water supplies, manage the quality of water resources and make independent volumetric allocations [16]. A regulator would also make decisions on the pricing of water to ensure the long term sustainability of supply. Experience from the UK and the USA, however, indicates that regulators tend to focus on water quality standards and the regulation of costs and that their effectiveness as independent adjudicators of water allocation has not been proven [36].

Ziervogel et al. argue that water resource planning in cities must now be informed by effective climate monitoring systems. These should provide early warnings to cities of impending water shortages based on factors such as successive dry seasons. This enables cities to prepare and put measures in place to address water scarcity well in advance.

Climate change scenarios based on the best available local climate change models need to be developed to inform planning. These should highlight climate change impacts such as the effect of changes in temperature, wind and precipitation patterns on water resources [50]. Global-scale climate change models are of limited use as these do not provide sufficient detail and downscaled local models are required [12]. However, the development of local weather, climate and water resource models and scenarios are complex and is likely to be beyond the competencies available within local water utility companies or municipalities. It, therefore, may be useful for authorities responsible for water resource planning to collaborate with local universities, research organisations or consultancies to build effective long-term capacity within this area.

Once developed, climate change guidance must be effectively translated into planning and management practices. This requires the provision of guidance in a useful and usable format that can be used to inform short term, seasonal and long term decision-making [50]. For instance, short-term forecasts can be used to help decide where water should be abstracted from and how water may be moved between dams to ensure its efficient use. Seasonal forecasts would inform the need for water restrictions and their type and level. Long-term forecasts would inform plans to increase water resources, such as the need for an additional dam, as well as requirements for the wide-scale application of more water-efficient fittings, such as toilets and showers, which could be addressed through city by-laws, where such mechanisms exist.

Municipal adaptation plans (MAPs) are local plans that aim to ensure that adaptations required for sustainability water supplies are integrated into development plans and policies [26]. They make sure that water supply resilience is addressed by all relevant government departments and at the required macro and micro levels. MAPs can also ensure early warning systems are in place to alert local government and communities of potential water shortages.

Building regulations and bylaws may be developed to reduce water consumption by banning wasteful water uses such as swimming and ornamental pools and certain types of car washing equipment and ensuring that water delivery fittings and appliances such as toilets and showers are efficient. Regulations can also require building owners to install rainwater harvesting and greywater systems to reduce the consumption of potable mains water. Tax incentives, tariffs and other financial subsidies help reduce the cost of installing and upgrading water infrastructure and fittings.

Water restrictions can be used by cities to manage water consumption. This is carried out by reducing supply to particular times or limiting flows. This approach was used with a significant impact in Cape Town, South Africa [51]. It is, however, a controversial measure as it may be seen to impinge on basic human rights. Therefore, it is likely only to be applied in extreme circumstances.

Pricing can be an effective mechanism to manage water consumption and may be regulated to match water resource availability. Thus, increases in the cost of water would be applied to reduce water use and wastage. Additional revenue generated can also be used to develop more reliable and resilient infrastructure [16]. Private finance may also be sought to ensure there is sufficient investment in water infrastructure. This, however, requires appropriate policies, regulations and contracting certainty [16].

Public awareness has been advocated as an effective way of responding to changing water supplies. Making communities aware of reducing water resources helps to ensure that people are supportive of initiatives that respond to water shortages when these are required [11].

Leakage management and water infrastructure upgrades can be used to reduce wastage and improve efficiencies [24]. In some areas, such as South Africa, studies indicate savings of up to 30 percent can be achieved [16]

Water catchment areas can be better managed to improve run-off and reduce erosion. This can be achieved by removing 'water-thirsty' alien tree species from catchments and installing erosion control measures. In South Africa, invasive alien plants have been found to reduce 7 per cent of the flow in South Africa's rivers and this is being addressed through the government's 'Working for Water' programme which uses unemployed people to clear alien vegetation such as Wattle and Pine [3].

Alternative water sources help to reduce the pressure on existing water resources; thereby ensuring that supplies are more resilient. Alternative water supplies include rainwater harvesting at a building level, street and neighborhood level [23]. Another source of water is through desalination which is often considered expensive. However, [53] shows that desalinated water can cost less than the penalty costs paid by users for excessive consumption during 2018 drought in Cape Town. Improved treatment of wastewater can supplement existing water supplies and cities such as Windhoek in Namibia have relied on this approach for many years with the first recycling plant being developed in 1968 [35]. Groundwater abstraction through wells and boreholes can also be used. This, however, should be regulated through, for instance, Water Use Licences (WULs) to ensure that supplies are sustainable. Continual monitoring of aquifers by municipalities enhance the understanding of their behavior under differing climate conditions and ensures that this contributes to the diversity and resilience of local water resources [54].

6.5 Water resilient cities

The preceding review of why cities run out of water indicated a wide range of interacting factors. The complex relationships between these factors and the diverse role-players involved are key challenges that cities face in developing more water resilient systems [30,47]. In this section, this complexity is addressed and a framework for more resilient water systems in cities is proposed. This consists of a) governance and collaboration for water resilience, b) planning and management for water resilience c) resilient water resources d) resilient infrastructure e) water resilient water use and f) resilient water treatment.

6.6 Governance and collaboration for water resilience

Water governance refers to the management framework within which decisions are made and shared. It refers to "the processes of interaction and decision-making among the actors involved in a collective problem that lead to the creation, reinforcement, or reproduction of social norms and institutions" [7]. A lack of governance can lead to dysfunctional water systems and Berg [4] attributes poor water resource management, limited coverage and inadequate water quality standards that characterizes water systems in many developing countries to poor governance.

In cities, water governance must provide leadership and a consistent, robust and collaborative system for planning and managing water systems. Effective water governance aims to ensure all parts of the water system in the city are integrated and coordinated. This is particularly important in water-scarce environments where a lack of communication and coordination between different parties can easily lead to water shortages.

Water governance structures that span different layers of government, the private sector and non-governmental organisations (NGOs) may be necessary to ensure effective accountability, coordination and integration of a city's water system. The Cape Town example shows the importance of being able to bring different stakeholders together to address water scarcity and to use their influence to drive change [51]. [55] show how informal, decentralized governance approaches work better in the early transformation stages of developing more water resilient systems in Australian cities, while formal and more centralized governance is more effective later. Appropriate governance structures build the public trust required to make behavioral changes when these are required.

Water governance and collaboration structures can be designed to support adaptive management and learning. These help to ensure that learning from successful practices are integrated into management processes and organisational structures to ensure that water resilience is responsive and does not rely on a "command and control" approach [51]. Relationships and networks built through effective governance are not only useful for addressing emergencies but also for creating a shared vision for a water resilient city and adopting practical initiatives like bylaws on water-efficient devices.

6.7 Planning and management for water resilience

Water planning and management refer to capabilities within the city to manage and plan water systems. This capability requires an understanding of all of the parts of the system to ensure these work together to maintain sustainable water supplies in the city. [56] shows how cities such as Rotterdam in the Netherlands address cross-scale water resilience issues by developing guidance and plans at regional, urban, neighborhood, and building scales. Similarly, regulations in the UK require water companies to produce a Water Resources Management Plan (WRMP) that must ensure there is a balance between supply and demand even in very dry years [57].

Water planning and management professionals must be able to understand current water demand patterns and accurately project these into the future based on data related to seasonal, climate change, economic activity and population immigration and growth. Similarly, the capabilities to understand the nature of current water resources and how this will behave in the future based on climate change, seasonal weather and water catchment area are important. Where this capability does not exist within municipalities, appropriate partnerships must be developed with universities and research organisations to ensure that appropriate guidance is developed and applied. In particular, specialist knowledge for understanding and projecting urban growth as well as for developing and interpreting local climate change projections for the city's water systems are valuable. Water engineering expertise, and the management support and administrative systems that ensure that these can be effectively applied, must, therefore, be in place [36]. An understanding of the funding implications of developing more water resilient systems and an effective means of accessing these funds are also required.

City water planning and management also need to identify, refine and apply the key mechanisms, or levers, to balance water resources and demand within the city to ensure sustainable supplies. These mechanisms include catchment area planning, water allocations and rationing, water supply pressure management, water bylaws, water pricing and awareness campaigns.

6.8 Water resilient resources

Water resources for cities include dams, rivers, groundwater and seawater (through desalination plants). Enhancing supply resilience requires a detailed understanding of these resources and their behavior under different conditions. For example, a good understanding of water catchment areas may result in management interventions such as the removal of alien vegetation and erosion control measures to increase run-off and avoid silt build-up in dams. It also includes identifying and monitoring alternative water resources within the city, such as water from rainwater and greywater systems, which can be used to supplement water resources within a city [22,45]. Initiatives to enhance rainwater harvesting in cities are increasingly widespread. Water restrictions, rebates and water pricing in Australia have led to about 1.7 million households fitting rainwater tanks providing a capacity of 156 GL of onsite stored water [9]. Large-scale GIS mapping surveys have also been carried out in Botswana, Ethiopia, Kenya, Malawi, Mozambique, Rwanda, Tanzania, Uganda, Zambia, and Zimbabwe which show significant potential for rainwater harvesting [23]. In Japan, the Great East Japan Earthquake and strong financial incentives from the government led to many households installing rainwater tanks as an emergency supply [59]. A pilot project in China has built 2 million rainwater tanks with a storage capacity of 73 million m^3, providing drinking water for 2 million people. This has been scaled up in a programme that has built more than 5.5 million rainwater harvesting tanks [20].

6.9 Water resilient infrastructure

Bulk infrastructure includes pipes, reservoirs and pumping stations used to transfer water from the source to where it is used in cities. This infrastructure can be planned and managed to support a water resilient supply to cities in several ways: Firstly, ensuring that high-quality materials and equipment are specified and maintained minimizes leaks and wastage within the system. Secondly, controlling water pressure and flows in the system can be used to reduce water consumption. Work by Nazif et al. [28] in Tehran suggests that pressure management can be used to reduce leakage by more than 30 percent a year.

Thirdly, in extreme conditions, parts of the system can be shut to minimize water consumption, while still allowing water flows to essential services such as hospitals and standpipes. This approach, however, is problematic and may be subject to legal challenges [34].

6.10 Water resilient usage

Water usage refers to how water is used in cities and includes industrial and agriculture uses, irrigation in parks and gardens, filling swimming pools, cleaning buildings and cars, as well water used for washing, drinking and flushing toilets. A range of mechanisms can be used to reduce water usage in these areas. These include bylaws and regulations that limit the installation of high water consumption activities such as swimming pools and car washes as well as detailing the efficiency requirements for fittings such as toilets and showers [5,31]. There is also an increasing number of standards and green building rating systems that provide guidance and water efficiency benchmarks [2,49]. Water restrictions can also be applied to limit water usage to defined amounts and this can be reinforced by penalties and water supply shut-off devices. Pricing can also be used to manage water usage and escalating tariffs applied to encourage reductions in consumption. A danger of this approach is that poor households can be unfairly penalized. This can be avoided by having a stepped tariff, where costs only escalate after a certain level of consumption. Alternatively, a 'free basic allocation' of water can also be used to avoid this. Both of these approaches were applied successfully in the Cape Town example discussed earlier.

6.11 Water resilient wastewater treatment

Wastewater systems treat water used in cities and direct this back into water resources such as rivers so it can be used again. Water resilient systems ensure that wastewater treatment systems are designed and managed effectively to ensure that treated water that is pumped back into rivers is of high-quality. Where feasible, wastewater can be treated to appropriate water standards for reuse, for instance for agricultural irrigation, or to drinking water standard for reuse within the city. In Windhoek, Namibia, for instance, water supplies are regularly supplemented with reclaimed wastewater [18]. Cities may require, or encourage, water users with appropriate facilities to recycle and reuse their water onsite. This is can be achieved through grey and black water systems that can cater to small sites, such as a single house as well as larger sites, such as student accommodation or hotels.

6.12 Conclusions

The chapter shows how the 'softer' issues of governance, coordination, and behavior, as well as the 'harder' technical aspects of appropriate water system infrastructure, are required to create resilient water systems. Effective governance, coordination and behavioral factors are crucial in making cities more water resilient. This ensures that different stakeholders understand current and potential future water scenarios and can contribute appropriately to creating and maintaining sustainable resilient water systems. Drawing on high-quality guidance, governance structures enable stakeholders to work together cohesively to respond dynamically to changing water resources. Strong water planning, infrastructure design and management capabilities within water companies and local authorities must be applied to create water systems that not only address existing demands but also take into account future requirements linked to city growth and climate change. These capabilities should draw on monitoring systems and scenario modeling to ensure that systems balance demand and supply. Finally, water users must be influenced to ensure that their behavior responds to changing water resource availability. Here, mechanisms such as pricing and penalties as well as regulations and awareness campaigns can be applied to reinforce the required changes in behavior.

A holistic, integrated approach is required to plan, design and manage the technical infrastructure for water resilient cities. This must span water resources, water infrastructure, water use and wastewater treatment. Water catchment areas, even if they are dams and rivers a considerable distance away, must be effectively managed. Allocations of water between competing users, such as agriculture, industry and residential areas, must also be negotiated. High-quality infrastructure in the form of reservoirs, pumping stations and distribution networks must minimize losses through leaks and evaporation. Water supply and pressure to different parts of the systems must be controllable to reduce leakage and pipe bursts and to balance demand with available water.

Potable water use for non-essential uses in cities, such as topping up ornamental ponds and irrigating ornamental gardens should be avoided. Regulations, bylaws and incentives should be in place to ensure water delivery devices and fittings like taps, showers, toilets, urinals and washing machines are efficient. Wastewater should be treated as a valuable resource and reused locally within buildings, neighborhoods and within the city as a whole.

References

[1] C. Alexander, Looking back on Cape Town's drought and 'day zero', citylab (2019). [ONLINE] Available at: https://www.citylab.com/environment/2019/04/cape-town-water-conservation-south-africa-drought/587011/ [Accessed 15/8/2019].

[2] O. Awadh, Sustainability and green building rating systems: LEED, BREEAM, GSAS and Estidama critical analysis, J. Build. Eng. 11 (2017) 25–29.

[3] D. Bek, E. Nel, T. Binns, Jobs, water or conservation? Deconstructing the green economy in South Africa's working for water programme, Environ. Dev. 24 (2017) 136–145.

[4] S.V. Berg, Seven elements affecting governance and performance in the water sector, Util. Policy 43 (2016) 4–13.

[5] E. Bertone, R.A. Stewart, O. Sahin, M. Alam, P.X. Zou, C. Buntine, C. Marshall, Guidelines, barriers and strategies for energy and water retrofits of public buildings, J. Cleaner Prod. 174 (2018) 1064–1078.

[6] N. Buxton, Eatout.co.za, (2019). SA's number 1 restaurant to launch The Drought Kitchen pop-up, Eat Out (2018). [Online] Available at http://www.eatout.co.za/article/sas-number-1-restaurant-launch-drought-kitchen-pop/ Accessed 30/7/2019.

[7] C40, C40, [Online] Available at: https://www.c40.org/awards/2015-awards/profiles Accessed 25/7/2019 2019.

[8] G.J. Carbone, K. Dow, Water resource management and drought forecasts in South Caroline, J. Am. Water Resour. Assoc. 41 (1) (2005) 145–155.

[9] A. Campisano, D. Butler, S. Ward, M.J. Burns, E. Friedler, K. DeBusk, L.N. Fisher-Jeffes, E. Ghisi, A. Rahman, H. Furumai, M. Han, Urban rainwater harvesting systems: research, implementation and future perspectives, Water Res. 115 (2017) 195–209.

[10] M. Charle, WATCH: #SaveWater ambassadors tackle 100 schools in 10 days, Cape Argus (2018). [Online] Available at https://www.iol.co.za/capeargus/news/watch-savewater-ambassadors-tackle-100-schools-in-10-days-13370509 Accessed 30/7/2019.

[11] I. Chirisa, E. Bandauko, African Cities and The Water-Food-Climate-Energy Nexus: An Agenda for Sustainability And Resilience at a Local Level, in Urban Forum, Springer, Netherlands, 26, 4, 2015, pp. 391–404.

[12] J.H. Christensen, B. Hewitson, A. Busuioc, A. Chen, X. Gao, I. Held, R. Jones, R.K. Kolli, W.T. Kwon, R. Laprise, V. Magaña Rueda, L. Mearns, C.G. Menéndez, J. Räisänen, A. Rinke, A. Sarr, P. Whetton, Regional climate projections, in: S. Solomon, D. Qin, M. Manning, Z. Chen, M. Marquis, K.B. Averyt, M. Tignor, H.L. Miller (Eds.), Climate Change. The Physical Science Basis. Contribution of Working Group I to the Fourth Assessment Report of the Intergovernmental Panel on Climate Change, Cambridge University Press, Cambridge, 2007.

[13] City of Cape Town, If it's yellow let it mellow." Flushing… - City of Cape Town, Facebook (2017). [Online] Available at https://www.facebook.com/CityofCT/posts/if-its-yellow-let-it-mellow-flushing-only-when-its-needed-can-help-save-up-to-9-/1361371987233050/ Accessed 30/7/2019.

[14] CIESIN, Columbia University, IFPRI, World Bank, CIAT Global Rural–Urban Mapping Project (GRUMP): Urban Extents , Center for International Earth Science Information Network, Palisades, NY., 2004. Available at http://sedac.ciesin.columbia.edu/gpw.

[15] City of Cape Town, Microsoft Word - Water Outlook 2018 (Rev 25) 20 May 2018 DRAFT01, 2018. [Online] Available at https://resource.capetown.gov.za/documentcentre/Documents/City%20research%20reports%20and%20review/Water%20Outlook%202018%20-%20Summary.pdf Accessed 30/7/2019.

[16] DBSA, New Opportunities for Set to Resolve Backlogs, Aid Delivery Funding for, 2019. [Online] Available at: https://www.dbsa.org/EN/DBSA-in-the-News/NEWS/Documents/FM%20Infrastructure%20Special%20Report%202019.pdf Accessed 5/8/2019.

[17] J. De Villiers, Watch: Residents Celebrate as Western Cape Municipal Dam Overflows, 2019. [Online] Available at: https://www.businessinsider.co.za/western-cape-ceres-dam-overflows-2018-9 Accessed 29/7/2019.

[18] P.L. Du Pisani, Direct reclamation of potable water at Windhoek's Goreangab reclamation plant, Desalination 188 (1–3) (2006) 79–88.

[19] B. Gasson, The ecological footprint of Cape Town: unsustainable resource use and planning implications Proc, SAPI Int. Conf., Durban, Planning Africa, 2002.

[20] J. Gould, Z. Qiang, L. Yuanhong, Using every last drop: rainwater harvesting and utilization in Gansu Province, China, Waterlines 33 (2) (2014) 107–119.

[21] L. Joubert, G. Ziervogel, Cities in the Global South can learn from how Cape Town., 2019. [Online] Available at: https://www.dailymaverick.co.za/article/2019-07-23-cities-in-the-global-south-can-learn-from-how-cape-town-survived-the-drought/ Accessed 25/7/2019.

[22] M.R. Karim, M.Z.I. Bashar, M.A. Imteaz, Reliability and economic analysis of urban rainwater harvesting in a megacity in Bangladesh, Resour. Conserv. Recycl. 104 (2015) 61–67.

[23] B. Mati, T. De Bock, M. Malesu, E. Khaka, A. Oduor, M. Meshack, V. Oduor, Mapping the potential of rainwater harvesting technologies in Africa. A GIS overview on development domains for the continent and ten selected countries, Tech. Manual 6 (2006) 126.

[24] R.I. McDonald, K. Weber, J. Padowski, M. Flörke, C. Schneider, P.A. Green, T. Gleeson, S. Eckman, B. Lehner, D. Balk, T. Boucher, Water on an urban planet: urbanization and the reach of urban water infrastructure, Global Environ. Change 27 (2014) 96–105.

[25] B. Moyo, E. Madamombe, D. Love, A model for reservoir yield under climate change scenarios for the water stressed City of Bulawayo, (2013).

[26] P. Mukheibir, G. Ziervogel, Developing a municipal adaptation plan (MAP) for climate change: the city of Cape Town, Environ. Urban. 19 (1) (2007) 143–158.

[27] P. Mukheibir, Qualitative assessment of municipal water resource management strategies under climate impacts: the case of the Northern Cape, South Africa, Water SA 33 (4) (2007) 575–581.

[28] S. Nazif, M. Karamouz, M. Tabesh, A. Moridi, Pressure management model for urban water distribution networks, Water Resour. Manage. 24 (3) (2010) 437–458.

[29] E. Oberholzer, SuzelleDIY and Helen Zille Hilariously Demonstrate How We Can Save Water, Thesouthafrican.com, 2019. [Online] Available at: https://www.thesouthafrican.com/lifestyle/watch-suzellediy-and-helen-zille-demonstrate-how-we-can-save-water-video/ Accessed 29/7/2019.

[30] S.J. Piketh, C. Vogel, S. Dunsmore, C. Culwick, F. Engelbrecht, I. Akoon, Climate change and urban development in southern Africa: the case of Ekurhuleni Municipality (EMM) in South Africa, in: Water SA 40 (4) (2014) 749–758.

[31] S.B. Schindler, Following industry's LEED: municipal adoption of private green building standards, Fla. L. Rev. 62 (2010) p.285.

[32] SCLI, Water Evaporation adds to Cape Town water shortage - Southern Cape Landowners Initiative, SCLI, 2019. [Online] Available at: https://www.scli.org.za/high-rates-water-evaporation-add-cape-town-water-shortage-woes/ Accessed 31/7/2019.

[33] K.B. Showers, Water scarcity and urban Africa: an overview of urban–rural water linkages, World Dev. 30 (4) (2002) 621–648.

[34] C.B. Soyapi, Water security and the right to water in Southern Africa: an overview, Potchefstroom Electron. Law J./Potchefstroomse Elektron. Regsblad 20 (1) (2017).

[35] P. Sorensen, The chronic water shortage in Cape Town and survival strategies, Int. J. Environ. Stud. 74 (4) (2017) 515–527.

[36] V.L. Speight, Innovation in the water industry: barriers and opportunities for US and UK utilities, Wiley Interdisciplinary Reviews, Water 2 (4) (2015) 301–313.

[37] T. Lancet, Water crisis in Cape Town: a failure in governance, Lancet Planet. Health 2 (3) (2018) p.e95.

[38] Treasury, Vote 36 Water and Sanitation, 2019. Available at http://www.treasury.gov.za/documents/National%20Budget/2019/enebooklets/Vote%2036%20Water%20and%20Sanitation.pdf Accessed 6/8/2019.

[39] B.W. van Wilgen, A. Wannenburgh, Co-facilitating invasive species control, water conservation and poverty relief: achievements and challenges in South Africa's Working for water programme, Curr. Opin. Environ. Sustain. 19 (2016) 7–17.

[40] C. Vogel, D. Olivier, *Re*-imagining the potential of effective drought responses in South Africa, Reg. Environ. Change (2018) https://doi.org/10.1007/s10113-018-1389-4.

[41] P. Wu, M. Tan, Challenges for sustainable urbanization: a case study of water shortage and water environment changes in Shandong, China, Procedia Environ. Sci. 13 (2012) 919–927.

[42] K. Wall, An investigation of the franchising option for water services operation in South Africa, Water SA 32 (2) (2006) 265–268.

[43] UN-Water, Scarcity | UN-Water, 2019. [Online] Available at https://www.unwater.org/water-facts/scarcity/ Accessed 25/7/2019.

[44] M. Visser, J. Brühl, Op-Ed: A drought-stricken Cape Town did come together t…, 2019. [Online] Available at https://www.dailymaverick.co.za/article/2018-03-01-op-ed-a-drought-stricken-cape-town-did-come-together-to-save-water/#.WpfFteeYNPZ Accessed 29/7/2019.

[45] E. Wanjiru, X. Xia, Optimal energy-water management in urban residential buildings through grey water recycling, Sustain. Cities Soc. 32 (2017) 654–668.
[46] P. Wolski, What cape town learned from its drought. April 16, 2018, Bull. Atom. Scientists (2018) https://thebulletin.org/2018/04/whatcape-town-learned-from-its-drought/.
[47] World Health Organisation, Summary and Policy Implications Vision 2030: The Resilience of Water Supply and Sanitation in the Face of Climate Change, WHO Press, 2009.
[48] WWF, WWFWaterFiles_3, 2018. [Online] Available at http://awsassets.wwf.org.za/downloads/WWFWaterFile_FINAL.pdf Accessed 30/7/2019.
[49] Y. Zhang, J. Wang, F. Hu, Y. Wang, Comparison of evaluation standards for green building in China, Britain, United States, Renew. Sust. Energ. Rev. 68 (2017) 262–271.
[50] G. Ziervogel, P. Johnston, M. Matthew, P. Mukheibir, Using climate information for supporting climate change adaptation in water resource management in South Africa, Clim. Change 103 (3–4) (2010) 537–554.
[51] G. Ziervogel, Unpacking the Cape Town Drought: Lessons Learned, Cities Support Programme Undertaken, by, African Centre for Cities, 2019.
[52] G. Köhlin, D. Whittington, M. Visser, Beyond day zero in cape town economic instruments for water-scarce cities, Environ. Develop. Init. (2018). Available from http://www.efdinitiative.org/blog/beyond-day-zero-cape-towneconomic-instruments-water-scarce-cities, accessed, 2.
[53] P. Sorensen, The chronic water shortage in Cape Town and survival strategies, Int. J. Environ. Studies 74 (4) (2017) 515–527.
[54] P. Mukheibir, Qualitative assessment of municipal water resource management strategies under climate impacts: the case of the Northern Cape, Water SA 33 (4) (2007) 575–581.
[55] J. Rijke, M. Farrelly, R. Brown, C. Zevenbergen, Configuring transformative governance to enhance resilient urban water systems, Environ. Sci. Policy 25 (2013) 62–72.
[56] D. Stead, Urban planning, water management and climate change strategies: adaptation, mitigation and resilience narratives in the Netherlands, Int. J. Sustain. Develop. World Ecology 21 (1) (2014) 15–27.
[57] Water UK, Water Resources Long Term Planning Framework., 2016 WaterUK-WRLTPF_Final-Report_FINAL-PUBLISHED-min.pdf [ONLINE] Available at: https://www.water.org.uk/wp-content/uploads/2018/11/WaterUK-WRLTPF_Final-Report_FINAL-PUBLISHED-min.pdf [Accessed 5/9/2019].
[58] M. Swilling, Sustainability and infrastructure planning in South Africa: a Cape Town case study, Environ. Urbanization 18 (1) (2006) 23–50.
[59] B. Mati, T. De Bock, M. Malesu, E. Khaka, A. Oduor, M. Meshack, V. Oduor, Mapping the potential of rainwater harvesting technologies in Africa. A GIS overview on development domains for the continent and ten selected countries, Tech. Manual 6 (2006) 126.

CHAPTER 7

Water, sanitation and hygiene (WASH) disease prevention and control in low resource countries

Mynepalli K.C. Sridhar[a,*], Mumuni Adejumo[a]

[a]*Department of Environmental Health Sciences, Faculty of Public Health, College of Medicine, University of Ibadan, Ibadan, Nigeria*
**Corresponding author*

7.1 Introduction

The water, sanitation and hygiene concept, or WASH represents three components, which are grouped together due to their interrelated deficiencies in each area. These deficiencies have to be corrected together in order to achieve a positive overall impact on public health. WASH was introduced for the first time in 1981. Over the years, however, the letter "H" was used for "Health" by USAID, "Health Education" in Zambia, "Hygiene", by the Water Supply and Sanitation Collaborative Council (WSSCC) and the International Water and Sanitation Centre (IRC) since 2001. "WATSAN" (water and sanitation) was also in vogue for a short period especially in emergency situations. WASH is a key public health issue and is also the focus of United Nations Sustainable Development Goal (SDG) 6 (https://www.un.org/sustainabledevelopment/sustainable-development-goals/), the "Water" SDG (https://www.un.org/sustainabledevelopment/water-and-sanitation/).

Water, the first component of WASH, is an essential aspect of human life because it is used for domestic, agricultural, and industrial purposes. Particularly at the domestic level, potable water is important to individuals and households to ensure healthy living, freedom from hazards and untimely deaths from water-related diseases. Lack of access to safe and adequate water supplies leads to the spread of disease and is a global public health threat. Unsafe water may lead to the risk of diarrhoea and other infectious diseases and illnesses due to toxic chemical constituents [8].

The second component of WASH, Sanitation, refers to public health conditions related to clean drinking water and the adequate treatment and disposal of human waste including wastewater [24]. Sanitation also includes prevention of human contact with faeces and thus

Sustainable Water Engineering. DOI: 10.1016/C2017-0-04301-X

promotes hand-washing practices, preferably with soap. Sanitation systems aim to protect human health by providing a clean environment that will stop the transmission of disease, especially through the faecal–oral route. For example, incidences of diarrhoea, a main cause of malnutrition and stunted growth in children, can be reduced through adequate sanitation [25]. There are a variety of other diseases which are easily transmitted in communities that have low levels of sanitation, such as ascariasis (or helminthiasis), cholera, hepatitis, polio, schistosomiasis, trachoma, to name just a few. In the African context, these include lymphatic filariasis, onchocerciasis, soil transmitted helminthiases, schistosomiasis, and trachoma. Over one billion people are affected by these preventable diseases, which are mostly prevalent among the poorest communities. WASH is essential in preventing and controlling these diseases.

Hygiene, the third component of WASH is a set of practices performed to preserve health and prevent the spread of disease [33]. Many people equate hygiene with "cleanliness," but it is a broad term to include such personal habit choices as how frequently to take a shower or bathe, wash hands, trim fingernails, and change and wash clothes. It also includes keeping surfaces in the home and workplace, including bathroom facilities clean and pathogen-free. Some regular hygiene practices are considered good habits, whilst the neglect of hygiene can be considered disgusting, disrespectful, or threatening.

7.2 History of the WASH concept

It is likely that the acronym was first coined in this specific context in a United States Agency for International Development (USAID) "WASH project report" dating to 1981. The acronym represented "Water and Sanitation for *Health*" and was used as such until 1988. In Zambia, the acronym WASHE was used in 1987, which stood for "Water Sanitation Health Education". From about 2001 onwards, international organisations active in the area of water supply and sanitation advocacy, such as the Water Supply and Sanitation Collaborative Council (WSSCC) founded in 1990 and the International Water and Sanitation Centre (IWRC) founded in the Netherlands in 1968 began to use "WASH" as an umbrella term for water, sanitation and hygiene. This has since been broadly adopted worldwide. The term "WatSan" was also used for a while, especially in the emergency response sector such as the IFRC, the International Red Cross and Red Crescent Movement along with the International Committee of the Red Cross (ICRC) and 190 National Societies and United Nations High Commissioner for Refugees (UNHCR). This term was however not as prevalent as WASH, which continues as a development priority for the United Nations and UNICEF. WASH underpins the Sustainable Development Goal (SDG) 6. UNICEFs declared strategy is "to achieve universal and equitable access to safe and affordable drinking water for all".[1] The World Health

[1] https://www.unicef.org/esa/water-and-environment

Organization (WHO) and United Nations Children's Fund (UNICEF) established the Joint Monitoring Programme (JMP) for Water Supply, Sanitation and Hygiene in 1990 to address these issues. The JMPs first global update was released in 2017 [39].

7.3 WASH and its relationship with the millennium and sustainable development goals (SDG)

The MDGs expired at the end of 2015, and were replaced with 17 SDGs, which extended the approach to 2030. The MDGs were set up to halve hunger and poverty, with similar proportional reductions in other targets. Properly addressed, WASH should reduce poverty, ill health, death, and improve the socioeconomic development of communities. Thus, the 2030 Agenda for Sustainable Development recognised "safe drinking water, effective sanitation, and good hygiene" as driving progress for many of the 17 SDGs. Health, nutrition, education, and gender equality are particularly important, with WASH standing out as a key factor, with trade-offs amongst many of them. The SDGs were designed to further reduce hunger, poverty, preventable child deaths and other targets, i.e. to achieve a statistical "zero" [12].

SDG 6 aims to "ensure availability and sustainable management of water and sanitation for all". UNICEFs targets are in close accord as eliminating inequality is a crucial purpose of Targets for Goal 6, which aims to achieve the following by 2030:

- Target 6.1: universal and equitable access to safe and affordable drinking water for all.
- Target 6.2: access to adequate and equitable sanitation and hygiene for all, end open defecation, meet the needs of women and girls and those in vulnerable situations.
- Target 6.3: improve water quality by reducing pollution, eliminating dumping, and minimizing release of hazardous chemicals and materials, halving the proportion of untreated wastewater, and substantially increasing recycling and safe reuse globally.

The situation of sanitation and hygiene as of 2017 are shown in Fig. 7.1.

WASH is also involved in several other SDGs with access to safe drinking water, safely managed sanitation and hygiene facilities part of Goal 2 (nutrition), Goal 3 (health), Goal 4 (education), Goal 5 (gender equality) and Goal 11 (cities and infrastructure).

7.4 Water supply, water quality and WASH-based diseases

Water supply is the provision of water by public utilities, commercial organisations, community endeavours or by individuals, usually through a system of pumps, pipes, and taps. A safe water supply - in particular, water that is not polluted with faecal matter from a lack of sanitation – is one of the most important determinants of public health. The destruction of water supply and/or sanitation infrastructure after major catastrophes (earthquakes, floods, war, etc.) poses an immediate threat of severe epidemics of waterborne diseases; which can be life

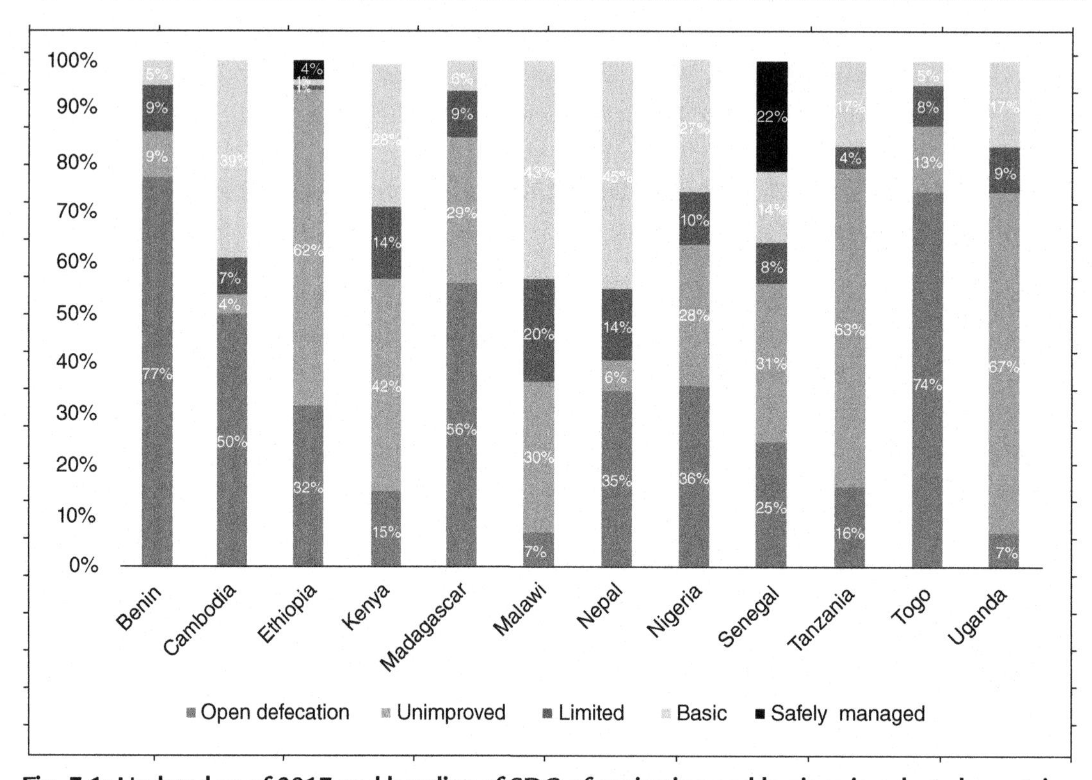

Fig. 7.1: Updated as of 2017 and baseline of SDG of sanitation and hygiene in selected countries. *Source*: Adapted from WHO/UNICEF, 2017. Progress in Drinking Water, Sanitation and Hygiene: 2017 update and SDG baselines, Geneva, pp. 1–108.

threatening. Water supply systems obtain water from a variety of sources such as groundwater (aquifers), surface water (lakes and rivers), and the sea through desalination. More recently, rainwater harvesting has gained traction [19]. Water from these sources sometimes require treatment, which may include simple purification, disinfection and sometimes fluoridation. Treated water then either flows by gravity or is pumped to reservoirs and finally distributed to end users. Examples of water supplies for some communities are shown in Fig. 7.2.

Water quality directly influences public health. Good microbiological quality is important in preventing ill health, a lack of which can sometimes lead to epidemics of cholera, typhoid, dysentery, hepatitis, giardiasis, guinea worm and schistosomiasis [35]. Many reported emerging diseases are also associated with improperly treated water, which may also promote vector breeding and the transmission of vector-borne infections. Certain chemicals when dissolved in water may lead to chronic long-term health effects, which may or may not be reversible. Acute effects may also be encountered because of major pollution events due to natural or anthropogenic catastrophes. Therefore, water has a role in promoting hygiene practices to control WASH related infections. Further detail on the place of WASH in specific health outcomes are given in Section 7.7.

Fig. 7.2: Various water supplies used in the Nigerian communities.

7.5 Components of WASH and their relevance

The WHO/UNICEF Joint Monitoring Programme (JMP)[2] is the custodian of global data for WASH. The JMP has been monitoring global WASH progress since 1990 and is responsible for reporting on the SDG WASH-related targets and indicators. A recent study by WHO[3] estimates that around 2 billion people worldwide use drinking water contaminated with human faeces. This is more than twice the official figure from the WHO/UNICEF JMP, which recorded that 748 million people lack access to an improved drinking water source. The MDG sanitation target, to halve the proportion of people lacking access to improved sanitation by 2015, has not been met. Currently, 2.5 billion people still lack access to basic sanitation. In many parts of sub-Saharan Africa, less than half of the population use a decent toilet. One

[2] https://www.unwater.org/publication_categories/whounicef-joint-monitoring-programme-for-water-supply-sanitation-hygiene-jmp

[3] https://www.who.int/news-room/fact-sheets/detail/drinking-water

billion people do not use a toilet at all [23]. In 2015, 750 million people lacked access to safe, clean drinking water and it was estimated that approximately 2,300 people died every day from diarrhoea [28]. WASH used in healthcare facilities prevent the spread of infections. However, WHO data from 54 countries in low- and middle-income settings representing 66,101 health facilities show that 38 percent of health care facilities lack improved water sources, 19 percent lack improved sanitation while 35 percent lack access to water and soap for handwashing [38]. The latter report found healthcare facilities, specifically in Africa, lacked adequate safe water supplies with only 42 percent being satisfactory. The complexity of some specific health issues which effective WASH strategies can resolve are shown in Fig. 7.3.

WASH emphasises "Improved Water Sources, Improved Sanitation, Sustainable Sanitation Systems and Basic Hygiene", as described below:

a) *Improved water sources* include piped water supplies in homes, yards, or plots; public standpipes and non-piped supplies such as boreholes, protected wells and springs, rainwater and packaged or delivered water. Between 2000 and 2015, the piped water usage by populations increased from 3.5 billion to 4.7 billion. Those using non-piped water supplies increased from 1.7 billion to 2.1 billion [39].
b) *Improved sanitation* refers to provisions designed to hygienically separate excreta from human contact. These include "wet" solutions such as flush and pour flush toilets connecting to sewers, septic tanks, or pit latrines. Dry sanitation technologies include ventilated improved pit latrines, pit latrines with slabs, or composting toilets. Improved sanitation is divided into three categories: limited (improved facilities shared with other households), basic and safely managed (facilities which are not shared).
c) *Sustainable sanitation systems* protect and promote human health, minimise environmental degradation and those preventing depletion of the resource base. They should be technically and institutionally appropriate, socially acceptable, and economically viable in the long run.
d) *Productive sanitation systems* include those making productive use of the nutrients, organic matter, water and energy content of human excreta and wastewater in crop and energy production. These systems have the potential to turn waste into valuable resources [2].
e) *Basic Hygiene* is encapsulated under SDG target 6.2 and represents the importance of hygiene and its close links with sanitation. Hygiene comprises much behaviour - handwashing, menstrual hygiene, food hygiene, amongst others. In the context of WASH, handwashing with water and soap is of paramount interest. The SDG indicator used is the "proportion of population with handwashing facilities with soap and water at home". For water, the acceptable sources are buckets with taps, tippy taps, and portable basins. Among soaps and detergents, bar soap, liquid soap, powder detergent and soapy water

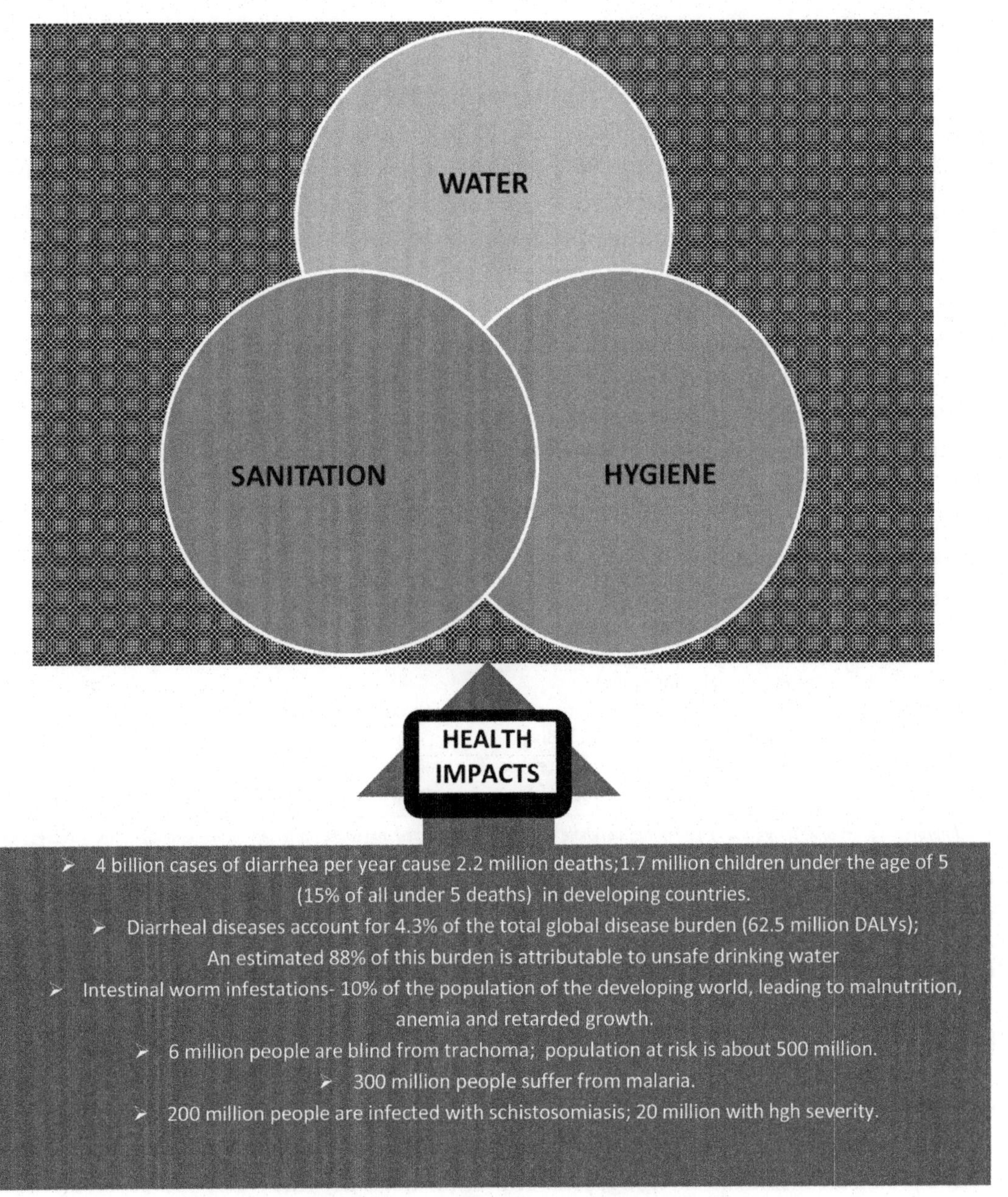

Fig. 7.3: Health impacts of WASH in developing countries.

are acceptable. The authors of this chapter propose that soap made from cocoa pod ash, banana or plantain peel is better than many of the commercial soaps available in Nigerian markets. They have developed a soap-coated paper to wash hands after using the "Pay and use toilets" in particular ([18]: Unpublished data).

7.5.1 The joint monitoring programme water ladder

The JMP ladder was developed in 2002 by the joint monitoring programme of UNCEF/WHO to classify water sources and sanitation facilities. Prior to this, a classification of improved and unimproved source type was used to compare the progress of countries in providing access to quality water sources [39]. The JMP ladder builds on the established classification of source type, thereby providing continuity with MDG monitoring. It also introduces additional criteria on the accessibility, availability, and quality of drinking water services (Table 7.1A). The rungs on the ladder are designed to enable countries at different stages of development to benchmark and compare progress over time. At the bottom of the ladder, the JMP identifies populations with "no service" as those using surface water such as rivers, lakes, and ponds. Populations with "unimproved" sources of water are those with no protection against contamination. During the SDG reporting period, populations using improved sources were further subdivided into three levels of service:

Table 7.1A: The new drinking water services ladder.

Service level	Definition
Safely managed	Sources of drinking water is improved, located on premises, available when needed and free of feacal and priority toxic
Basic	Drinking water from an improved source with collection time not more than 30 minutes for roundtrip including queuing
Limited	Drinking water from an improved source with collection time of more than 30 minutes for roundtrip to fetch water including
Unimproved	Drinking water from an unprotected dug well or spring
Surface water	Drinking water collected directly from a river, dam, lake, pond, stream, canal or irrigated channel

Note: Improved sources of water include: piped water, borehole or tubewells, protected dug wells, protected spring, rainwater, and packaged or delivered water.
Source: Adapted from WHO/UNICEF, 2017. Progress in Drinking Water, Sanitation and Hygiene: 2017 update and SDG baselines, Geneva, pp. 1–108.

i) Limited - if the improved source is not readily accessible, where the round trip to collect water, including queuing, is more than 30 min.
ii) Basic - if the improved source is readily accessible, close to home where a round trip takes 30 min or less.
iii) Safely managed - the improved source has to satisfy three conditions in order to meet the threshold of "safely managed", that is accessibility (the source should be located on the premises, inside the dwelling, yard or plot); availability (the water source should be available when needed); and quality (the water supplied should be free from faecal and priority chemical contamination).

Table 7.1B details the JMP WASH Ladder, addressing safe water provision, available sanitation, and handwashing facilities in the home in terms of the categories given above.

Whilst hygiene has long-established links with public health, it was not included in any of the MDG targets or indicators. However, the explicit reference to hygiene in SDG 6.2 shows increasing recognition of the importance of hygiene and its close links with sanitation.

International consultation among WASH sector professionals identified handwashing with soap and water as a top priority in all settings. It is also as a suitable indicator for national and global monitoring. Handwashing facilities may be fixed or mobile and include a sink with tap water, buckets with taps, tippy-taps, and jugs or basins designated for handwashing. The JMP hygiene ladder, which classifies the facilities available, is shown in Table 7.1C.

Table 7.1B: The new sanitation services ladder.

Sanitation service level	Definition
Safely managed	Use of improved facilities which are not shared with other households and where excreta are safely disposed in situ or transported and treated off-site.
Basic	Use of improved facilities which are not shared with other household
Limited	Use of improved facilities shared between two or more households
Unimproved	Use of pit latrines without a slab or platform, hanging latrine or bucket latrine
Open defecation	Disposal of human faeces in fields, forests, bushes, open bodies or water, beaches and other open spaces or with solid waste

Note: Improved facilities include: flush/pour flush to piped sewer systems, septic tanks or pit latrines, ventilated improved pit latrines, composting toilet or pit latrine with slabs.
Source: Adapted from WHO/UNICEF, 2017. Progress in Drinking Water, Sanitation and Hygiene: 2017 update and SDG baselines, Geneva, pp. 1–108.

Table 7.1C: The new hygiene service ladder.

Service level	Definition
Basic	Availability of a handwashing facility on premises with soap and water
Limited	Availability of a handwashing facility on premises without soap and water
No facility	No handwashing facility on premises

Note: Handwashing facilities may be fixed or mobile and include a sink with tap water, buckets with taps, tippy-taps, and jugs or basins designated for handwashing. Soap includes bar soap, liquid soap, powder detergent, and soapy water but does not include ash, soil, sand or other handwashing agents.
Source: Adapted from WHO/UNICEF, 2017. Progress in Drinking Water, Sanitation and Hygiene: 2017 update and SDG baselines, Geneva, pp. 1–108.

7.6 Sanitation options and the management of waste

Sanitation is important for all, in maintaining health and increasing lifespan. It involves safe collection, transportation, and disposal of human excreta (faeces and urine), and the safe management of municipal and other wastes. These wastes may be sullage or greywater (i.e. from food preparation in the kitchen, personal bathing, and clothes washing), sewage or blackwater (i.e. liquid from the toilet), faecal sludge, solid wastes, their leachates among others.

A sanitation system includes the collection, storage, transport, treatment and disposal or reuse of human excreta and wastewater [10]. A range of sanitation technologies and approaches exists, including container-based sanitation, ecological sanitation, emergency sanitation, environmental sanitation, onsite sanitation, community-led total sanitation, and sustainable sanitation. There are several sanitation technologies available, which can be either wet or dry. Most urban sanitation is "wet", involving some form of flush toilet connected to a soak-away pit, septic tank, or sewer. Dry technologies do not use water for flushing; examples include traditional pit latrines, ventilated improved pits, and contemporary designs that promote the safe reuse of excreta (Table 7.1C). Although pit latrines are commonly used in some rural areas, this option is rarely used in urban communities. In recent years, however, some small-scale initiatives have promoted ecological sanitation (known as Ecosan), a form of dry sanitation that involves the separation of faeces and urine at source and the reuse of treated excreta. The advantages of Ecosan include reduced water demand for flushing; reduced wastewater management problems; no blackwater generation; and improved nutrient recycling, particularly that from urine [27].

Sanitation systems may also be classified as On-site (retaining wastes in the vicinity of the toilet in a pit, tank or vault), Off-site (removing wastes from the vicinity of the toilet for disposal elsewhere) and Hybrid (retaining solids close to the latrine but removing liquids for off-site disposal away from generation site). In urban areas, on-site sanitation systems usually require periodic removal of the faecal sludge and septage from the storage pits, tanks, and vaults. As a result, no urban sanitation system is completely self-contained. There must be provision made at a designated area to enable the removal and treatment of the faecal sludge [31]. To achieve total sanitation in an urban area, consideration must be given to the way in which household services are linked with higher level transport and disposal facilities.

Every sanitation system should include some form of toilet for the management of excreta. Most consist of a water-sealed pan but a hole in a pit latrine cover is also a basic form of toilet. The toilet type is important because it will determine whether the sanitation system is wet or dry. This in turn will influence choices relating to other components of the sanitation system. On-site and hybrid systems require storage of the sludge in pits, tanks or vaults before desludging takes place. Table 7.2 summarises these systems.

Table 7.2: Commonly used on-site sanitation systems.

Sanitation facility options	Some of their feature
Open defecation	Dry facility with no needs for faecal sludge management. It has a very low human dignity and pose a very high public health risk.
Flying toilet (or pack and throw)	Dry facility with no needs for faecal sludge management. It has a very low human dignity and pose a very high public health risk.
Bucket latrine	Dry facility with provision for faecal sludge management. It has a low human dignity and pose a high public health risk.
Simple pit latrine	Dry facility with provision for faecal sludge management. It has a fair human dignity and pose a medium public health risk.
Ventilated improved pit (VIP) latrine	Semi wet facility with provision for faecal sludge management. It has a good human dignity and pose a low public health risk.
Pour-flush latrine with pit, aqua privy	Wet facility with provision for faecal sludge management. It has a good human dignity and pose a low public health risk.
Water-flush or pour-flush toilet with septic tank	Wet facility with provision for faecal sludge management. It has a good human dignity and pose a low public health risk.
Water-flush toilet with holding tanks	Wet facility with provision for faecal sludge management. It has a good human dignity and pose a low public health risk.

Source: Adapted from https://ocw.un-ihe.org/pluginfile.php/621/mod_folder/content/0/C2U2_Conventional_onsite_sanitation.ppt

7.6.1 Faecal sludge management

It is essential to treat wastewater and faecal sludge prior to discharge into the environment. This is especially important in situations where sources of drinking water are at risk from contamination, where residents use rivers or drainage channels for bathing or washing or where the wastewater is reused for irrigating vegetables or horticultural crops. The purpose of treatment is to reduce the concentration of potentially harmful substances to levels that will not cause harm to either the environment or the people who might be exposed to them. While some degradation of the waste can occur on-site in the vaults, leach pits or septic tanks, it is usually necessary for further treatment of these materials. While it is possible to provide this additional treatment on-site, the more common arrangement is to provide it off-site or "end-of-pipe" (i.e. at the end of the sewerage system) or to where vehicles carrying the faecal sludge discharge the wastes. Due to the high concentration of organic matter and pathogens, in a relatively low volume, faecal sludge should normally be dealt with separately in a stand-alone management method. Unfortunately, there are still places in the world where safe disposal is not used, examples of which are shown in Fig. 7.4.

Faecal sludge management in selected West African countries
Left – Unhygienic practice of dumping on farmlands untreated (Kano, Nigeria);
Right - Discharging into 8 Treatment lagoons before sending into Ocean (Banjul, The Gambia)

Fig. 7.4: Faecal sludge management practices in West Africa.

Sullage normally requires a much lower level of treatment as it is less polluting and potentially less harmful than sewage or blackwater. It is normally advisable to treat faecal sludge and wastewater separately, although they can be combined in a wastewater treatment plant if the sludge loads are relatively small. Several options are available for separate sludge treatment. These include [4,5]:

a) Solids–liquid separation in batch-operated settling-thickening tanks
b) Primary sedimentation/anaerobic stabilization ponds

c) Sludge drying beds (unplanted; planted)
d) Combined composting with organic solid waste (co-composting); and
e) Anaerobic digestion (potentially with biogas utilization).

7.6.2 The JMP sanitation ladder and management of excreta

There are three main ways to meet the criteria for having a safely managed sanitation service (SDG 6.2). People should use improved sanitation facilities that are not shared with other households, and any excreta produced should be:

- treated and disposed in situ;
- stored temporarily and then emptied and transported to treatment off-site; or
- transported through a sewer with wastewater and then treated off-site.

If the excreta from improved sanitation facilities are not safely managed, people using those facilities will be classed as having a basic sanitation service (SDG 1.4). People using improved facilities, which are shared with other households, will be classified as having a limited service (Table 7.1B). The JMP will also continue to monitor the population practicing open defecation, the reduction of which is an explicit focus of SDG target 6.2 [39].

Treatment separates the wastewater into solid and liquid fractions. The solid fraction (bio-solids) can be of variable consistency. Hence, it requires additional concentration (thickening), drying or extra treatment until disposal to landfill, or its potential reuse in agriculture as a soil-conditioner and fertiliser. Polishing might be necessary for the liquid fraction, to satisfy criteria for discharge to surface waters or to avoid long-term impacts on groundwater quality [16].

7.7 WASH and health outcomes

UNICEF 2016 highlighted the influence WASH has on child development, particularly in respect of nutrition, food hygiene, female psychosocial stress, menstrual hygiene management, violence, maternal and newborn health, school attendance, oral vaccine performance, and neglected tropical diseases. Some associated common diseases include diarrhoea, dysentery, cholera, infectious skin and eye diseases, malaria, dengue, river blindness, schistosomiasis and guinea worm. Their occurrence is enhanced due to water shortages, contaminated water, and poor sanitation.

Globally, diarrhoea (largely caused by poor WASH) is a leading cause of death in children under-five [11], and its constant presence in low-income settings may contribute significantly to under-nutrition. Similarly, parasitic infections, such as soil-transmitted helminths (worms), caused by a lack of sanitation and hygiene, infect around two billion people globally [3], while an estimated four and a half billion people are at risk of infection [42]. Such infections can lead to anaemia, reduced physical development, and inhibited cognitive development. Furthermore, countries with high maternal mortality rates are those where the burden of infectious diseases remains high, and health information and primary healthcare are difficult to access. Improving access to WASH and providing expectant mothers with basic services and accurate hygiene information, is vital in reducing maternal mortality rates and meeting global goals to end preventable child mortality [30]. Collecting and carrying water while pregnant can cause difficulties in pregnancy and other reproductive health consequences, such as uterine prolapse [20]. Women who lack safe water are more prone to WASH-related illnesses, such as hookworm infestation, which, when occurring during pregnancy, are linked to low birth weight, slow child growth and hepatitis [36]. Emerging evidence suggests that giving birth in a setting without safe drinking water or sanitation has a negative impact on the health and survival of both mother and baby [1]. Hygiene promotion and supplies are key to safe delivery and breastfeeding. A lack of safe drinking water can be fatal for babies who must have infant feeding formula to prevent the transmission of HIV. Lack of safe WASH causes up to 50 percent of undernutrition worldwide [34]. Therefore, improved access to safe WASH is pivotal for good nutrition during the first 1,000 days of life.

Disease outbreaks such as the 2020 coronavirus pandemic occur sporadically, and WASH plays an important role through interventions. An outbreak is the occurrence of disease in excess of the normal baseline (two times the baseline); a sudden spike in cases (two times the incidence of new cases); single case of a communicable disease long absent from a population; caused by a pathogen not previously recognised; emergence of a previously unknown disease; or a single case of particular diseases of interest (cholera, ebola and hepatitis E). Yates [41] gave a comprehensive review of various WASH interventions covering disease outbreaks in 51 humanitarian contexts in 19 low and middle-income countries. WASH interventions consistently reduce both the risk of disease and the risk of transmission in outbreak contexts. A significant outcome was a long-running Community-Led Total Sanitation (CLTS) intervention implemented before and during an Ebola outbreak which recorded a large and significant reduction in disease risk.

The impact of WASH interventions on risk of transmission was more common than disease risk evaluations. These may include well disinfection, chlorine dispensers and household water treatment (liquid chlorine, chlorine tablets and flocculants/ disinfectants). WASH components act as barriers to disease transmission and hence disease risk. Some evaluations also demonstrated reduced short-term transmission risk with environmental hygiene interventions. This is because WASH interventions are user-friendly due to their simplicity, timing, community-based and involvement, and multisectoral linkages with relief, rehabilitation, and development [41].

7.8 WASH and nutrition

Approximately one third of all child deaths are attributable to nutrition-related factors, such as low birth weight, stunted growth (low height for age) and severe wasting, all of which are closely linked to a lack of access to water and particularly sanitation and hygiene [47]. Many children in developing regions suffer stunted growth, which reflects chronic nutritional deficiencies, and repeated ingestion of animal and human faeces due to poor waste management and a lack of sanitation. If the only supply of safe water must be purchased, this limited quantity leaves little for good hygiene practices. The UN post-2015 global thematic consultation on food security and nutrition (https://www.who.int/nutrition/events/2013_consultation_hungerfoodsecuritynutrition_2015dev_4apr13/en/) directly highlighted the role of safe drinking water, sanitation and hygiene in enabling good nutrition to eliminate stunted growth in children under two. It called for access to adequate WASH to be combined with other measures, such as ensuring a diversified diet.

7.9 WASH and hygiene

Hygiene is associated with a set of practices, which preserve health and prevent the spread of diseases, and is closely related to cleanliness, personal and professional care practices. It is an important barrier to many infectious diseases, including faecal–oral diseases. It not only includes personal hygiene, but keeping surfaces in the home and workplace clean and pathogen-free [39].

7.9.1 Components of hygiene

These include personal, domestic, community and food hygiene. Whilst community education is key, engineering approaches can be supportive in enabling the promotion of hygiene practices. These components outlined in Box 7.1 are engineering-focused from a design, siting of facilities and infrastructure point of view.

7.10 WASH, gender and education

Lack of access to WASH at home and school has a negative impact on children's education, particularly for girls; effective education programmes need to be supported by fully accessible, child-friendly and gender-segregated WASH facilities [6]. Opportunities for learning are lost when children must spend time collecting water or finding a safe place to defecate or urinate in the open. This is a problem particularly for girls due to their additional burden of MHM. In particular, adolescent girls are disinclined to use school toilets that are dirty or lack privacy, especially when they are menstruating, and this affects their attendance. Women and girls need to be equipped with the knowledge and means to manage their menstruation

Box 7.1 Components of Hygiene

A) Personal and domestic hygiene [32,39]

- Handwashing e.g. using soap and wood ash, prevents diarrhoeal diseases. Thus WASH facilities must be located near where they are to be used. If running water is available, facilities should include soap, a tap and sink. If not, large oilcans, or buckets with taps are adequate.
- Some traditional bathing practices may not be water efficient but can be modified using water containers with taps.
- "Pay and Use" community showers should provide separate facilities for men and women. They require careful maintenance and must be conveniently located.
- Laundry of clothes and bedding prevents scabies, ringworm, trachoma, conjunctivitis, and louse-borne typhus. To promote this practice, laundry slabs or sinks should be constructed near water points.
- Sinks should be large enough, draining water away, but not in or near natural water bodies, streams, or irrigation canals to prevent schistosomiasis and other transmissible diseases.

B) Community hygiene [32]

- Includes the maintenance, protection, and upkeep of water sources.
- Management of solid waste and excreta, wastewater management and drainage, control of the rearing of livestock and market hygiene.
- Availability and accessibility of functional sanitation facilities.
- Promotion of cleanliness and household maintenance by community leaders.

C) Market and Food hygiene [32]

- To reduce health hazards of markets, cafes, restaurants etc., a safe water supply, sanitation arrangements appropriate for the number of people, separate for male and female, personal cleanliness, the means to dispose of solid waste and adequate drainage need to be provided.
- Ideally, markets should have several taps for traders and customers to wash and drink safely. Solid waste, requires proper, and frequent disposal to prevent rodents and insects from feeding and breeding.
- Market stalls should be laid out in such a way as to allow easy access for the vehicles collecting waste and for efficient cleansing of the area.

D) Animal rearing

- Animals should always be kept and slaughtered away from households, (at least 100 m from water sources and 10 m from dwellings), to avoid contamination with pathogens.
- Animal waste should be stored and disposed of properly.
- This also avoids disease vectors, which can feed on livestock [32].

E) Menstrual Hygiene

- WASH also covers menstrual hygiene, whereby inadequate facilities, particularly in schools, workplaces, or health centres can pose major problems for women and girls in carrying out effective Menstrual Hygiene Management (MHM).
- Separate and usable toilets for girls is necessary, with doors that close safely.
- It is challenging to maintain MHM in a private, safe and dignified manner if the means to dispose of used sanitary pads is not available or water to wash hands ([26,37]; The [22,29]).

Source: Adapted from WHO/UNICEF, 2017. Progress in Drinking Water, Sanitation and Hygiene: 2017 update and SDG baselines, Geneva, pp. 1–108.

hygienically and with dignity and be provided with the means for the safe disposal of menstrual waste [7].

A lack of access to water, sanitation, and hygiene (WASH) affects women and girls disproportionally, due both to biological and cultural factors. In addition to meeting women's needs across menstrual, sexual, and reproductive health, WASH is also essential for their social and economic development, contributing towards gender equity and realisation of their rights. In any post-2015 framework, decision-makers must address the persisting inequalities between women and men, embracing the human rights principles of equality and non-discrimination to ensure universal access to water and sanitation for all women and girls everywhere [17]. Globally, women and girls are still the primary water collectors and the main carers when children or others get sick with diarrhoeal diseases. Women are often vulnerable to harassment or violence when they travel long distances to fetch water, use shared toilets, or practice open defecation, often waiting until nightfall, which increases the risk of assault. The shame and indignity of defecating in the open also affects women's self-esteem, as does a lack of water for washing clothes and personal hygiene [9]. Women and girls perform most of the unpaid labour associated with WASH in households and communities. This reduces the time they have available for education, economic activities, and leisure. A lack of economic independence compromises their empowerment and perpetuates gender inequality. With improved access to WASH, women will have more time to undertake income-generating activities. WASH programmes also provide women with the water needed to carry out economic activities and can create opportunities for paid work. For example, setting up and watering a kitchen garden, to improve her family's food security and potentially an opportunity to earn money by selling the surplus.

7.11 Stakeholders in WASH programmes

Several stakeholders are involved in WASH programmes. These can be individuals, organizations or groups and include:

- public-sector agencies involved in water resources (for example, departments of agriculture, of industry, of transportation, or of recreation);
- various levels of public-sector agencies in the water sector (state, regional or local);
- private-sector organizations and companies with WASH interests;
- environmental and professional NGOs; and
- representatives of those people likely to be affected, specifically including people who may have little knowledge of the effects of the strategy and who may lack the means to participate. May be represented by a Community Based organisation (CBO).

FGN 2015 includes a useful and detailed table of stakeholders in the Nigerian context, suggesting that engagement with them should be "underpinned by strong Government leadership at various levels; collaborative partnerships; private sector engagement and community led processes". These various stakeholders need to be engaged when planning the design and installation or improvement of a WASH programme of facilities or infrastructure. There are a variety of methods which can be used to identify stakeholders. Three of the simplest are self-identification, third-party identification, and identification by the strategy team.

Self-identification simply means that individuals or groups step forward and indicate an interest in participating. Third-party identification uses knowledgeable parties, such as existing advisory committees, informal or formal community leaders, and representatives of known interests, to suggest people or organisations that could be included. Identification by the strategy team relies on systematic identification followed by approaching potential stakeholders. Social and environmental impact assessments, financial and gender analysis can all assist in identifying stakeholders. The team should identify those parties essential to implementing projects, those who are benefitting or will benefit from WASH projects, and those who are bearing project costs and impacts. Most importantly, people who would be affected by such strategies, but may not yet know that they will be affected, should be identified. Stakeholder participation is a means of giving them a voice.

7.12 WASH achievements and benefits

The many achievements of WASH globally are shown in Fig. 7.5. This figure shows that WASH facilities for the poor are generally half of those provided for the riches. Also, whilst the programme has delivered many benefits, there are still areas of concern regarding the incidences of disease, the continued prevalence of open defecation and lack of access to a safe source of water.

7.12.1 Sustaining WASH programmes

The sustainable delivery of WASH services remains a challenge since there is insufficient emphasis on their maintenance and operation. This results in the failure of water supply and sanitation infrastructure in low economy countries where major constraints are the low priority of WASH in terms of governance, lack of commitment and poor funding. As World Vision International [40] state, sustainability is dependent on technological, institutional, social, environmental, and financial systems working together to ensure longevity. A 5-country evaluation of the development of sustainable WASH programmes in Afghanistan, Iraq, Jordan, Lebanon and Syria carried out by [40] identified the following successes:

i) Successful community engagement due to targeted demand and community-led interventions.
ii) Knowledge and skills improved; thus WASH programmes became more effective.
iii) Development of key indicators across the region to enable an evaluation to be made of intervention effectiveness.
iv) Wider application of lessons learnt was made possible due to collaboration across user groups, partners and interested parties.

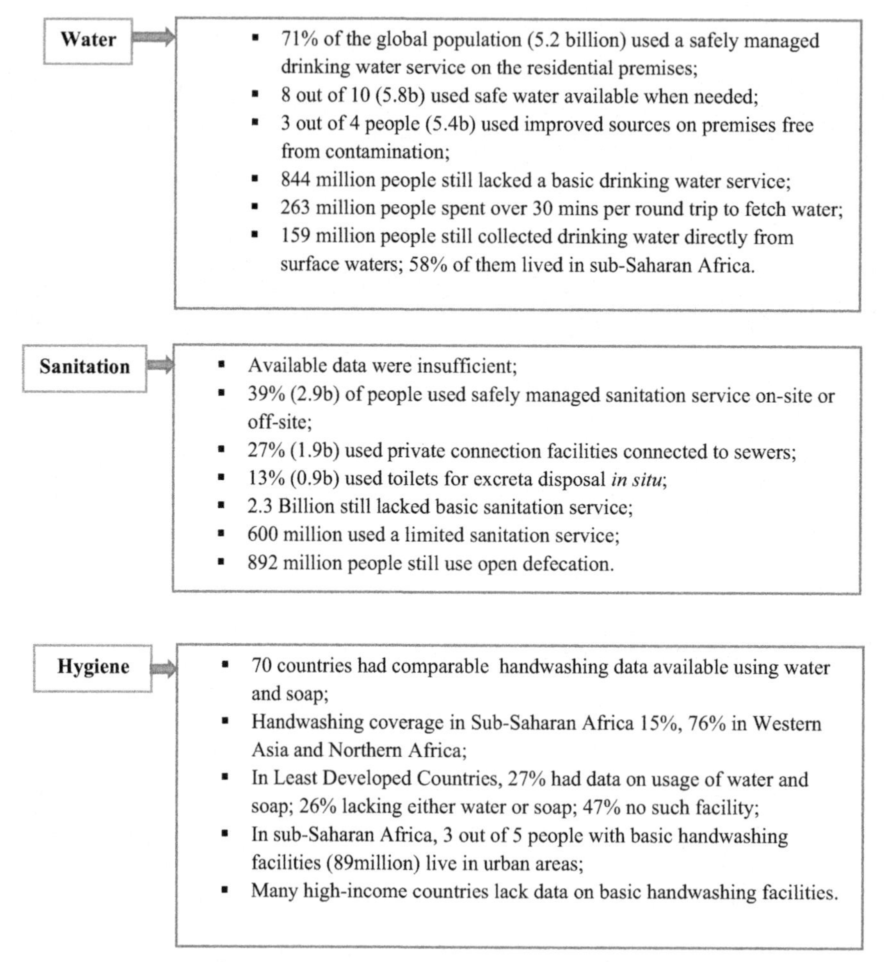

Fig. 7.5: WASH achievements and benefits.

7.13 Conclusions

The importance of WASH in general in promoting human and environmental health and the role of engineers in design and construction cannot be overstated. Underlying this is the need to achieve SDG 6.1 and 6.2 by 2030. As stated by [15]:

"It is clear that if we fail to achieve SDG 6, we undermine progress towards other goals on poverty, inequality, health and wellbeing, economic growth, sustainable cities and communities, gender equality, zero hunger and quality education – almost all of the other SDGs in fact."

In its 2018 report on progress in achieving SDG 6, the United Nations emphasised in what it calls "A Snapshot" of its strategy, [21], identified three categories of focus up to 2030. These are: to continue to learn and to adapt; aspects that could be done better; and new areas to explore (Fig. 7.6).

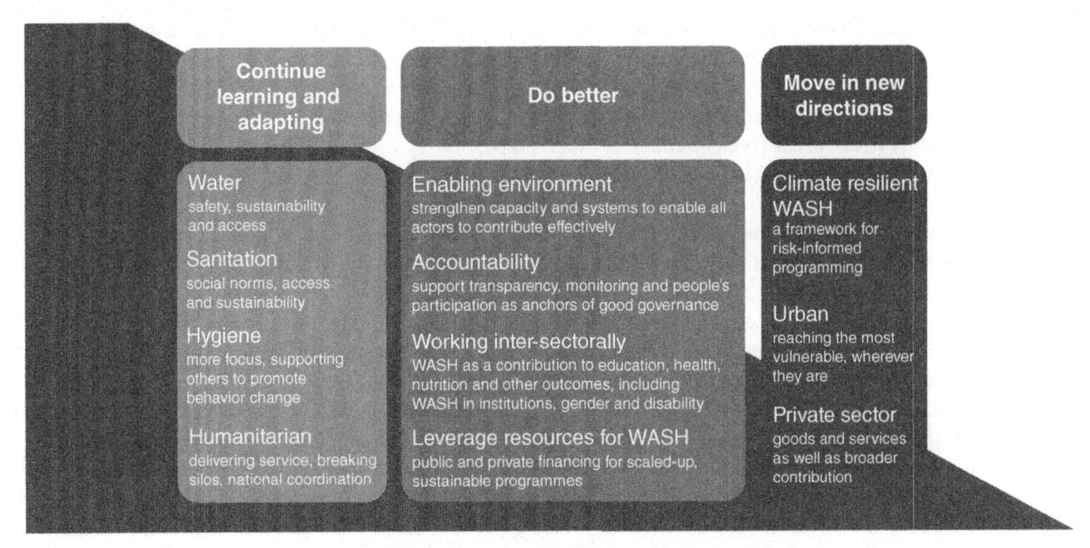

Fig. 7.6: UNICEFs Strategy for WASH: A Snapshot.

Mason et al. [13] list the 10 priorities for WASH, which need to be addressed by 2030. These include considering impacts on managing water resources, and the provision of sanitation, due to climate change, migration and urbanisation, and the development and use of new technology. All of which can provide opportunities for growth, but can also test programmes in communities and user groups as well as challenging governments and whole economic structures.

The central role of engineering is summarised in Mihelcic et al. [14], as follows:

> "the skills and expertise that reside within the environmental engineering community are fundamental to achieving a much broader range of SDGs, including those related to health; water, energy, and food security; economic development; and reduction of poverty and social inequalities".

References

[1] T. Ali, F. Fikree, M. Rahbar, S. Mahmud, Frequency and determinants of vaginal infection in postpartum period: a cross-sectional survey from low socioeconomic settlements, Karachi, Pakistan, J. Pak. Med. Assoc. 56 (2006) 99–103.

[2] K. Andersson, A. Rosemarin, B. Lamizana, E. Kvarnstrom, J. McConville, R. Seidu, S. Dickin, C. Trimmer, Sanitation, Wastewater Management and Sustainability From Waste Disposal to Resource Recovery, SEI and UN Environment, Nairobi and Stockholm, 2016, bit.ly/2dsgnA8.

[3] S. Brooker, A.C. Clements, D.A. Bundy, Global epidemiology, ecology and control of soil transmitted helminth infections, Adv. Parasitol. 62 (2006) 221–261.

[4] EAWAG, Solids Separation and Pond Systems for the Treatment of Faecal Sludges in the Tropics Lessons Learnt and Recommendations for Preliminary Design. SANDEC Report No. 05/98, 1998. Available at: https://tinyurl.com/y8pq5skg.

[5] EAWAG, Faecal Sludge Treatment, 2002. Available at: https://www.pseau.org/outils/ouvrages/eawag_faecal_sludge_treatment_2002.pdf.

[6] J. Fisher, For her it's the big issue – putting women at the centre of water supply, sanitation and hygiene, WSSCC and WEDC. (2006). Available at: https://tinyurl.com/y9p8qqr7.
[7] S. House, T. Mahon, S. Cavill, Menstrual hygiene matters – a resource for improving menstrual hygiene around the world, WaterAid Report., 2012. Available at: https://www.wsscc.org/wp-content/uploads/2017/11/Menstrual-Hygiene-Matters-WaterAid.pdf.
[8] J.M. Hughes, J.P. Koplan, Saving lives through global safe water, J. Emerg. Infect. Dis. 11 (10) (2005) 1636–1637.
[9] International Women's Development Agency and WaterAid, Now we feel like respected adults – positive change in gender roles and relations in a Timor L'este WASH programme, ACFID. (2012). Available at: https://tinyurl.com/yd75uqtz.
[10] D. Jong, Advocacy for water, environmental sanitation and hygiene – Thematic overview paper, IRC, The Netherlands, 2003. Available at: https://www.susana.org/_resources/documents/default/2-439-jong-de-et-al-2003-advocacy-water-sanitation-health-irc-en.pdf.
[11] L. Liu, H.L. Johnson, S. Cousens, J. Perin, S. Scott, J.E. Lawn, I. Rudan, H. Campbell, R. Cibulskis, M. Li, C. Mathers, R.E. Black, Global, regional, and national causes of child mortality: an updated systematic analysis for 2010 with time trends since 2000, Lancet North Am. Ed. 379 (9832) (2012) 2151–2161 .
[12] N. Malik, WASH and the MDGs: the Ripple Effect, (2015), Available at: https://www.huffpost.com.
[13] N. Mason, M.D. Le Sève, R. Calow, G. Jobbins, I. Langdown, E. Ludi, P. Newborne, F. Pichon, 10 Things to Know About the Future of Water and Sanitation. Briefing Papers, Overseas Development Institute (ODI), 2017. Available at: https://www.odi.org/sites/odi.org.uk/files/resource-documents/11720.pdf.
[14] J.R. Mihelcic, C.C. Naughton, M.E. Verbyla, Q. Zhang, R.W. Schweitzer, S.M. Oakley, E.C. Wells, L.M. Whiteford, The grandest challenge of all: the role of environmental engineering to achieve sustainability in the world's developing regions, Environ. Eng. Sci. 34 (1) (2017) 16–43, 2.
[15] B. Mosello, M. Matoso, Obstacles to WASH, UNA, UK, 2017. Available at: https://www.sustainablegoals.org.uk/obstacles-to-wash/.
[16] E.A. Oluwadamisi, M.K.C. Sridhar, A.O. Coker, S.O. Jacob, Septage purification and nutrient uptake potential of water hyacinth (Eichhornia crassipes) in a batch system, Int. J. Eng. Res. Manage. 2349-2058 4 (2) (2017) 49–56 .
[17] M. Satterthwaite, I. Winkler, JMP Working Group on Equity and Non-discrimination final report, WHO/UNICEF Joint Monitoring Programme, 2012. Available at: www.wssinfo.org/fileadmin/user_upload/resources/JMP-END-WG-FinalReport-20120821.pdf.
[18] M.K.C. Sridhar, E.O. Oloruntoba, H.B. Taiwo, O. J., Effectiveness of locally available hand washing materials on pathogen removal from hands of selected market traders, (2016) (Unpublished data).
[19] M.K.C. Sridhar, A.O. Coker, O. Shittu, Rain- A Resource Untapped, Ecosolutions Ltd, Nigeria, 2019, pp. 1–420, ISBN.
[20] F. Sultana, B. Crow, Water concerns in rural Bangladesh: a gendered perspective, in: J. Pickford (Ed.), 26th WEDC Conference – Water, Sanitation and Hygiene: Challenges of the Millennium, 2000, pp. 416–419.
[21] UNICEF, Review of evidence by Joanna Esteves Mills and Oliver Cumming, the impact of water, sanitation and hygiene on key health and social outcomes: review of evidence, (2016) pp. 28.
[22] UNICEF, Guidance on menstrual health and hygiene, (2019), Available at: https://www.unicef.org/wash/files/UNICEF-Guidance-menstrual-health-hygiene-2019.pdf.
[23] UN-Water Global Analysis and Assessment of Sanitation and Drinking Water, Investing in Water and Sanitation: Increasing Access, Reducing Inequalities, World Health Organization, 2014 p. iv, ISBN 978 92 4 150808 7.
[24] USAID, Guidelines for institutional assessment: water and wastewater institutions, (1988) WASH Technical Report No 37. Arlington, Va., WASH Project. Available at: https://niwe.org.ng/wp-content/uploads/2017/10/WATER-Institutional-Assessment.pdf.
[25] USAID/WASH, WASH Technical Report No 7 (1981). Facilitation of community organization: an approach to water and sanitation programs in developing countries (WASH Task No 94): prepared for USAID, USAID/WASH Washington, DC, USA, 1981.

[26] WASH Advocates, Water, sanitation, and hygiene and menstrual hygiene management: a resource guide, (2015). Available at: https://menstrualhygieneday.org/wp-content/uploads/2016/04/WASH-MHM-Resource-Guide-2015.pdf.
[27] Water Affairs and Forestry Department, Sanitation for healthy nation, in: Sanitation Technology Options, Department: Water Affairs and Forestry, Health, Education, Provincial and Local Government, Housing, Environmental Affairs and Tourism, Public Works, Treasury, South Africa, 2002.
[28] Water Crisis, Global water crisis fact sheet, (2017). Available at: https://watermission.org/wp-content/uploads/2017/05/WaterCrisis_FactSheet_2017.pdf.
[29] World Bank, Menstrual hygiene management enables women and girls to reach their full potential, (2018). Available at: https://www.worldbank.org/en/news/feature/2018/05/25/menstrual-hygiene-management.
[30] World Bank, World development report on gender equality and development, (2012). Available at: https://tinyurl.com/yddydcxc.
[31] World Bank, A Guide to Decision making: Technology Options for Urban Sanitation in India, Water and Sanitation Program-South Asia, 2008, pp. 72253. Available at: http://documents1.worldbank.org/curated/en/772471468307155976/pdf/722530WSP0Box30IC00Urban0Sanitation.pdf.
[32] WHO, Healthy Villages: A Guide for Communities and Community Health Workers, World Health Organization, Geneva, Switzerland, 2002.
[33] WHO, Sanitation and hygiene promotion. In: Programming Guidance, Water Supply and Sanitation Collaborative \council and World Health Organisation, Geneva, Switzerland, 2005. Available at: https://www.who.int/water_sanitation_health/hygiene/sanhygpromo.pdf?ua=1.
[34] WHO, Safer water, better health: costs, benefits and sustainability of interventions to protect and promote health, 2008, Available at: http://whqlibdoc.who.int/publications/2008/9789241596435_eng.pdf.
[35] WHO, Drinking Water Fact Sheet, World Health Organization, Geneva, 2019, https://www.who.int/news-room/fact-sheets/detail/drinking-water.
[36] WHO/UNICEF, Water for Life: Making it Happen, WHO/UNICEF, Geneva, 2005.
[37] WHO/UNICEF, Joint monitoring programme (JMP) for water supply and sanitation. Consultation on draft long list of goal, target and indicator options for future global monitoring of water, Sanit. Hyg. 2012, 1–19. Available at: https://washdata.org/sites/default/files/documents/reports/2017-06/JMP-2012-post2015-consultation.pdf.
[38] WHO/UNICEF, Water, sanitation and hygiene in health care facilities: status in low and middle income countries and way forward, (2015). Available at: https://apps.who.int/iris/bitstream/handle/10665/154588/9789241508476_eng.pdf?sequence=1.
[39] WHO/UNICEF, Progress in Drinking Water, Sanitation and Hygiene: 2017 update and SDG baselines, Geneva, Licence:CC BY-NC-SA 3.0 IGO, 2017, pp. 1–108.
[40] World Vision International, The five principles of sustainable WASH, (2018) Available at: https://www.wvi.org/sites/default/files/Sustainable%20WASH%20programmes_lessons%20learnt.pdf.
[41] T. Yates, J. Allen, J.M. Leandre, D. Lantagne, WASH interventions in disease outbreak response. In: Humanitarian Evidence Programme, Oxfam GB, Oxford, 2017.
[42] K. Ziegelbauer, B. Speich, D. Mäusezahl, R. Bos, J. Keiser, J. Utzinger, Effect of sanitation on soil transmitted helminth infection: systematic review and meta-analysis, (2012) PLoS Med, 9.1: e1001162, doi: 10.1371/journal.pmed.1001162.
[43] D. Spears, How much international variation in child height can sanitation explain, working paper 1436, Princeton University, Woodrow Wilson School of Public and International Affairs, Research Program in Development Studies. https://rpds.princeton.edu/sites/rpds/files/media/spears_how_much_international_variation.pdf. (Accessed on 28 September, 2020).

CHAPTER 8

Modelling of a rainwater harvesting system: Case studies of university college hospital, residential apartment and office block in Ibadan city, Nigeria

Omolara Lade[a], David Oloke[b,*]

[a]*Department of Civil Engineering, University of Ibadan, Nigeria* [b]*Faculty of Science and Engineering, University of Wolverhampton, UK*

**Corresponding author.*

8.1 Introduction

Urban water systems are globally under recurring and increasing water stress due to growth in demand, ageing infrastructure, and the uncertainty and variability of climate change. Attention is thus focused on the need to manage demand for potable water rather than investing in large civil engineering projects that would further cause a greater distress to existing water systems [1].

In Nigeria, the Water Corporation for each state is the sole organisation responsible for the treatment and distribution of potable water to households and industries. However, the cost of supplying water to government organisations such as hospitals are significant, and the supply is intermittent and unreliable [2]. Hence, alternative sources of supply such as rainwater harvesting (RWH) can be explored to reduce the cost and other challenges of producing and supplying water.

Rainwater can be a supplementary source of water supply for various non-potable purposes in the home, workplace and garden. RWH is an option where conventional water supply systems have failed to satisfy demand [3]. This technology can serve as part of an integrated water supply system where the city supply is undependable or where local water sources dry up during the year. In developed countries, rainwater is used to complement non-potable purposes, such as clothes washing, toilet flushing, irrigation and outside washes [6]. In developing countries, rainwater is used for potable and non-potable purposes to prevent water shortages [7]. RWH can also help reduce surface runoff. For instance, rainwater was used for recharging the groundwater in a domestic well in Ibadan city, Nigeria. The study revealed

Sustainable Water Engineering. DOI: 10.1016/C2017-0-04301-X

that there was water conservation through reduced evaporation. The experimental well yields water all year round compared to the control well that dries up during the dry season [4]. A typical RWHS consist of the basin surface, the transportation, the storage and dissemination systems. Although watertight areas, such as roads, car parks and pavements can be used for runoff collection [8], the basin surfaces are commonly roofs [9]. Rooftop water harvesting is the collection of rainwater from a roof for potable and non-potable use [5]. Roofs are constructed of various materials such as corrugated cement and clay tiles, corrugated plastic and metal sheets. The rainwater quality and quantity are affected by the catchment material. After collection, it goes through the transportation system to treatment. There are more pollutants in the initial (first) surface runoff compared to successive flows. There is an exponential reduction in the quantity of contaminants integrated with a given rainfall event [10]. Thus, there is a need to divert the initial surface runoff away from the storage device to enhance the standard of water entering storage to reduce or eliminate successive treatment [11]. As rainfall events are unpredictable compared to system demand, a reservoir is needed to collect and hold basin runoff [9].

The use of RWHS in urban areas is still limited due to economic reasons such as, long payback periods which are common in smaller domestic systems. Payback periods depend on factors such as rainfall pattern, maintenance and installation costs of the RWHS, cost of energy, water supply and workmanship. Non-potable water demand was stimulated to size storage tanks in rainwater harvesting system using a dwelling and public building in Brazil. A payback period of 200 years has been found for the typology of a dwelling [12]. Payback periods of more than 75 years were revealed for scenarios simulated in a university accommodation building [13]. The modelling of a RWHS in a residential apartment in Nigeria found a payback period of 21 years and savings of $259 [14].

In Nigeria, safe drinking water is available to less than 30 percent of the population. In 2007, water from improved sources is accessible to only 47 percent of the total population [15]. Several billions of dollars have been spent by Nigerian governments to provide safe drinking water, but most of these projects failed due to fraud. This has led to people drinking contaminated water resulting in water-borne diseases. The demand for water supply is very high in Ibadan City due to inadequate supply by the Water Corporation in the state. Alternative water supply in form of RWHS would alleviate the challenge faced by the city. This study evaluates the potential for improved water supply and savings if adopted in Ibadan. The evaluation was carried out using the following data: precipitation; roof area and average daily water demand of the three case studies.

8.1.1 Study area

The study was based on three case studies: Children out-patient ward of the University College hospital, a residential apartment and an office block in Ibadan, Oyo state. Oyo State is the largest City in the south-west, south of the Sahara Africa (longitude 3°45′–4°00′E, latitude 7°15′–7°30′N). Ibadan is the second largest city in Nigeria with a population of 2,559,853 in 2007 [16] and land area of 400 km^2 [17]. Ibadan is the capital of Oyo state consisting of 11 Local Government areas. In these case studies, there is inadequate water supply for the daily needs of: staff and patients, household occupants and the staff in the office block.

In Nigeria, precipitation is consistent for six months of the year, with a mean annual intensity of 1200–2250 mm. Therefore, rainwater can be viably collected in the southern parts of the country [18]. Depending on the rainfall pattern each year, the rainy season is from May/June to September/October, while November- April are dry. In Ibadan, the highest rainfall occurs in June and has a mean value of 188 mm while the lowest rainfall is in January with a value of 3.7 mm [18].

8.2 Performance of system component

In this section, the following components are constituted within a conceptual RWHS hydrological model.

8.2.1 Precipitation

Factors such as location, weather and year have a significant impact on precipitation. The variance of precipitation is influenced by distance from the coast and the local topology [19]. In Nigeria, the annual rainfall intensity is 0–2400 mm, with the bulk of population residing in locations receiving 0–1350 mm [20]. The North receives less rainfall (~800 mm), than the south. Rainfall data of 30 years was collected from sources such as the Meteorological office and Nigerian Airport Authority. Average monthly rainfall was input into the Rain Cycle model while the rainfall wizard was used to define the rainfall pattern. The annual and average monthly precipitation contained within the data set is presented in Figs. 8.1 and 8.2 respectively.

8.2.2 Catchment surface

Runoff can be harvested from roads, pavements and car parks. However, in urban areas, rainwater is typically collected from roof catchments. Thus, this study is based on roofs rainfall harvesting only.

8.2.3 Runoff coefficient

Runoff coefficient is the ratio of the volume of water that runs off a surface to the total volume of precipitation falling on it [21]. Data are gathered from multiple years to calculate the coefficient, including many storm events. For each storm event, the combination of the runoff

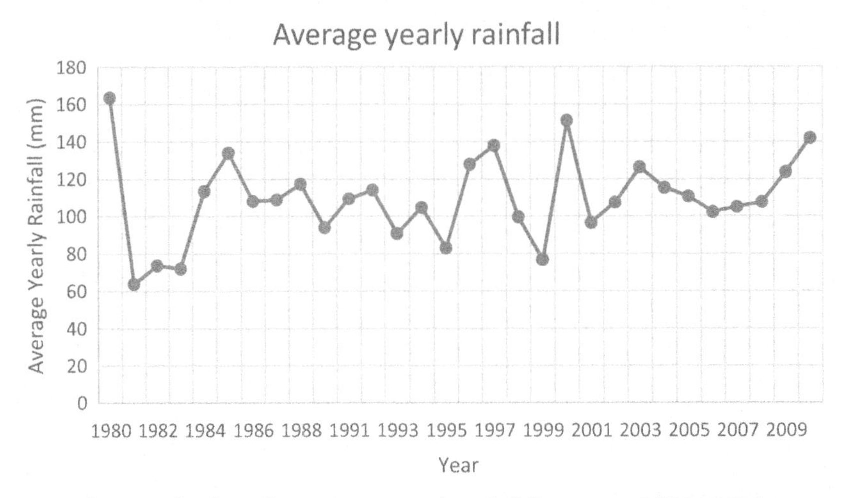

Fig. 8.1: Ibadan City average yearly rainfall pattern 1980–2009. Source: [18]

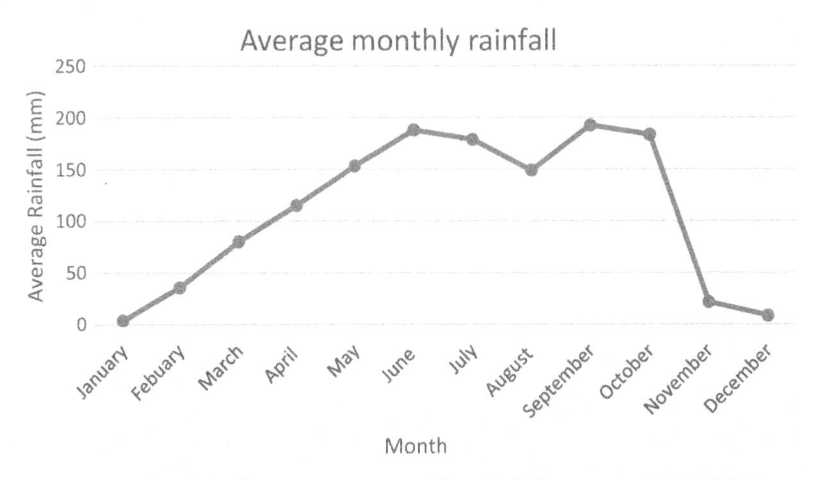

Fig. 8.2: Ibadan City Average monthly rainfall pattern 1980–2009. Source: [18]

coefficients gives the mean value. The runoff coefficient (C_R) can be determined using (Eq. (8.1)) [21].

$$C_R = \text{Runoff volume in } t/\text{rainfall volume} \tag{8.1}$$

where t is the time of measurement.

The amount of precipitation on a catchment surface in time t is given by multiplying the intensity of precipitation in time t by the effective basin area, which is estimated by multiplying the catchment length by the width. Precipitation is assumed to fall vertically onto the roof surface.

$$\text{Effective catchment area} = \text{length} \times \text{width} \tag{8.2}$$

After calculating the effective area of the catchment (Eq. (8.2)), an acceptable runoff coefficient should be determined. Then, the volume of runoff occurring in time t can be calculated using Eq. (8.3).

$$\mathrm{ER}_t = R_t A \times C_\mathrm{R} \tag{8.3}$$

where:

ER_t = effective runoff in time t (m^3)

R_t = rainfall depth in time t (m)

A = effective catchment area (m^2)

C_R = catchment runoff coefficient

8.2.4 Roof areas

Since the level of occupancy strongly influences total water demand within a dwelling, roof areas as a function of occupancy is needed to conduct simulations of water harvesting system installed for the three case studies [22].

8.2.5 Pump

A pump can be modelled hydraulically using the quantity of water requiring pumping per unit time and the rate of pumping. The operating period can be calculated, from which the energy usage of the pump can be determined (Eq. (8.4)). The operating cost per unit time can be determined by the product of pump energy usage and the unit cost of electricity depending on the amount charged by the relevant energy utility.

$$C = \mathrm{Pu}_\mathrm{POW} \times \mathrm{Pu}_\mathrm{TIME} \tag{8.4}$$

where:

Pu_POW = pump hydraulic power (kW)

$\mathrm{Pu}_\mathrm{TIME}$ = pump operating period t (h)

C = operating cost per unit time

8.2.6 Storage tanks

There exists a relationship between the performance of storage capacity, rainfall pattern and demand on the system [9]. Therefore, a rainwater tank is sized to satisfy system demand by considering it to be a reservoir that receives surface runoff over time [9].

The mass curve method has formed the basis of many adaptations [21], for example, sizing fresh water supply reservoirs. The specific periods when the difference between cumulative inflows (precipitation) and cumulative outflows (demand) are at a maximum are identified. This difference represents the maximum volume for the future and maximising the storage capacity for optimum supply. For the storage reservoir in a water harvesting system to be effective, the relationship shown in Eq. (8.5) must be satisfied [9].

$$S \geq \mathrm{Max}\left(\int_{t1}^{t2} [Dt - Qt]\partial t\right) \tag{8.5}$$

$t_1 < t_2$ and:

S = storage volume (m^3)

Dt = water demand during time interval t (m^3)

Qt = precipitation during time interval t (m^3)

t = time of measurement

8.3 Methodology

The economic viability of installing a RWHS can be evaluated by estimating the return period using a Rain-cycle model to maximise storage capacity and water reduction. Sensitivity examination and Monte Carlo simulation were also carried out. A lower return period forms a more attractive investment. The return period is determined by pairing the expenditure with water savings. Water savings resulting in a decrease in potable water cost is considered as it affects total charges. The market value of the RWH components was used to account for direct cost as expenditures are associated with the investment and operational costs of the RWH system. The methodology to calculate the water savings and costs are presented in the following sections.

8.3.1 Water savings

Water savings were achieved by the balance between daily water consumption and harvested rainwater. For this study, a behavioural theory was used and Yield After Spillage (YAS) was approved [23] due to the conservativeness of the estimate given on system accomplishment. However, time accuracy was estimated using Yield Before Spillage (YBS) regulations in preference to YAS. The rain-cycle model adopted in this work incorporates YAS/YBS algorithm with the storage operating variable θ set to zero (YAS) as the default approach. However, investigation proposed that YAS models can model system performance within 10 percent of that envisaged by an hourly time-step model which was an acceptable limit of error if certain constraints regarding the chosen time-step are engaged [23].

8.3.1.1 Water availability

Factors such as precipitation pattern, catchment surface and water losses determine the available rainfall. A continuous 30-year daily rainfall record (1980–2009) was obtained from the City's Meteorological stations [18,24]. The monthly rainfall contained within the data set is presented in Fig. 8.2.

8.3.1.2 Water demand

Potable and non-potable uses (toilet flushing and clothes washing) were examined in this study. In Nigeria, it is difficult to gather viable information on water usage. Water demand was estimated by determining the amount of water used based on the number of 8 L buckets consumed per day [25]. The standard water consumption is approximately 50 Ls per person per day [26].

8.3.1.3 Estimating non-potable domestic demand

Factors such as household size, the season of the year, type of property and ages of household occupants have an impact on per capita consumption [27]. To attain the lowest level of disposition, a minimum per capita usage of 120 Ls per day is required. Thus, a daily per capita usage of 120 Ls was assumed in this work.

8.3.1.4 Water closet demand

A maximum flush volume of 6 Ls is recommended for single flush WCs. Available data on previous monitoring studies was used as an indicator of future behaviour as it was not possible for WC usage frequency to increase or decrease significantly. The mean value equals 4.59 flushes per person per day (Table 8.1). However, a per capita usage of 6 times/day deduced for weekends (Saturday and Sunday) was used for case studies 1 and 2. A progressive relationship exists between household residents and the rate of WC flushes [28]. Household usage was calculated by multiplying the household occupancy rate by the frequency of capita usage.

Table 8.1: Scope of domestic WC usage frequencies.

Use/person/day	References
3.3	[29]
3.7	[28]
5.25	[30]
6–8*	[31]
4.3	[32]
4.8	[33]
4.8	[34]
4.59	Mean (of above)

8.3.1.5 Washing machine demand

There is an association between household occupancy and washing machine usage [29]. Predicted future per capita use will not differ much from those occurring at present. The mean value of 0.21 usage per person was used as the standard value for domestic simulations. The family usage was determined by simply multiplying the household occupancy rate by the per capita frequency.

8.3.2 Costs

The cost of water supply was derived from Water Corporation of Oyo State (WCOS) while the cost of the rainwater catchment components was obtained from a market survey. The data on economic details were processed into the model (Table 8.2).

Table 8.2: Economic details.

Variable	Hospital	Residential	Office
Total cost	$1277.00	$496.00	$1047.00
Low decommissioning rate	$0.000	$0.00	$0.00
Interest rate	3.5 percent	3.5 percent	3.5 percent
Energy	0.1 c/kWh	0.1 c/kWh	0.1 c/kWh
Main water cost	$0.83/m^3$	$0.83/m^3$	$0.83/m^3$

8.3.3 Return period

The payback period is the amount of time a project is expected to take to earn net revenue equal to the capital cost of the project. It is measured as the ratio between total capital costs and the difference between annual revenue and annual expenditures, considering the discount rate. Data on existing water charges by the water industry are used to compare the payback period.

8.3.4 The analysis of the rain cycle model

A succession of analytical steps is followed to increase the likelihood of creating a successful RWHS design. The steps are as follows (Fig. 8.3):

- Estimate reservoir sizes
- Estimate savings of reservoir
- Gather data required for comprehensive analysis
- Carry out a comprehensive analysis and appraise results

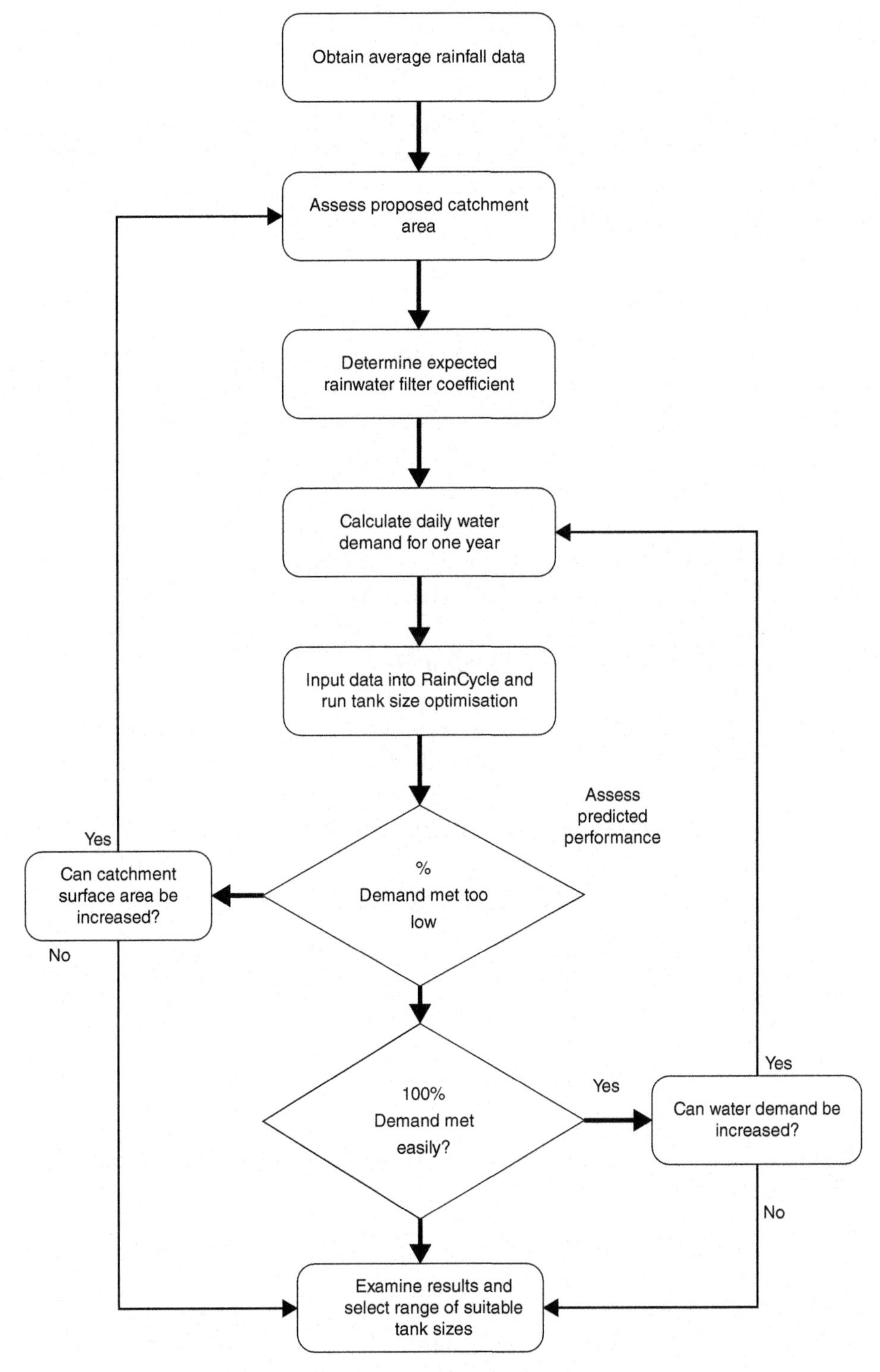

Fig. 8.3: Determining range of suitable tank sizes.
Source: [35]

8.4 Results and discussion

The proposed system was assessed using seven fixed (UV hydraulic power, operating time, basin area, pump hydraulic power, storage capacity, pump capacity and initial runoff volume) and eleven variable parameters (disposal and asset disposal obligation, runoff coefficient, main water supply, total cost, filter coefficient, rainfall profiles, energy cost, discount rates and water demand).

8.4.1 Monte Carlo simulation

In this simulation method, random numbers and probability distributions were used in solving problems. New values for the variable parameters were randomly generated and a system analysis was run using the new values. The results of thousands of simulations were used to evaluate the RWHS response to a wide range of conditions; each parameter involves three values: the highest most probable value, most probable and lowest value. For each iteration, a new set of variables were generated by the random process of set binomial dissemination. The results of such iteration were used to predict the binomial that long-term savings are equal or more than a specified amount or that system payback takes place within a given period.

8.4.2 System detail (case study 1): a hospital block

The hospital ward provides facilities for 203 patients and nurses. Direct measurement was completed to determine the length and width of the building as a detail roof plan was not available. The hydraulic elements of the three case studies are presented in Table 8.3. The time frame for the analysis was 50 years which is the expected operational life of the building.

Table 8.3: Hydraulic details.

Parameter	Hospital	Residential	Office
Rainfall intensity	1,311 mm/year	1311 mm/year	1311 mm/year
Catchment area	8132 m^2	196 m^2	566 m^2
Runoff coefficient	0.85	0.85	0.85
Filter coefficient	0.90	0.90	0.90
Reservoir capacity	12 m^3	4 m^3	10 m^3
Pump hydraulic power	1.4 kW	0.80 kW	1.2 kW
Pump capacity	60 Ls/min	60 Ls/min	65 Ls/min
UV hydraulic power	0 W	0 W	0 W
Water demand	4704 m^3/year	258 m^3/year	485 m^3/year

The results of optimizing storage capacity show that 78.1 percent of the maximum demand could be met with a tank size of 12 m^3. The quantity of water available was the restraining factor, hence, a storage capacity greater than 12 m^3 will not be beneficial to the system.

In optimizing water saving, seven reservoirs with potential long-term profit were revealed. A storage capacity of 12 m^3 estimated to save \$51,072 over 50 years with a pay-back period of 1 year was found to be the best. 78.1 percent of the predicted demand was met which was very good for a commercial system. The water supply value for the three case studies is presented in Table 8.4. A storage tank of 12 m^3 gave an acceptable result. The data was input into the Storage Tank Module and WLC Details module, the result in the Analysis System module was examined.

Table 8.4: Water supply value.

Parameter	Hospital	Residential	Office
Main water price/m^3	\$0.40	\$0.40	\$0.40
Harvested water price/m^3	\$0.14	\$0.37	\$0.29

8.4.3 System details (case study 2): a residential apartment

This building is a twin semi-detached bungalow designed and built to provide facilities for a household of 12 people. It has a roof area of 195.69 m^2. The data on the hydraulic detail of the system is presented in Table 8.3. The expected operational life of the building is 50 years, and this was used as the time frame for the analysis.

The tank size optimisation and analysis reveal that the maximum percentage of demand that could be met by harvested rainwater was 70.6 percent with a tank size of 4 m^3. Therefore, the limiting factor was the amount of water available. Thus, increasing the tank size above 4 m^3 would have little benefit.

The analysis on optimising saving showed the suitability of four storage tank sizes with a potential long-term profit. The best size was a 4 m^3 tank which was predicted to save \$259 over 50 years and had a payback period of 21 years, which is typical for a current domestic system at 70.6 percent of predicted demand. The 4 m^3 gave acceptable results and so the data for this tank was input into the storage tank module (tank size) and whole life cycle (WLC) details module and then the result in the analysis system module were examined.

8.4.4 System details (case study 3): an office block

Case study 3 is a two-storey office block with a roof area of 566 m^2. Data on the system's hydraulic detail is presented in Table 8.3. The operational life of the storey building is 50 years which was the time frame used for the analysis.

The tank size optimisation and analysis reveal that the maximum percentage of demand that could be met by harvested rainwater was 75.0 percent with a tank size of 10 m^3. The limiting factor was the amount of water available. Thus, increasing the tank size above 10 m^3 would have little benefit.

The analysis on optimising saving showed the suitability of six storage tank sizes with a potential long-term profit. The best was the 10 m^3 tank which was predicted to save $2,564 over 50 years and had a pay-back period of 8 years. This is typical of a commercial system. The 10 m^3 gave acceptable results and the data was input into the Storage Tank module and WLC Detail module. The result in the Analysis System module was then examined.

8.4.5 Payback period

The prospective water savings from rainwater usage was evaluated in 195 towns of South-eastern Brazil [36]. The result revealed potable water savings of 12–79 percent per year for the towns and dwelling. An ideal tank capacity of 3–7 m^3 is required for a high potable demand while 2–20 m^3 for low demand. Another study examines the feasibility of RWH for roof catchments in Australia [37], a model was developed to simulate the performance of a RWH system. The findings revealed that the reliability of a RWH system depends on the mean annual rainfall in which 20 kL tank can provide reliability of 61–97%percent for toilet and laundry usage depending on the location in Australia. In another study, rainwater tank was evaluated and model for large roof areas in Australia [40]. Decision support tool was used in carrying out behavioural analysis and model reservoir capacity. Analysis revealed the effectiveness of both tanks in wet season while it becomes less effective in dry seasons. A return period of 15–21 years was revealed. The financial variability of a RWHS in single and multi-buildings was investigated in Spain [39]. Return periods were between 30 and 60 years.

This study examined the economic variability of a hospital, residential apartment and an office block in Nigeria with a payback period of 1, 21 and 8 years respectively. The results were similar to the study on the economic variability of domestic RWHS in high rise buildings in four towns in Australia and Sidney [38], where the shortest return period (about 9 years) was found.

8.5 Conclusion

This chapter presented the hydraulic and financial modelling of a RWHS using three case studies: a bungalow, an office with two floors and a hospital with five floors. The water usage and precipitation pattern of the three case studies were monitored, and the result indicated both monetary and water savings are possible in the long-term. A decision support tool (rain cycle) was used to maximise storage capacity and water reduction. Sensitivity examination and Monte Carlo testing were also carried out.

The proposed system was assessed using seven fixed and eleven variable parameters. The fixed criteria were: UV hydraulic power, operating time, basin area, pump hydraulic power, storage capacity, pump capacity and initial runoff volume, while the variable criteria were: disposal and asset disposal obligation, runoff coefficient, main water supply, total cost, filter coefficient, rainfall profiles, energy cost, discount rates and water demand.

Analysis of Case Study 1 reveals that the maximum percentage of annual demand that could be met was 78.1 percent with a tank size of 12 m^3. A savings of $51,072 over 50 years and a payback period of 1 year were predicted, which is very good for a current commercial system.

Analysis of Case Study 2 reveals that the maximum percentage of annual demand that could be met was 70.6 percent with a tank size of 4 m^3. A savings of $259 over 50 years and a payback period of 21 years were predicted, which is typical for a current domestic system.

Analysis of Case Study 3 reveals that the maximum percentage of annual demand that could be met was 75.0 percent with a tank size of 10 m^3. A savings of $2,564 over 50 years and a payback period of 8 years were predicted, which is good for a current commercial system.

Acknowledgement

This research was funded under the ETF programme, Nigeria

References

[1] C.M. Rui, S.P. Francisco, M. João, Redrafting, water governance, guiding the way to improve the status quo, Utilities. Policy 43 (2016) 1–3.

[2] B. Sudeshna, F. Vivien, Y. Yvonne, S. Heather, W. Quentin, Cost recovery, equity, and efficiency in water tariffs: evidence from African utilities, African Infrastructure Country Diagnostic, World Bank, 2008, pp. 5.

[3] M.A. Alam, Technical and social assessment of alterative water supply options in Arsenic affected areas, M.Sc. thesis. Civil Engineering of Bangladesh University of Engineering and Technology, 2006 (BUE 1).

[4] O.O. Lade, A.O. Coker, M.K.C. Scridhar, Sustainable water supply for domestic use: application of roof-harvested water for groundwater recharge, J. Environ. Sci. Eng. A 1 (5) (2012) 581–588.

[5] C.B.J. Mendez, B. Klenzendorf, B.R. Afshar, M.T. Simmons, M.E. Barrett, K.A. Kinney, M.J. Kirisits, The effect of roofing material on the quality of harvested rainwater, Water. Resour. 4 (5) (2011) 2049–2059.

[6] T. Hermann, U. Schimida, Rainwater utilisation in Germany: efficiency, dimensioning, hydraulic and environmental aspects, Urban. Water 1 (4) (1991) 307–316.

[7] V. Meera, M.M. Ahammed, Water quality of rooftop rainwater harvesting system: a review, J. Water Supply Res. Technol.-AQUA 55 (2006) 257–268.

[8] Environment Agency, Saving water: on the right track, A Summary of Current Water Conservation Initiatives in the UK, March 1999, Environment Agency, National Water Demand Management Center, Worthing, West Sussex, UK, 1999.

[9] A. Fewkes, The technology, design and utility of rainwater catchment systems, in: D. Butler, F.A. Memon (Eds.), Water Demand Management, IWA Publishing, London, 2006.

[10] J.H. Lee, K.W. Bang, L.H. Ketchum, J.S. Choe, M.J. Yu, First flush analysis of urban storm runoff, Sci. Total Environ. 293 (1–3) (2002) 163–175.

[11] C. Wu, L. Junqi, L. Yan, W. Wenhai, First flush control for urban rainwater harvest systems, Proceedings of 11th International Rainwater Catchment Systems Conference, Texcoco, Mexico, 2003.

[12] C. Santos, F. Taveira-Pinto, Analysis of different criteria to size rainwater storage tanks using detailed methods, Resour. Conserv. Recycl. 71 (2013) 1e6.

[13] J. Devkota, H. Schlachter, D. Apul, Life cycle-based evaluation of harvested rainwater use in toilets and for irrigation, J. Cleaner. Prod. 95 (2015) 311e321.

[14] O.O. Lade, D.A. Oloke, Modelling rainwater system harvesting in Ibadan, Nigeria: application to a residential apartment, Am. J. Civil Eng. Architect. 3, (3), (2015), pp. 86–100. Online: http://pubs.sciepub.com/ajcea/3/3/5 © Science and Education Publishing, doi:10.12691/ajcea-3-3-5.

[15] O.O. Aladenola, O.B. Adeboye, Assessing the potential of rainwater harvesting, Water Resour. Manage. 24, (2010) 2129–2137.
[16] SSN, Socio-economic survey of Nigeria, Social Statistics in Nigeria, Abuja, Federal Republic of Nigeria, 2007.
[17] A. Onibokun, A. Faniran, Women in Urban Land Development in Africa: Nigeria, Ghana and Tanzania, Centre for African Settlement Studies and Development (CASSAD), Nigeria, 1995, pp. 75.
[18] DMS, Weather Observation Station, Department of Meteorological Services, Samonda, Ibadan, 2010.
[19] T. Thomas, RWH performance predictor for use with coarse (i.e. monthly) rainfall data, Domestic Roof Water Harvesting Research Programme Report RN-RWH04, Development Technology Unit, University of Warwick, 2002.
[20] NPC, Nigeria Population Commission Official Result for 2006 House and Population Census Figures, Bureau for National Statistics Abuja, Nigeria, 2006. Online: http://placng.org/Legal%20Notice%20on%20Publication%20of%202006%20Census%20Final%20Results.pdf (accessed 22.07.19).
[21] J. Gould, E. Nissen-Peterson, Rainwater Catchment Systems for Domestic Supply: Design, Construction and Implementation, Intermediate Technology Publications, London, 1999, pp. 335.
[22] F.A. Memon, L. Ton-That, D. Butler, An investigation of domestic water consumption through taps and its interaction with urban water flows, Water Sci. Technol. Water Supply 5–6 (7) (2007) 69–76.
[23] A. Fewkes, D. Butler, Simulating the performance of rainwater collection and reuse system using behavioural models, Build. Serv. Eng. Res. Technol. 21 (2000) 99–106.
[24] IITA, Weather Observation Station, International Institute of Tropical Agriculture, 2010.
[25] K.O. Adekalu, J.A. Osunbitan, O.E. Ojo, Water sources and demand in South Western Nigeria: implications for water development planners and scientists, Technovation 22 (12) (2002) 799–805.
[26] UN, Water supply and waste disposal poverty and basic need series, september warm glacial climate, Science (1997) 1257–1266.
[27] F.A. Memon, D. Butler, Water consumption trends and demand forecasting techniques. In: Water Demand Management, IWA publishing, London, 2006, pp. 1–26.
[28] D. Butler, The influence of dwelling occupancy and day of the week on domestic appliance wastewater discharges, Build. Environ. 28 (1) (1991) 73–79.
[29] J.E. Thackray, V. Crocker, G. Archibald, The Malvern and Mansfield studies of domestic water usage, Proc. Inst. Civil Eng. 64 (1978) 37–61.
[30] SODCON Survey of domestic consumption, in: Water Demand Management, Anglian Water, Normich, 1994, pp. 361.
[31] A. Fewkes, The use of rainwater for WC flushing: the field testing of a collection system, Build. Environ. 34 (1999) 765–772.
[32] Environment Agency, A Scenario Approach to Water Demand Forecasting, National Water Demand Management Center, Environment Agency, West Sussex, UK, 2001.
[33] V.K. Chambers, J.D. Creasey, E.B. Glennie, M. Kowalski, D. Marshallsay, Increasing the value of domestic water use data for demand management- summary report, WRC collaborative Project CP 187, Report no. P6805, Swindon, Wiltshire, UK, 2005.
[34] DCLG, Code for sustainable homes: technical guide, Department for Communities and Local Government, Communities and Local Government Publications, HMSO, London, 2007.
[35] R.M. Roebuck, R.M. Ashley, Predicting the hydraulic and life cycle cost performance of rainwater harvesting systems using a computer-based modelling tool, Proceedings of 7th International Conference on Urban Drainage, Melbourne, Australia, 2006.
[36] E. Ghis, D.L. Bressan, M. Martina, Rainwater tank capacity and potential for potable water savings through using rainwater in the residential sector of South-eastern Brazil, Build. Environ. 42 (2007) 1654–1666.
[37] E. Hajani, A. Rahman, Rainwater utilization from roof catchments in arid regions: a case study from Australia, J. Arid. Environ. 111 (2014) 35–41.
[38] Y. Zang, D. Chen, L. Chen, S. Ashbolt, Potential for rainwater use in high-rise buildings in Australian cities, J. Environ. Manage. 91 (1) (2009) 222–226.

[39] L. Domenech, D.A. Sauri, Comparative appraisal of the use of rainwater harvesting in single and multi-family buildings of the metropolitan area of Barcelona (Spain): social experience, drinking water savings and economic costs, J. Cleaner. Prod. 11 (2010) 1–11.

[40] M. Alam Imteaz, S. Abdallah, R. Ataur, A. Amimul, Optimisation of rainwater tank design from large roofs: a case study in Melbourne, Australia, Resour. Conserv. Recycl. 55 (2011) 1022–1029.

CHAPTER 9

Phytotechnologies in wastewater treatment: A low-cost option for developing countries

Mynepalli K.C. Sridhar[a], Akinwale O. Coker[b,*], Olalekan I. Shittu[b], Temitope A. Laniyan[a], Chibueze G. Achi[b]

[a]*Department of Environmental Health Sciences, Faculty of Public Health, College of Medicine, University of Ibadan, Ibadan, Nigeria* [b]*Department of Civil Engineering, Faculty of Technology, University of Ibadan, Ibadan, Nigeria*
**Corresponding author.*

9.1 Introduction

Phytoremediation is a term for the group of environmentally friendly technologies that use plants (phyto) for remediating contaminated water. Phytoremediation is the inherent ability of certain plants known as hyperaccumulators to bio-accumulate, degrade or render harmless, contaminants in soils, water or air [39]. Phytoremediation is defined as the use of green plants, including grasses and woody species, to address environmental contamination by heavy metals, metalloids, trace elements, organic compounds and radioactive compounds in soil or water [24]. They make use of science and engineering principles. There are three types:

(a) ecotechnology involving ecological systems;
(b) phytotechnologies involving plants and
(c) biotechnologies involving living beings.

They take up the pollutants through a series of processes such as absorption, degradation, accumulation, sequestration, volatilisation or stabilisation of organic and inorganic materials from the environment. About 300 years ago, the idea of using plants for treatment of wastewater was mooted [22]. A recent survey by Marmiroli et al. [32] showed that there are about 29 countries and 350 groups who are actively involved in research or the application of phytotechnologies.

Wetlands, an example of natural phytotechnology, existed in several parts of the globe. They are transitional between terrestrial and aquatic systems where the water table is usually at or near the surface or the land is covered by shallow water. It is a generic term and

Sustainable Water Engineering. DOI: 10.1016/C2017-0-04301-X

is known for supporting aquatic vegetation and providing a bio-filtration capability in the removal of water pollutants. They may be natural marsh and swamp environments but also artificially constructed storage basins or ponds. They were also called as 'the kidneys of the landscape' [34]. Natural wetlands reduce diffuse pollution within a definite range of operational conditions. When hydrologic or pollutant loadings exceed their assimilative capacity, they rapidly become stressed and degraded. Artificially constructed wetlands (storage basins or ponds) create 'generic' wetland habitats, and play a role in the control of flooding and pollution.

Constructed wetlands are artificial, engineered, designed and constructed to treat wastewater similar to a natural wetland. It is a practical alternative to the functional component of conventional wastewater treatment plants. The technology of constructed wetlands for wastewater treatment has been in use for over 40 years. Constructed wetlands (or basins) normally have non-soil substrates and a permanent (but normally shallow) water volume that can be almost entirely covered in aquatic vegetation. These wetlands may contain marsh, swamp and pond (lagoon) elements and may trap sediment. The dominant feature of the system is the macrophyte zone containing emergent and/or floating vegetation that requires (or can withstand) wetting and drying cycles. The constructed wetland may be classified as either dependent on surface flows (i.e. SFs constructed wetland, SFCW) or subsurface flows (i.e. SSF constructed wetland SSFCW). The SSFCW has been further classified either as a horizontal flow (HF) constructed wetland (HFCW) or vertical flow (VF) constructed wetland (VFCW). They can be used as part of decentralised wastewater treatment systems, due to their 'robust', 'low-tech' nature with none or few moving parts such as pumps and they work on relatively low operational requirements. Phytotechnologies can be used for the treatment of domestic and municipal wastewater, greywater and selected industrial and commercial wastewaters.

9.2 Wastewaters

Wastewater is any water that has been affected by human use. Wastewater is 'used water from any combination of domestic, industrial, commercial or agricultural activities, surface runoff or stormwater and any sewer inflow or sewer infiltration' [56]. Wastewaters may be classified into several following categories:

a) Greywater: Generated in a house such as the bathroom, kitchen, laundry but excludes faecal matter.
b) Black water: Greywater plus human faeces arising from toilets.
c) Stormwater: Rain runoff that may pick up various contaminants from drains or washings. Stormwater can either be mixed with greywater and black water in a combined system, or be disposed of in a separate system. Older built-up areas tend to have combined systems, newer ones tend to be separated.

d) Industrial wastewaters: Specific to a particular industry, some may contain toxic and hazardous chemicals. In Nigeria and other African countries these may also include small-scale industries such as food processing and agro-based.
e) Trade and commercial wastewaters: Emanate from trading and business outfits.
f) Infectious wastewaters: Generated from various categories of healthcare facilities such as hospitals, wards, clinics or primary healthcare centres. These are suspected to contain hazardous chemicals, drugs and pathogenic organisms.

9.3 The origins of phytotechnologies

The history of the use of plants in the treatment of wastewater may go back to antiquity, as shown in Table 9.1, their design and use has developed rapidly over the last few decades.

Table 9.1: History of constructed wetlands using phytotechnologies.

Year	Development	Author
Early period	Drainage in every human settlement, ponds were common. Constructed (or 'artificial') wetlands developed to manage drainage until 1980s.	[61]
1952	First experiments to assess the potential of wastewater treatment using wetland plants at the Max Planck Institute, Germany by Dr Seidel.	[62]
Mid-1960s	Seidel and Kickuth (Institute of Bodenkunde, University of Gottingen) introduced the root zone method, which was a rectangular bed planted with reeds (*Phragmites australis*) and horizontal subsurface flow. In this method, organic matter decomposed and nitrogen nitrified and then denitrified.	[12]
1967	Based on Seidel's ideas, a large-scale treatment system, the Lelystad Process, developed by the Ijsselmeer polders Development Authority in the Netherlands. The first treatment facility constructed treated wastewater from a campsite near Elburg catering for 6000 people per day during summer. This system was star-shaped, with a free water flow over 1 ha and a depth of 0.4 m. Subsequent systems consisted of shallow ditches up to 400 m long, which were mechanically maintained. Reeds beds were also used across the Netherlands.	[26]
1970s	Experiments with different designs began in North USA.	[12]
1980s	Austria began to experiment with horizontal beds. National regulations for the design of horizontal constructed wetlands released in 1997.	[12]
1990s	Major increase in constructed wetlands in the 1990s to treat different kinds of wastewater e.g., industrial wastewater and storm water.	[12]
Current	Use of constructed wetlands for wastewater treatment more popular worldwide. Subsurface flow constructed wetlands common in many developed countries e.g., Germany, UK, France, Denmark, Austria, Poland and Italy.	[12]

Constructed wetlands may be classified based on the following:

i) The type of macrophytic growth (emergent, submerged, free floating and rooted with floating leaves); or
ii) The water flow regime (SF, subsurface, vertical or HF). VF systems (VFS) consist of intermittent batch-fed wastewater, in which the wastewater infiltrates down through the media bed.

Phytotechnologies are also classified according to the mechanism involved. They may be applied in the treatment of wastewater or contaminated soils.

a) Phytosequestration. This is also called 'phytostabilisation', which involves absorption of pollutants by roots, adsorption to the surface of roots or the production of biochemicals by the plant released into the medium in the immediate vicinity of the roots that can sequester, precipitate or otherwise immobilise contaminants.
b) Rhizodegradation. This process takes place immediately surrounding the plant roots. Exudates from plants stimulate rhizosphere microbes to enhance biodegradation of contaminants.
c) Phytohydraulics. This is carried out by use of deep-rooted plants, usually trees—to contain, sequester or degrade contaminants that come into contact with their roots. For example, poplar trees have been used in the treatment of methyl-tert-butyl-ether [25,30].
d) Phytoextraction. Also known as 'phytoaccumulation' where plants take up or hyperaccumulate contaminants through their roots and store them in the tissues of the stem or leaves. The contaminants are not necessarily degraded but are removed from the environment when the plants are harvested. This is particularly useful for removing metals from soil. In some cases, metals can be recovered for reuse by incinerating the plants in a process called phytomining.
e) Phytovolatilisation. Plants take up volatile compounds through their roots, transpire them or their metabolites through the leaves and release them into the atmosphere.
f) Phytodegradation. Contaminants are taken up into the plant tissues where they are metabolised, or bio-transformed. The nature and degree of transformation depends on the type of plant and may occur in any of the plant parts.

Broadly these technologies are grouped as SF, SSF and VF types as illustrated in Fig. 9.1.

The area required for using phytotechnologies is proportional to the size of the residential population and is generally around 5 m^2 per person. According to [13], constructed wetlands may be of three types:

(i) Type A can be single- or two-stage form; the latter is intended to utilise the sediment adsorptive capacity to degrade nutrients.
(ii) Type B is intended to provide a secondary biological treatment for surface runoff; low in solids but carrying high levels of organics and dissolved pollutants.
(iii) Type C provides a hybrid tertiary form of treatment for low flow volumes that may be source: associated with high levels of micro-pollutants, e.g., metals or pesticides.

Constructed wetlands normally have non-soil substrates and a permanent shallow water column that may be entirely covered in aquatic vegetation. They may contain marsh, swamp and pond (lagoon) elements; the inlet zone, for example, can resemble a pond and be used as a sediment trap. The dominant feature of the system is the macrophyte zone containing

Surface Flow Wetlands

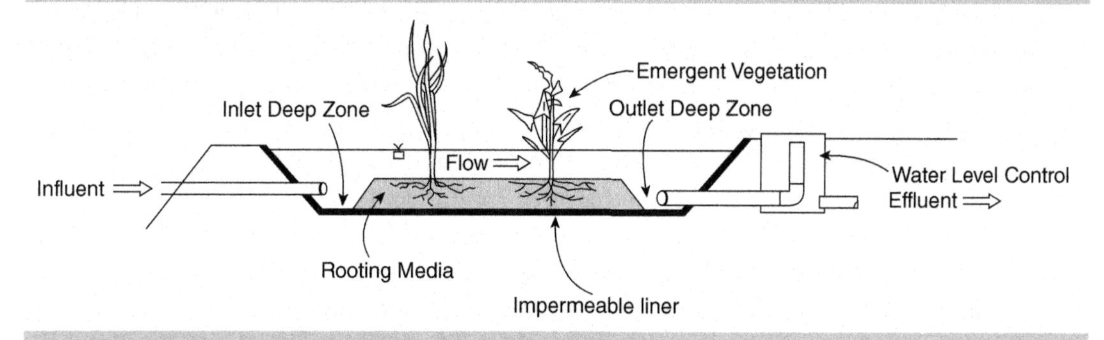

Subsurface horizontal flow constructed wetlands (CWM1)

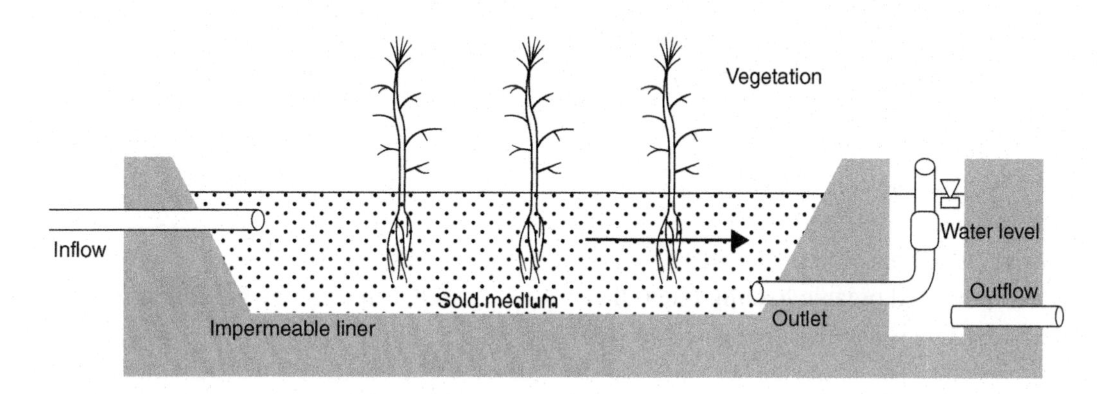

Subsurface vertical flow constructed wetlands (CW2D)

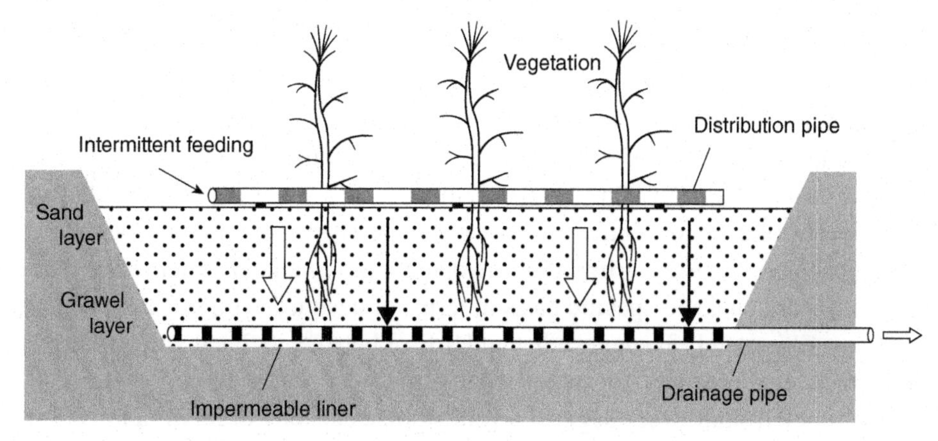

Fig. 9.1: Types of phytotechnologies in wastewater treatment.

emergent and/or floating vegetation; they can withstand wetting and drying cycles. A few terminologies are in vogue:

- Free water surface (FWS). May be SF or SSF types. SF systems contain aquatic plants rooted into the soil layer of the wetland; water flows through their leaves and stems above the substrate. They are usually designed to tolerate a permanent depth of the treated water, which flows horizontally across the wetland bed.
- Vegetated submerged bed (VSB) systems. These contain aquatic plants in a media such as crushed rock, small stones, gravel, sand or soil. Wastewater to be treated flows horizontally or vertically through a selected bed medium and the root zone. These cost more to construct than FWS wetlands due to cost of the media.
- HF systems. Here the effluent is fed into the inlet and flows slowly through the porous medium (normally gravel) horizontally under the surface of the bed to the outlet. To design these systems, for example in a domestic setting, the following equations are generally used [53]:

$$\text{Bed design equation: } Ah = Qd\,(\text{Cin} - \text{Cout})/KBOD$$

where Ah = Surface Area of the bed (m^2)
Qd = average flow (m^3/day)
Cin = influent BOD_5 (mg/L),
Cout = effluent BOD_5 (mg/L)
KBOD = rate constant (m/day)
Specific area: 5-10 m^2/PE
where PE = Population equivalent
Recommended organic loading rate (OLR): 8 – 12 g BOD_5 m^2/day; 25 g COD m^2/day (where COD is chemical oxygen demand)
- VF systems. These may have a sand cap overlying the graded gravel/rock substrate, and are intermittently dosed from top to flood the surface of the bed; the effluent drains vertically down through the bed and collected at the base.

$$\text{Bed design equation: } A1 = 5.25P0.35 + 0.9P$$

where
Specific area = 1–2 m^2/PE^2
Recommended OLR: 8–12 g BOD_5 m^2/day; 25 g COD m^2/day
- Hydraulic loading rate (HLR) refers to the loading on a water volume per plan area basis. Loading = (parameter concentration) (water volume/area). Water depths and duration of flooding become important criteria for the operation of wetland systems and these need to be considered on a site-specific basis in terms of design storm, substrate and vegetation conditions. Ellis et al.[13] suggest that an HLR of between 0.2 and 1m^3/m^2/day (wetland surface area) provides the maximum treatment efficiency. For a 5 mm and 10 mm effective runoff volume, they suggest a void storage capacity of 50 m^3 and 100 m^3 per

impervious hectare, respectively. To estimate HLR (m/d), the following equation is generally used [13]:

$$HLR = Qin / As$$

Where Qin = inflow rate (m^3/d)
As = wetland surface area (m^2)
During storm events, when high rates of stormwater runoff may discharge onto constructed wetlands, optimal HLRs should not exceed $1 m^3/m^2/d$ to achieve a satisfactory treatment.

- Hydraulic residence time (HRT) is the average time that water remains in the wetland, expressed as mean volume divided by mean outflow rate. If short-circuiting develops, effective residence time may differ significantly from the calculated residence time.
- Hydraulic retention rate is the most important factor influencing the treatment mechanism and is the average time that water remains in the wetland. This can be expressed as the ratio of the mean wetland volume to mean outflow (or inflow) rate.
- HRT, usually measured in days, this is the volume (LWD) of free water in the wetland divided by the volumetric inflow rate (Qin; m^3/day):

$$HRT = LWD/Qin\left(\text{or } D/Qin\right)$$

Where L and W = length and width (m);
D is free water depth (expressed as: porosity × water depth).

The mean retention time can also be determined by undertaking an accurate tracer study. Wetlands should have a minimum retention time of at least 10–15 h for the design storm event or alternatively retain the HLR (m/d):

$$Qin/As$$

where Qin = inflow rate; m^3/d;
As = wetland surface area; m^2

An average annual storm volume of at least 5–10 h should be considered to achieve a high level of removal efficiency.

- Flow velocity. This should not exceed 0.3–0.5 m/s at the inlet zone if effective sedimentation is to be achieved. At velocities greater than 0.7 m/s, high flow may damage the plants physically and cause a decline in system efficiency.
- Inlet. The inlet pipe should be constructed in such a way that influent flow is evenly distributed across the width of the bed. This may be achieved using slotted inlet pipes or a notched gutter (slots should be large enough to prevent clogging by algae).
- Outlet. The level at which the outlet is set is determined by the lowest water level required in the constructed wetland. Until further information is available, it is considered

[13] that the lowest level in the wetland should be 300 mm below the substrate surface dependent on plant type.

- Aspect ratio. An aspect ratio (AR) (length: width) of 4:1 for SSF wetlands and 10:1 or higher for SF wetlands has been recommended for domestic wastewater treatment wetlands. Studies have indicated that this can vary if the distribution is well designed [17,18].
- Aeration. A grid of slotted plastic pipes (dia 100 mm) could be installed vertically in the substrate, 100 mm protruding above the surface, and penetrating the full depth of the substrate at 5 m centres to serve as static ventilation tubes for aeration of the root zone.
- Substrate structure. Horizontal SF wetlands utilise a natural soil substrate to provide organics and nutrients to maintain plant growth, whereas SSF wetland substrates should primarily provide a good hydraulic conductivity. Gravel provides the most suitable substrate for SSF emergent plants, supporting adequate root growth, high conductivity and superior permeability.
- Optional structures. These may be installed depending on the need and may include oil separators, pre-treatment, sedimentation tank and final settlement tanks.

9.4 Mechanisms of phytoremediation

Phytoremediation uses trees, shrubs, grasses and aquatic plants and their associated microorganisms to remove, degrade or isolate toxic substances from the environment. It is a passive technique based on the ability of plants to clean and restore the contaminated media. Green plants are particularly endowed with great metabolic and absorption capabilities, coupled with a good transport system that has the ability to accumulate both food and pollutants from any media. They also have genetic ability to adapt. Accumulation of pollutants occurs essentially through the root system, where the principal mechanisms for preventing pollution are found. The root system provides an enormous surface area that absorbs, adsorbs and accumulates the water and nutrients essential for growth, and other non-essential contaminants.

More than 400 plant species belonging to 45 plant families have been identified and reported from temperate to tropical regions with the ability to tolerate and hyper-accumulate heavy metals. These plants have been considered suitable for soil stabilisation and extraction of heavy metals [45]. Plants used in phytoremediation are grouped into C3 and C4 plants depending on their photosynthetic mechanisms. The main difference between C3 and C4 plants is that single fixation of carbon dioxide is observed in C3 plants and double fixation of carbon dioxide is observed in C4 plants. Around 95% of plants on the Earth are C3 plants. Sugarcane, sorghum, maize and grasses are C4 plants capable of photosynthesising even in low concentrations of carbon dioxide as well as in hot and dry conditions. In view of this, grasses, with characteristics of rapid growth, large amount of biomass, strong resistance, effective stabilisation to soils and ability to remediate different types of soils, are widely used in phytotechnologies. However, trees with a tap root system have an advantage over others as the roots can penetrate deeper into the soil. Hence, they are more useful in deep contaminated ground water remediation where the groundwater is pumped and treated with the chosen plant. Plant roots also cause changes at the soil–root interface as they release inorganic and organic compounds (root exudates) in the rhizosphere. These root exudates affect the rhizosphere microorganisms through enhancing the availability of the contaminants. Root exudates can also mobilise or immobilise the availability of the contaminants in the root zone. To remove pollutants from soil, sediment and/or water, plants can break down, or degrade organic pollutants or contain and stabilise metal contaminants by acting as filters or traps (Tangahu et al., 2011).

9.5 Merits and demerits of phytotechnologies

Greipsson [21] identified several merits and demerits of using phytotechnologies for pollution control:

Merits

- Cost of the phytoremediation is lower than that of conventional processes both in situ and ex situ.
- Plants can be easily monitored.
- Possibility of the recovery and reuse of valuable metals through 'phyto mining'.
- Potentially, the least harmful to the environment because it uses environmentally friendly organisms and hence less environmental degradation.
- Preserves the topsoil, maintaining the fertility of the soil when used for soil remediation.
- Increase soil health, yield and plant phytochemicals.
- Use of plants also reduces erosion and metal leaching in the soil.

Demerits

- Phytoremediation is limited to the surface area and depth occupied by the roots.
- Slow growth and low biomass require a long-term commitment.
- Not possible to completely prevent the leaching of contaminants into the groundwater (without the complete removal of the contaminated ground).
- The survival of the plants is affected by the toxicity of the contaminated land and the general condition of the soil.
- Contaminants, especially metals, bio-accumulate into the plants which are then pass into the food chain, from primary-level consumers upwards or require the safe disposal of the affected plant material.
- When taking up heavy metals, sometimes the metal is bound to the soil organic matter, which makes it unavailable for the plant to extract.

9.6 Commonly used plants

The plants used for phytoremediation should have certain features, e.g. economic benefits, ease of harvesting management and by-product utilisation [35]. The common plants used in phytotechnologies are as follows (Fig. 9.2):

Fig. 9.2: Various reeds reported using in wastewater treatment.

- **Trees:** Poplars (hybrids), Cottonwoods, Willows, *Nauclea diderrichii*
- **Grasses:** Prairie grasses, Fescue, Elephant grass, *Chrysopogon Zizanioides* (*Vetiver*)
- **Legumes:** Alfalfa
- **Metal accumulators:** Hyperaccumulators *Thlaspi caerulescens*, *Brassica juncea*, sunflower *Helianthus annuus* and Mexican sunflower *Tithonia diversifolia* (Hemsl).
- **Aquatic plants:** Parrot feather, Phragmites reeds, Typha or cattails or *Typha latifolia*, water hyacinth, *Pistia stratiotes*, duckweed.
- **Cacti:** A thorny succulent plant (e.g. *Opuntia* sp.) common in semi-arid and arid tropical regions, which is found to be good for the treatment of wastewater. This plant is emerging for many uses—as coagulant/flocculant, biosorbent and packed material for biofilter.

Some of the enzymes in the plant are shown to be active against transformation of toxic textile dyes. A review prepared by Rebah and Siddeeg [40] revealed that cactus preparations showed very high and promising pollutant removal efficiency as evident from the significant reduction in turbidity, COD, heavy metals, conductivity and salinity. A cacti polyculture in Ibadan, Nigeria is found to absorb nutrients from municipal solid wastes from an illegal dump and flourish well (Observation from Sridhar, 2018 unpublished) which was used as a measure to prevent illegal dumping of wastes in communities, which also helps preventing grazing animals destroying farms.

There are more plants that are site and country specific. Some of the plants are those which are found in the contaminated areas. Fig. 9.2 shows common reeds used in wastewater treatment.

In phytotechnologies, biomass production is an added advantage. Reddy and DeBusk, [41] give an indication of the biomass that can be produced by some common floating macrophytes such as *Eichhornia crassipes* (water hyacinth), Pistia stratiotes (water lettuce) and *Hydrocotyle* spp. (pennywort): 20.0–24.0, 6.0–10.5 and 7.0–11.0 t (dry weight) per hectare, respectively. The figures for emergent macrophytes vary from as much as 35 t (dry weight) per hectare for Phragmites (reed) to 4.3 and 4.5 t (dry weight) per hectare for Typha (cattail) and *Saururus cernuus* (lizard's tail), respectively.

9.7 Examples of phytotechnologies in the treatment of various wastewaters in Nigeria

9.7.1 Wastewater from a tertiary hospital

For the possible utilisation of aquatic plants in treating polluted waters and industrial effluents, Sridhar [48,49] adopted water lettuce, *P. stratiotes* L obtained from a eutrophic lake in Ibadan, Nigeria. The study indicated that the shoot system accumulated more potassium, calcium and magnesium, whereas the root system accumulated significantly more cadmium, chromium, cobalt, copper, iron, lead, manganese, mercury, nickel, sodium and zinc. Water hyacinth entered Nigerian waters in 1984 and proved to be a stubborn weed and also has ability to remove pollutants; it became a chosen plant in many studies since 2000 to date. Noah [36] and Sridhar et al. [50] carried out a series of experiments on municipal wastewater from a tertiary hospital using a variety of macrophytes available locally and optimised their utilisation. They obtained acceptable quality effluents through the phytotreatments (Fig. 9.3).

Continuing the work by Noah [36], Badejo et al. [9,10] designed a phyto-based treatment facility for a tertiary hospital in Ibadan and used *Phragmites karka* and *Vetiveria nigritana.* The average flow rate of wastewater during the dry and wet seasons were (m^3/day), 131.5±51.9 and 262.2±56.2, respectively. The beds have a uniform depth of 0.7 m, and the two cells used were

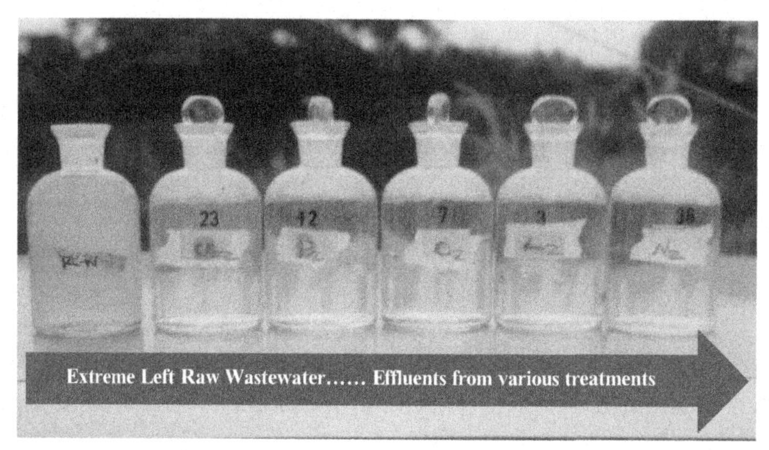

Fig. 9.3: Effluents obtained from various phytotechnologies used in wastewater treatment.

310 × 340 cm separated by sandcrete blocks. The beds had a slope of 2% and the substrate incorporated was washed granite and sand. The beds were irrigated uniformly with a pump connected to a 150 mm diameter (perforated at 100 mm intervals) PVC pipe intermittently. The influent fed into the plant had a biochemical oxygen demand (BOD) of 293.5± 20.43 mg/L, which was reduced to 53.83 ± 16.2 mg/L and 44.03 ± 17.5 mg/L for the cells with *V. nigritana* and *P. karka,* respectively. Similarly, total suspended solids (TSS) were reduced from 213.5 ± 9.24 mg/L to 59.78 ± 10.15 mg/L and 57.64± 8.23 mg/L; nitrate was reduced from 0.141 ± 0.003 mg/L to 0.055 ± 0.011 mg/L and 0.049 ± 0.014 mg/L whilst the phosphate level was reduced from 2.36 ± 0.05 mg/L to 0.52 ± 0.21 mg/L and 0.45 ± 0.65mg/L.

9.7.2 Treatment of septic tank effluent

In Nigeria, there are no large-scale municipal wastewater treatment plants that serve large urban areas. There are 26 wastewater treatment plants (biological such as trickling filters, activated sludge, aerated lagoons and imported packaged plants) identified but they cater to confined areas such as institutions and selected industries. Septic tank system is, therefore, popular in all cities and towns owned and managed by individuals. Managing septic tank effluent is a serious problem as they are scattered and dispose indiscriminately into water bodies and open drains. Therefore, we tried to apply the phytotechnologies to treat septic effluents. A design was used to treat a block of students' hostel at Balewa hall, University of Ibadan, Oyo State, Nigeria. The hall with a population of 191 produced 933.2–1066.25m^3/day of wastewater generating at the rate of 76.25 L/head/day. Performance of two-layered reed bed consisting of sandcrete blocks (150m × 225m × 450m) and granite was carried out. The influent and effluent samples were analysed for physicochemical characteristics namely, pH value, total dissolved solids, BOD, phosphate, ammonia and nitrate using Standard methods [7].

The design adopted was a VF reed bed consisting of coarse to medium sand on coarse granite, *P. karka,* was planted in the sand layer and flooded by influent unto the surface by means of perforated 0.1-m-diameter PVC pipe and hose at an average flow rate of 0.1×10^{-3} m^3/s. A retention period of between 14 and 24 h was used. For the experimental design 0.36 m^3 of influent representing a population equivalent of seven people was allowed to pass down through the sand layer by gravity (Fig. 9.4).

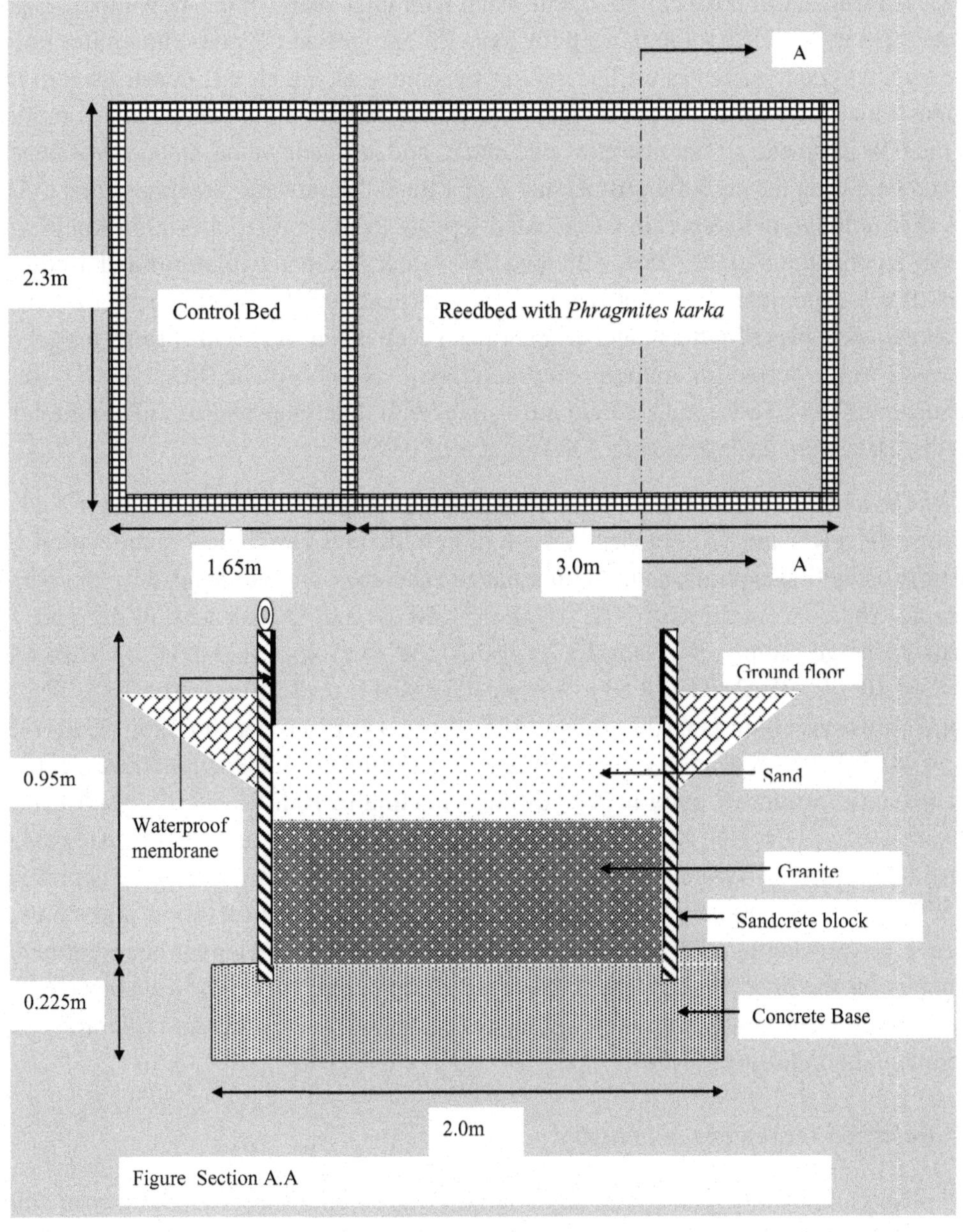

Fig. 9.4: Plan view of reed bed and control bed.
Source: [9].

9.7.3 Water hyacinth, Typha and Phragmites polyculture for septic effluents and wastewaters in Abuja

A treatment plant was designed using three aquatic plants: water hyacinth, *E. crassipes*; Typha, *T. latifolia* and Phragmites, *Phragmites australis* obtained from Jabi Lake in Abuja. Pilot Constructed Wetland (CW) using Phragmites, Typha–Phragmites polyculture and water hyacinth sequentially in three compartments each with inlet chamber 0.5-m width, 0.29 m^3/h flow rate; 5.38 m × 2.25 m × 0.6 m deep for lone–Phragmites and Typha–Phragmites polyculture each was constructed as well as a water hyacinth tank which was constructed in two flow-lines with 24 Queen of the night (*Epiphyllum oxypetalum*) stem planted at a 1 m interval along the CW perimeter to stabilise flow rate and to add aesthetic value. In one treatment, wastewater entering the 42,000 kg/BOD/day Wupa Basin Wastewater Treatment Plant, Abuja was used as influent. In another line, evacuated septage from six districts within Abuja were collected, reconstituted to 0%, 25%, 50% and 75% water to septage dilutions and used as influent. Batch experiments were carried out using 10 healthy offshoots of each of the three reeds chosen. Samples of influent and effluent from each batch and control were analysed daily over 2 weeks period for ammonia nitrogen (NH_4^+–N), phosphate (PO_4^{2-}),BOD_5, dissolved oxygen (DO), TSS, total kjeldahl nitrogen (TKN), total phosphorus (TP) and total coliforms. Data were analysed using ANOVA at α of 0.05.

The CW showed 93.0%, 94.0%, 82.0%, 69.0% and 94.0% reductions in BOD_5, total coliforms, TP, TKN and TSS, respectively, within 4.88 days HRT. Flow stabilisation, uniformity as well as operation and maintenance efficiency were enhanced by the perimeter plants. The concentration of NH_4^+–N, PO_4^{2-}, BOD_5 and DO for wastewater and evacuated septage samples was (mg/L) 2.14±0.37, 28.13±1.46, 2.13±0.04, 24.57±6.14, 172.50±60.10, 483.33±219.61, 2.83±1.09 and 1.19±0.47, respectively. The TSS, TN, TP and total coliforms count per 100 mL was 185.00±14.14, 2234.17±1009.26; 10.00±4.24, 46.42±4.41; 2.32±0.62, 13.50±3.79, 2650.00±919.24 and 4333.33±1032.80, respectively. Water hyacinth favourably grew in 75% dilution leading to 100.0%, 95.0%, 99.0% and 91.0% removal of NH_4^+–N, PO_4^{2-}, TKN and TP, respectively. In undiluted wastewater, increase of biomass of Typha, Phragmites and Typha–Phragmites polyculture is observed to be 57% and 33%, 50% and 67%, 58%–44% and 55%–65% for offshoots and stems, respectively, resulting in a 77% nutrient removal efficiency. Variations in the treatment efficiencies for the different aquatic plants were significant. *Typha latifolia* showed a high capacity for nutrient removal under a highly concentrated wastewater and tolerating harsh environmental conditions [37,50].

9.7.4 Reed bed technology in industry

A steel processing factory manufactures galvanised iron roofing sheets and long span colour-coated roofing sheets. The basic feedstocks for the processes are cold rolled sheets, zinc and colour-coated galvanised sheets. The factory produces between 6000 and 8000 L of effluent per day in a 24-h cycle, with all the drainage systems centralised and well controlled. The effluent generated from the factory has pH = 8.29, Total Solids = 1650 mg/L, electrical conductivity = 980 μS, BOD = 5.80 mg/L, P= 2.50 mg/L, NO_3 = 5.80 mg/L, Mn = 6.30 mg/L, Zn = 0.68 mg/L, Pb = 0.01 mg/L and Cr = 0.127 mg/L. The effluent is usually mixed with other wastewater flows and hence diluted.

The design of the Reed bed consists of three units, which include a well-rendered equalisation basin, two Reed beds placed parallel to one another and a storage tank for effluent collection. The Reed bed is made up of a trapezoidal basin, 10.4 × 4.4 × 1 m in section, lined with 300 mm well-compacted clay. It consisted of three sections; the inlet and outlet zone (20–30 mm rounded granite), and the treatment zone (150 mm deep 20–30 mm overlaid by 200 mm deep 10–15 mm granite). The macrophytes are locally available *P. karka, V. nigritana* and *Cana* lily planted 400–500 mm (centre to centre) using organic fertiliser as a nutrient supplier. The settling tank connects directly to an inspection chamber that is connected to a bypass and the two reed beds linked with the concrete-filled sandcrete blocks storage tank (Fig. 9.5). The results showed a significant removal of heavy metals: zinc was reduced by 55.7%, lead 15.4% and chromium by 97.9% and the quality of effluent complied with the national guidelines. The design and the beds are shown in Fig. 9.5 [50].

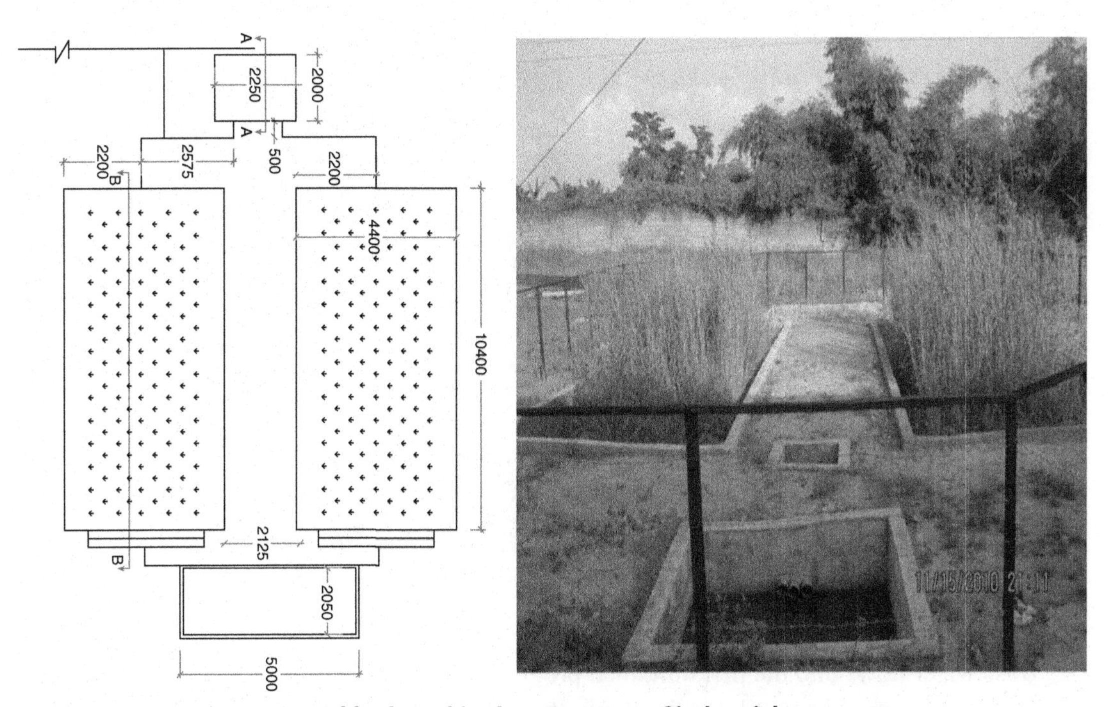

Fig. 9.5: Reed bed used in the treatment of industrial wastewater.
Source: [50].

9.7.5 Examples of water hyacinth in conjunction with other reeds as a wastewater treatment

(A) Water hyacinth based wastewater treatment

Three halls of residence for students with a total population of 3644 generate the wastewater being treated at the water hyacinth wastewater treatment facility. These are Obafemi Awolowo Hall (1700 students), Queen Idia Hall (1244 students) and Abdulsalami Abubakar Hall (700 students). The estimated volume of wastewater flow is 255,080 L/day, which enters the treatment plant by gravity flow. The wastewater flows through underground drainage pipes and travels about 100–200 m before discharging into the treatment plant. Fig. 9.6 shows the wastewater treatment plant using the water hyacinth plants. The combined flow from the three halls enters into a grit chamber where debris and other solid materials are removed. From this unit, the wastewater flows into the main plant. The treatment plant consists of two sets of tanks in series (four compartments each), one on the right and the other on the left. There are two separate control valves, which direct the flow of wastewater to either of the series. All the tank units have dimensions 9 × 9 × 1.2 m deep. In each unit, water hyacinth was planted evenly to aid in treatment. A depth of 1.2 m is just sufficient to allow maximising root growth and the absorption of nutrients and heavy metals. The design also favours closer contact of the roots with the wastewater flow [1].

Fig. 9.6: Water hyacinth wastewater treatment plant.

Wastewater flows into the first unit, then progressively flows into the second unit, then the third and fourth units, until it is finally discharged and flows into Awba lake which is 500 m away. The quality of effluent is shown in Table 9.2.

Table 9.2: Characteristics of the wastewater and effluent obtained from the water hyacinth treatment plant.

Parameters	Influent	Effluent	Percent change	FEPA (2009) guidelines
Temperature, °C	24	24	-	<40
pH	8.80	7.80	−11.4	6.0-9.0
Turbidity, Formazin Turbidity Unit (FTU)	29.00	8.00	−72.4	-
Total suspended solids, mg/L	463.00	65.031	−85.95	30
Total dissolved solids, mg/L	955.50	266.50	−72.1	2000
Dissolved oxygen, mg/L	0.00	2.03	Increased	-
Chemical oxygen demand, mg/L	2160.00	1040.00	−51,85	-
Biochemical oxygen demand, mg/L	686.10	388.60	−43.36	50.0
Ammonia, NH_3, mg/L	168.00	98.00	−41.67	-
Nitrate, mg/L	7.80	4.80	−38.46	20
Phosphate, mg/L	0.33	0.04	21.21	5

FEPA Standard = Nigeria Federal Environmental Protection Agency

(B) Phragmites and water hyacinth in sequence for treating institutional wastewater

A treatment plant was constructed to treat wastewater from University of Ibadan using Phragmites and water hyacinth plants [47]. An area of 1–2 m²/person is recommended for a warm country like Nigeria [8]. The available land area of 150 m² was used for the pilot plant. The influent was drawn from the main hole nearest to the treatment plant. The treatment system consisted of a balancing tank, *P. karka* tank and water hyacinth tank, all arranged in parallel to ensure gravity flow according to the topography of the site. The plant was designed to closely conform to plug flow hydraulics to minimise short circuit. The AR or length-to-width ratio (L:W) fell within the 1:1 and 90:1 limits (Environmental Protection Agency minimum). For Phragmites tank (L:W) is 2.5:1, and 3:1 for water hyacinth tank. The tanks were rectangular in shape, constructed with 0.225 m hallow sandcrete wall, filled with 1:2:4 mix mass concrete and rendered internally with ratio 1:4 cement and sand mortar. The floor was made of same concrete mix to prevent seepage of wastewater. A slope of 1.3%, using a dumpy level, was adopted for both common reed and water hyacinth tanks. The balancing tank was 0.6 m below and 0.3 m above the ground. The Phragmites and water hyacinth tank were both 0.4 m below and 0.3 m above the ground. The height above the ground prevented any external water from entering the tanks from the natural ground surface.

The filter media used in the Phragmites tank was obtained from the excavated material during the construction. The granular material is a mixture of sand and gravel. The material was sieved to separate sand and gravel, and the gravel was then graded. The bottom of the tank was covered with polyethylene sheets to prevent leakages before the media were placed. The biggest size was placed at the bottom of the tank followed by the smallest and then sand. Water hyacinth tank operated on SF system or FWS, while the Phragmites tank worked on SSF system or VSB. The design features are shown in the Box 9.1 and Fig. 9.7.

Box 9.1

Design parameters of the tanks

Phragmites tank

Design features: Volume of empty Phragmites tank = $1.95 \times 5.1 \times 0.7$ m = 6.96 m^3
Volume of media in Phragmites tank = $1.95 \times 5.1 \times 0.6$ m = 5.97 m^3
Volume available space for water = (6.96–5.97) m^3 = 0.99 m^3
With 9.23% water absorption of the media, total volume of water in the tank,
Total volume of water in Phragmites tank = 1.0923×0.99 m^3 = 1.08 m^3
Detention time = 3 days

Water hyacinth tank

Volume of water hyacinth tank = $1.95 \times 5.85 \times 0.7$ m = 7.98 m^3
Volume of water in the water hyacinth tank = $1.95 \times 5.85 \times 0.47$ m = 5.33 m^3
Total volume of wastewater in the two tanks = (1.08 + 5.33) m^3 = 6.41 m^3
Volume of balancing tank = $6 \times 2.4 \times 0.9$ m = 12.96 m^3
This volume is more than double the wastewater to be treated in both tanks.

9.7.6 Treatment of leachates from municipal solid wastes

Leachate is the fluid percolating through the landfill, or solid waste dumpsite and is generated from liquids present in the water and outside pool, including rainwater, filtering through the waste. Globally, the uptake of pollutants from leachates in the environment has resulted in environmental pollution particularly, surface and groundwater pollution and developmental disorders in humans such as congenital disabilities [3]. In Nigeria, sanitary landfills are few or none and privately established and operated by multinational corporations, while open dumpsites dot the landscape in urban and rural settlements in Nigeria. In landfills and open dumpsites, leachate generation is inevitable, resulting in environmental pollution, especially where there are no plans or schemes for onsite leachates treatment or evacuation of leachates from leachate dumps for co-treatment with domestic wastewater. Leachates are hazardous and highly polluted wastewaters with considerable seasonal variations in chemical compositions and volumetric flow [4,5]. The leachate composition and the quality of effluents obtained by phytotechnologies using *Ipomoea aquatica forsk* are shown in Table 9.3. The results showed the treated effluents meet the required standards by the regulatory bodies.

9.7.7 Removal of arsenic

Arsenic, a toxic metal, can enter the environment through geological weathering of rocks, industrial wastes and processing effluents, leachates from solid waste dumps and litter, animal and human excreta [15]. It may lead to different diseases such as heart diseases, skin cancer (arsenicosis) and even death. Several efforts have been made to remove arsenic from water and wastewater in the environment using various adsorbents [38]. In the 19th century, Bauman

Fig. 9.7: Treatment plant design.

[11] identified plants capable of accumulating uncommonly high Zn levels. In 1935, Byers documented the accumulation of selenium in *Astragatus* sp. and a decade later, Minguzzi and Vergnano [33] identified plants capable of hyper accumulating up to 1% of Ni in shoots.

Table 9.3: Leachate characteristics and the quality of effluents obtained from application of phytotechnologies.

Parameter (unit)	Influent[a]	Effluent[a]	Percent reduction	P-value	FEPA standard[b]
Temperature (°C)	25.2±0.3	25.7±0.4	–	–	<40
pH	8.4±0.1	7.5±0.1	–	–	6-7
Colour, Hazen Units (HU)	456.5±9.2	17.5±3.5	96.2	.0004	7
Turbidity (FTU)	132.0±15.6	6.0±0.0	95.5	.005	–
Conductivity (μS/cm)	4515.0±148.5	1393.5±217.1	69.2	.002	–
Total solids (mg/L)	6062.5±381.1	899.0±144.3	85.2	.002	–
Total suspended solids (mg/L)	197.5±21.9	37.5±6.4	81.0	.007	30
Total dissolved solids (mg/L)	5865.0±403.1	861.5±150.6	85.3	.003	2000
Alkalinity (mg/L)	1444.0±560.0	307.0±216.4	78.7	.116	–
Chloride (mg/L)	1034.0±159.8	356.0±32.5	65.6	.024	600
Sulphate (mg/L)	66.5±13.9	57.3±20.1	13.8	.648	500
Dissolved oxygen (mg/L)	1.9±0.0	5.4±0.2	184.2[c]	.004	–
Biochemical oxygen demand (mg/L)	712.0±62.2	99.5±5.7	86.0	.012	50
Chemical oxygen demand (mg/L)	3365.0±31.1	560.5±84.2	83.3	0.002	–
Ammonia (mg/L)	610.9±365.4	13.6±3.8	97.8	0.148	–
Nitrate (mg/L)	1.1±0.3	3.7±0.2	246.2[c]	0.008	20
Phosphate (mg/L)	0.6±0.0	0.1±0.1	76.5	0.008	5
Lead (mg/L)	1.6±1.9	0.0±0.0	97.9	0.121	<1
Nickel (mg/L)	0.9±0.1	0.2±0.1	81.6	0.006	<1
Cadmium (mg/L)	0.2±0.1	0.0±0.0	80.5	0.011	<1
Iron (mg/L)	198.1±129.4	3.6±0.2	98.2	0.121	20
Manganese (mg/L)	23.2±12.2	0.1±0.1	99.6	0.121	5
Zinc (mg/L)	1.2±0.1	0.4±0.4	72.2	0.085	<1

[a]Values are of 12 individual and separate samples obtained from the same sampling point and treatment system for influent and effluent samples, respectively.
[b]FEPA Standard = Nigeria Federal Environmental Protection Agency standard for discharge into surface waters (1991).
[c]Percentage increase.
Source: [4]

Greenhouse experiments carried out to assess the removal of lead, copper and cadmium from polluted soils showed *T. diversifolia,* a common road side weed in Nigeria, remediated better than *Helianthus annuus*; while, *E. crassipes* was found to remove arsenic (As) from polluted water [2,59]. Laniyan et al. [28,29] studied arsenic removal through use of filtration and water hyacinth based phytotechnology. Four different geo-materials (marble, activated charcoal, filtration carbon and clay) were placed on layers of sand in glass filtration tanks; while phytoremediation method was carried out by cultivation of matured water hyacinth (*E. crassipes* Mart. Solms) in arsenic acid solution of equal concentration (100 mg/L). Arsenic concentration in filtrates showed no reduction, indicating poor absorption capacity of the geo-materials. Highest arsenic bio-accumulation was found at 100 mg/L in mature water hyacinth. Remediation of arsenic using water hyacinth proved to be a better method for arsenic removal compared to filtration.

9.7.8 Treatment of miscellaneous wastewaters

Badejo et al. [60] carried out experiments to treat dye wastes. The pilot-scale Reed bed consisted of two 1200 × 1000 × 1000 mm plastic tanks with 500 mm deep 10–15 mm granite overlain with 150-mm-thick sand (Cu = 1.15 and Cc = 6.8) substrate, and planted with *P. Karka* at 200 mm c/c. Intermittent irrigation was undertaken with a 6-dayretention period with 0.05 m^3 indigo dye rich wastewater from the local tie and dye textile industries in Abeokuta, Ogun state in Nigeria. Plant growth was monitored along with physicochemical parameters and heavy metals such as Cr, Pb, Cu, Zn and Fe removal. The study revealed a 24% growth rate reduction in the plants irrigated with indigo dye rich wastewater. The reduction of TDS by 50%, TSS 66%, EC 46% was also observed and the heavy metals uptake was 64%, 68%, 78%, 5 % and 68 % for Cr, Pb, Zn, Cu and Fe, respectively, by the Reed bed. Similarly, *P. karka* was found useful in reducing the pollutants from abattoir wastewaters [60].

9.8 Successful global phytotechnology applications

Application of phytotechnologies is rather more recent, say less than two decades. Several countries have been using phytotechnologies to varying extent either at research or pilot-scale level. The applications have been on wastewaters contaminated with heavy metals and other organic matter, soils used for waste disposal, land treatment of wastewaters, mining areas, etc. Some of the relatively large-scale ongoing applications are given in Table 9.4.

Table 9.4: Selected global phytotechnology applications in wastewater treatment.

Location	Facility role	Capacity	Plant used	Observations	Reference
Putrajaya, Malaysia, Federal Government Administrative Centre.	Stormwater remediation	400 hectare artificial lake; catchment area 50.9 km^2	12.3 million wetland plants	One of the largest fully constructed tropical freshwater wetlands.	[52]
Nakivubo Wetland, Kampala, Uganda	Tertiary effluent treatment, Kampala Sewage Works	5.29 km^2 wetland drains 50 km^2 from Kampala city	Papyrus and Coco Yam. These can be harvested	Floating papyrus mats with strong rhizomes sediment out suspended matter	[58]
South Kalimantan, Indonesia	Acid mine waters	Multilevel methods more effective to lower acid levels	*Eleocharis dulcis, Cyperus odoratus, Hydrilla verticillata, Ipomea aquatic, Pistia stratiotes*	Acid pH neutralised, reduced Fe, Mn.	[23]

Table 9.4: (*cont'd*)

Location	Facility role	Capacity	Plant used	Observations	Reference
AngloGold Ashanti's West Wits and Vaal River, 2009-2012	Remediation of groundwater and soils	Experimental	Indigenous trees	Wetlands and riverside woodlands treat water	[45]
Chinna Kalapet, Pondicherry, India	Designed to treat 10,000 L greywater from 38 houses	Pilot scale	Leaf clover and Water Hyacinth	Sedimentation tank and channels constructed using sand bags in pits. Covered with impermeable sheet so wastewater does not percolate into the ground. Aquatic plants grown on top. Within 2 h effluent was clean.	[63]
Pariyej Wetland, Gujarat, India	Heavy metal contamination	Wetland field investigation	*Ipomoea aquatica Forsk, Eichhornia crassipes, (Mart.) Solms, Typha angustata Bory, Chaub, Echinochloa colonum (L.) Link, Hydrilla verticillata (L.f.) Royle, Nelumbo nucifera Gaerth. and Vallisneria spiralis* L	To assess accumulation of: Cd, Co, Cu, Ni, Pb and Zn in aquatic macrophytes using them as bio-monitors. Compared with water and sediments	[27]
Bangladesh, tannery waste polluted water with chromium		Pilot scale	Several plants; *Urtica dioica* efficient treatment.	K addition was required	[46]
China, Dianchi Lake	Pollution control		Water hyacinth	Biomass was harvested, biogas generated and GHG reduced; 11,004.2 tonnes of water hyacinth, generated 5.3 trillion joules.	[54]
Western Treatment Plant (WTP), Victoria, Australia	Open cast mining, contaminated sites with phytoremediation of biosolids	2 million ha	Land and grass filtration	WTP covers 11000 ha, discharging treated sewage into Port Philip Bay. Extensive lagoon systems for both peak daily and year round wet weather flows	[42]

9.9 Lifecycle of phytotechnologies

Phytotechnologies have ancillary positive impacts on the surrounding environment, providing ecosystem services with tangible, quantifiable value for public health and social welfare. These technologies may have strong community acceptance, in part, because covering a contaminated site with vegetation creates an open green space and such spaces have been shown to reduce stress, particularly in urban environments. Sometimes the treated effluent may be useful in increasing farm outputs [20]. This added psychological benefit reinforces the public health value of phytotechnologies, in view of the World Health Organisation (WHO) definition of health: "*a state of complete physical, mental and social well-being and not merely the absence of disease or infirmity*" [55]. Garfi et al. [19] carried out a Lifecycle Analysis (LCA) of the conventional wastewater treatment (activated sludge process) and two nature-based technologies (i.e. hybrid constructed wetland and high rate algal pond systems). These systems served a Population Equivalent (PE) of 1500 PE using a functional unit of 1 m^3 of water. System boundaries comprised input and output flows of material and energy resources for system construction and operation. A software SimaPro 8, with the ReCiPe midpoint method was used. The results showed the nature-based solutions were the most environmentally friendly, while the conventional wastewater treatment plant presented the worst results due to the high energy and chemicals consumption. A study by Machado et al. [31] showed that the LCA quantification identified the constructed wetland and the slow rate infiltration systems as appropriate technologies in rural areas. A key factor was the reduction of global warming impact due to carbon sequestration, as opposed to the activated sludge processes, which require a high energy input and present a negative carbon balance. In yet another LCA study, different performance scenarios are used to compare the environmental impacts of VFCW and HFCW and that of conventional treatment. The LCAs include greenhouse gas (N_2O, CO_2 and CH_4) emissions. The LCAs suggest that constructed wetlands have less environmental impact, in terms of resource consumption and greenhouse gas emissions. The VFCW is a less impactful configuration for removing total nitrogen from domestic wastewater. Both wetland designs have negligible impacts on respiratory organics, radiation and ozone [16,43].

9.10 Common mistakes in the application of phytotechnologies

In developing countries some of the most common mistakes are improper designs and overloading of the plant. Here, we cite two examples where the treatment failed (Fig. 9.8).

For the treatment shown in Fig. 9.8, pumping was eliminated as the system used gravity flow. At a design level, it was intended for a limited population but later it was connected to two other hostels thereby increasing organic loading and reducing retention time. Problems with the treatment plant included silted sewers, broken chambers, overflow of

A= A water hyacinth treatment plant with excessive loading with organic matter with less retention time;

B= A treatment plant used for brewery waste with very high BOD (about 1000mg/l) led to

Fig. 9.8: Poorly managed water hyacinth treatment plants, (A) A water hyacinth treatment plant with excessive loading with organic matter with less retention time; (B) A treatment plant used for brewery waste with very high BOD (about 1000 mg/L) led to anaerobic conditions with hydrogen sulphide emissions.

wastewater causing pollution, thereby reducing hydraulic conductivity, non-functional valves and inadequate personnel to monitor and maintain the plant. It was therefore not functioning well as a result of high organic loading, causing the water hyacinth in the first chamber to wilt and later die. Keeping wastewater in a retention tank for 24 h reduces BOD by about 40%.

9.11 Phytotechnologies and public health

Phytotechnologies with natural treatment systems concurrently provide valuable ecosystem services; the integration and coordination of phytotechnology activities with public health will allow technology development that focuses on the prevention of environmental exposures, reducing the burden of disease. Constructed wetlands that receive primary effluent or partially treated effluent are potentially hazardous with pathogens. In addition, mosquito populations may increase and, as a result their bites, there is risk of certain mosquito-borne infections particularly in tropical developing countries where Culex and other mosquitoes breed for extended periods. Operators working in the plant are prone to more health problems, where dermal contact and other health risks are of greater concern at the inlet of the system. When the effluents are reused for irrigation and other purposes, health should be considered and appropriate disinfection should be incorporated.

9.12 The Future of phytotechnologies

In the last 20 years or more, many researchers favour the use of phytotechnologies in the treatment of municipal and other community wastewaters. Water hyacinth, Phragmites, sunflower, Typha, amongst other indigenes species of plants, are finding their place in managing local wastewater problems. In developing countries, there is still a belief that it is better using established technologies rather than taking chances [51]. It seems governments are still not fully aware of all the available choices. Phytotechnologies are most appropriate for small wastewater flows with low organic matter. The literature to date suggests they are good for removing heavy metals through bio-concentration and managing other hazardous wastes along with the recovery of water.

The benefits of these technologies can be appreciated in the management of greywater, stormwater, leachates and in secondary treatment of wastes. These technologies make use of the principles of reuse and recycle, with added returns of by-products, such as biomass and other secondary resources. However, there needs to be a review of existing environmental regulations, targeted towards adaptation in specific locations. Alongside this, planners need to allocate available land for wastewater management in towns or cities.

There is a need for improved awareness of phytotechnologies. Demonstration models could be built to educate decision makers to understand the principles of the technology and to evaluate their cost effectiveness. After all—seeing is believing! This would need to be supported by media sources to promote them amongst communities and establishments. The economics of phytotechnologies is promising, as they do not need any expensive equipment (which would also need periodic maintenance to prevent breakdowns). Therefore, as the future lies in promoting these technologies at household and community levels, it needs to become an established part of the planning process across developing countries.

References

[1] C.G. Achi, M.K.C. Sridhar, A.O. Coker, Performance evaluation a water hyacinth based institutional waste treatment plant to mitigate aquatic macrophyte growths at Ibadan, Nigeria, Int. J. Appl. Sci. Technol. 4 (3) (2014) 117–124.

[2] M.B. Adewole, M.K.C. Sridhar, G.O. Adeoye, Removal of heavy metals from soil polluted with effluents from a paint industry using *helianthus annuus* l. and *tithonia diversifolia* (hemsl.) as influenced by fertilizer applications, Bioremediat. J. 14 (4) (2010) 169–179, doi:10.1080/10889868.2010.514872.

[3] O.O. Aluko, M.K.C. Sridhar, P.A. Oluwande, Characterization of leachates from a municipal solid waste landfill site in Ibadan, Nigeria, Int. J. Environ. Health Res. 2 (1) (2003) 32–37.

[4] O.O. Aluko, M.K.C. Sridhar, Application of constructed wetlands to the treatment of leachates from a municipal solid waste landfill in Ibadan, Nigeria, J. Environ. Health 67 (10) (2005) 58–62.

[5] O.O. Aluko, M.K.C. Sridhar, Evaluation of leachate treatment by trickling filter and sequencing batch reactor processes in Ibadan, Nigeria, Waste. Manag. Res. 31 (7) (2013) 700–705, doi:10.1177/0734242X13485867.

[6] O.O. Aluko, M. Sridhar, Evaluation of effluents from bench-scale treatment combinations for landfill leachate in Ibadan, Nigeria, Waste. Manag. Res. 32 (1) (2014) 70–78, doi:10.1177/0734242X13514624.
[7] APHA, Standard Methods for the Examination of Water and Waste water, 19th edition, American Public Health Association;, Washington DC:, 1997.
[8] S.J. Arceivala, S.R. Asolekar, Wastewater Treatment for Pollution Control and Reuse, Tata McGraw-Hill Publishing Company Limited, New Delhi:, 2008.
[9] A.A. Badejo, A.O Coker, M.K.C. Sridhar, Treatment of tertiary hospital wastewater in a pilot-scale natural treatment system (Reedbed technology), Res. J. Eng. Appl. Sci. 1 (5) (2012) 274–277.
[10] A.A. Badejo, M.K.C. Sridhar, A.O. Coker, J.M. Ndambuki, W.K. Kupolat, Phytoremediation of water using *Phragmites karka* and *Veteveria nigritana* in constructed wetland, Int. J. Phytoremediation 17 (9) (2015) 847–852, doi:10.1080/15226514.2014.964849.
[11] A. Baumann, Das verhalten von zinksatzen gegen pflanzen und imboden. Landwirtsch.Verss 3 (1885) 1–53.
[12] H. Brix, Do macrophytes play a role in constructed treatment wetlands?, Water Sci. Technol. 35 (5) (1997) 11–17.
[13] J.B. Ellis, R.B.E. Shutes, D.M. Revitt, Guidance manual for constructed wetlands, R&D Technical Report P2-159/TR2, Bristol: Environment Agency, 2003, pp. 1–69.
[14] EPA, Constructed Wetlands Treatment of Municipal Wastewaters- Manual, EPA/625/R-99/010 September 1999, National Risk Management Research Laboratory Office of Research and Development U.S. Environmental Protection Agency, Cincinnati, Ohio, vol 45268, 1999, pp. 1–165.
[15] U. Forstner, G.T.W. Wittmann, Metal Pollution in the Aquatic environment 486 p. Environmental Health Perspectives, Springer, Berlin, vol. 19, 1981, pp. 67–71.
[16] V.J. Fuchs, J.R. Mihelcic, J.S. Gierke, Life cycle assessment of vertical and horizontal flow constructed wetlands for wastewater treatment considering nitrogen and carbon greenhouse gas emissions, Water Res. 45 (5) (2011) 2073–2081, doi:10.1016/j.watres.2010.12.021.
[17] J. Garcia, J. Chiva, P. Aguirre, E. Alvarez, J.P. Sierra, R. Mujeriego, Hydraulic behaviour of horizontal subsurface flow constructed wetlands with different aspect ratio and granular medium size, Ecol. Eng. 23 (2004) 177–187.
[18] J. Garcia, P. Aguirre, J. Barragan, R. Mujeriego, V. Matamoros, J.M. Bayon, Effect of key design parameters on the efficiency of horizontal subsurface flow constructed wetlands, Ecol Eng. 25 (2005) 405–418.
[19] M. Garfí, L. Flores, I. Ferrer, Life cycle assessment of wastewater treatment systems for small communities: activated sludge, constructed wetlands and high rate algal ponds, J. Clean. Prod. 161 (2017) 211–219, doi:10.1016/j.jclepro.2017.05.116.
[20] P. Grahna, U. A. Stigsdotte, Landscape Planning and Stress, Urban Forestry & Urban Greening, Elsevier GmbH, 2003, pp. 1–18.
[21] S. Greipsson, Phytoremediation, Nature Education Knowledge, vol. 3 (10), 2011, pp. 7.
[22] W.J. Hartman, An Evaluation of Land Treatment of Municipal Wastewater and Physical Siting of Facility Installations, US Department of Army, Washington DC, 1975.
[23] P.J. Herniwanti, Y.B. Soemarno, Water plants characteristic for phytoremediation of acid mine drainage passive treatment, Int. J. Basic Appl. Sci. 13 (06) (2013).
[24] H. Hoffmann, C. Platzer, E. von Münch, M. Winker, Technology review: constructed wetlands. In: Deutsche Gesellschaft für Technische Zusammenarbeit GmbH (GTZ), Sustainable sanitation - Ecosan Program Postfach 5180, 65726, Eschborn, Germany, 2010, pp. 1–35.
[25] M.S. Hong, W.F. Farmayan, I.J. Dortch, C.Y. Chiang, S.K. McMillan, J.L. Schnoor, Phytoremediation of MTBE from a groundwater plume, Environ. Sci. Technol. 15 35 (6) (2001) 1231–1239.
[26] J.De Jong, The purification of wastewater with the aid of rush or reed ponds, in: J. Tourbier, R.W. Pierson Jr. (Eds.), Biological Control of Water Pollution, University of Pennsylvania Press, Philadelphia, 1976, pp. 133–139.
[27] J.I.N. Kumar, H. Soni, R.N. Kumar, I. Bhatt, Macrophytes in phytoremediation of heavy metal contaminated water and sediments in Pariyej Community Reserve, Gujarat, India, Turk. J. Fish. Aquat. Sci. 8 (2008) 193–200.

[28] T.A Laniyan, A.F. Abimbola, M.K.C. Sridhar, Experimental comparison of phytoremediation and filtration methods in the remediation of water contaminated with arsenic, Int. J. Phytomedicine 7 (1) (2015) 08–17 http://www.arjournals.org/index.php/ijpm/index.

[29] T.A. Laniyan, A.F. Abimbola, M.K.C. Sridhar, Comparison of phytoremediation and filtration in remediation of contaminated waters, Ife J. Sci. 17 (1) (2015a) 183–188.

[30] X. Ma, A.R. Richter, S. Albers, J.G. Burken, Phytoremediation of MTBE with hybrid poplar trees, Int. J. Phytoremediation 6 (2) (2004) 157–167.

[31] A.P. Machado, L. Urbano, A.G. Brito, P. Janknecht, J.J. Salas, R. Nogueira, Life cycle assessment of wastewater treatment options for small and decentralized communities, Water Science &Technology, Q IWA Publishing, 2 Clove Crescent London, E14 2BE, UK, 2007, pp. 15–22.

[32] N. Marmiroli, B. Samotokin, M. Marmiroli, E. Maestri, V. Yanchuk, Capacity building in phytotechnologies, in: P.A. Kulakow, V.V. Pidisnyuk (Eds.), Application of Phytotechnologies for Cleanup of Industrial, Agricultural and Wastewater Contamination, Springer science+ Business Media B. V., 2010.

[33] C. Minguzzi, O. Vergnano, II contento di nichel nelli ceneri di Alyssum bertlonii Desv. Atti della Societa, Toscana di science Naturali, Mem Ser A 55 (1948) 49–77.

[34] W.J. Mitsch, J.G. Gosselink, Wetlands, 2nd ed., Van Nostrand Reinhold, New York, 1993, pp. 1–722.

[35] W. Nakbanpote, O. Meesungnoen, M.N.V. Prasad, Potential of ornamental plants for phytoremediation of heavy metals and income generation. Bioremediation and Bioeconomy, 2016, pp. 179–217.

[36] Noah Okunromade Olumade, A comparative assessment of plants used for treatment of domestic waste water using root zone technology, Unpub Master of Public Health Thesis, Faculty of Public Health, University of Ibadan, Ibadan, Nigeria, 2002.

[37] E.A Oluwadamisi, M.K.C Sridhar, A.O Coker, S.O. Jacob, Septage purification and nutrient uptake potential of water HYACINTH (*Eichhornia crassipes*) in a batch system, Int. J. Eng. Res. Manag. (IJERM) 4 (2) (2017) 49–56 ISSN: 2349–2058.

[38] I. Polowczyk, A. Bastrzyk, T. Kozlecki, W. Sawinski, P. Rudnicki, A. Sokolowski, Z. Sadowski, Use of fly ash agglomerates for removal of arsenic, J. Environ. Geochem. Health 32 (2010) 361–366.

[39] R. Razzaq, Phytoremediation: an environmental friendly technique - a review, J. Environ. Anal. Chem. 4 (2017) 195, doi:10.41722380-2391.1000195.

[40] F.B. Rebah, S.E. Siddeeg, Cactus an eco-friendly material for wastewater treatment: a review, J. Mater. Environ. Sci. 8 (5) (2017) 1770–1782.

[41] K.R. Reddy, W.F. DeBusk, Nutrient storage capabilities of aquatic and wetland plants, in: K.R. Reddy, W.H. Smith (Eds.), Aquatic Plants for Water Treatment and Resource Recovery, Magnolia Publishers, Orlando, Florida, 1987, 337–357.

[42] B. Robinson, C. Anderson, Phytoremediation in New Zealand and Australia, methods in biotechnology, Phytoremediation: Methods and Reviews, N. Willey, Vol. 23, 2003.

[43] P. Roux, C. Boutin, E. Risch, A. Heduit, Life cycle environmental assessment (LCA) of sanitation systems including sewerage: case of vertical flow constructed wetlands versus activated sludge, 12th IWA International Conference on Wetland Systems for Water Pollution Control, Oct 2010, 2010 879–887.

[44] K. Seidel, Reinigung von Gewassern durch hohere Pflanzen, Naturwiss 53 (1966) 289–297.

[45] P. Sharma, S. Pandey, Status of phytoremediation in world scenario, Int. J. Environ. Bioremediat. Biodegradation 2 (4) (2014) 178–191, doi:10.12691/ijebb-2-4-5.

[46] K.M. Shams, G. Tichy, A. Fischer, K. Filip, M. Sager, A. Bashar, T. Peer, M. Jozic, Chromium contamination from tannery wastes in the soils of Hazaribagh area in Dhaka City, Bangladesh, and aspects of its phytoremediation, Geophys. Res. Abstr. 10 (2008)1–2.

[47] O.I. Shittu, Evaluation of an Institutional Wastewater Treatment Facility in Ibadan, Nigeria and Development of a Low-Cost Treatment Plant, Using Reedbed Technology. Unpub. PhD thesis, Faculty of Technology, University of Ibadan, Ibadan, Nigeria, 2019.

[48] M.K.C. Sridhar, Trace element composition of *Pistia stratiotes* L. in a polluted lake in Nigeria, J. Hydrobiology 131 (1986) 273–276.

[49] M.K.C. Sridhar, Uptake of trace elements by water lettuce (*Pistia stratiotes* L.), Acta Hydrochimica et Hydrobiologica DDR. (*Journal of Hydrobiology*) 16 (1988) 293–297.

[50] M.K.C. Sridhar, A.A. Badejo, A.O. Coker, O.O. Noah, Reed bed, root zone or constructed wetlands technology to treat domestic wastewater, septic tank effluent and industrial waste waters, (2013) Registered Number NG/PT/NC/ 2013/36 Application Date 2013-11-11.

[51] E. Tilley, L. Ulrich, C. Lüthi, Ph. Reymond, C. Zurbrügg, Compendium of Sanitation Systems and Technologies – (2nd Revised Edition), Swiss Federal Institute of Aquatic Science and Technology (Eawag), Duebendorf, Switzerland, 2014 ISBN 978-3-906484-57-0.

[52] Y. Trihadiningrum, B. Hassan, M. Muhammad, L. Denny, N. 'Ain bt Ab. Jalil, citing Phytotechnology, a nature-based approach for sustainable water sanitation and conservation, cited, in: K.C. Huat (Ed.), 2002. Development of Putrajaya Wetland for Stormwater Pollution Control, 2008, https://www.academia.edu/2688345/Phytotechnology_a_Nature-Based_Approach_for_Sustainable_Water_Sanitation_and_Conservation.

[53] J. Vymazal, Horizontal sub-surface flow and hybrid constructed wetlands systems for wastewater treatment, Ecol. Eng. 25 (2005) 478–490.

[54] Z. Wang, J. Wan, An Economic Analysis of the Use of Water Hyacinth for Phytoremediation and Biogas Production in Dianchi Lake, China, WorldFish (ICLARM) - Economy and Environment Program for Southeast Asia (EEPSEA), 2013.

[55] WHO, Constitution of the World Health Organization, 1948, http://www.who.int/governance/eb/who_constitution_en.pdf.

[56] Wikipedia, Wastewater, 2019, https://en.wikipedia.org/wiki/Wastewater Accessed June 2, 2019.

[57] Wikipedia, Constructed Wetlands, 2019, https://en.wikipedia.org/wiki/Constructed_wetland Accessed June 2, 2019.

[58] WLE Nile and East Africa Regional programme, CGIAR, https://www.nile-eco-vwu.net/ecosystems, Accessed June 2, 2019.

[59] M.B. Adewole, G.O. Adeoye, M.K.C. Sridhar, Effect of inorganic and organo-mineral fertilizers on the uptake of selected heavy metals by Helianthus annuus L. and Tithonia diversifolia (Hemsl.) under greenhouse conditions, Toxicol. Environ. Chem. 91 (5) (2009) 963–970. doi:10.1080/02772240802614705.

[60] A.A. Badejo, A.O. Coker, V.O. Oyedele, Subsurface flow constructed wetland system vegetated with phragmites karka in the treatment of dye-rich wastewater, J. Nat. Sci., Eng. Technol. 112 (2) (2012) 97–103.

[61] D.A. Hammer, Creating freshwater wetlands, Lewis Publishers, Chelsea, MI, 1992, pp. 1–298.

[62] K Seidel, Abbau von bacteriumcoli durch hohere wasserpflanzen., Naturwiss 51 (1964) 395.

[63] Sanchari Pal, 2017. https://www.thebetterindia.com/89733/shefrol-wastewater-treatment-chinna-kalapet-puducherry-university/.

CHAPTER 10

Sustainable drainage systems in highway drainage

Alireza Fathollahi[a], Stephen J. Coupe[a], Luis A. Sañudo-Fontaneda[a,b,*]

[a]*Centre for Agroecology, Water and Resilience (CAWR), Coventry University, UK* [b]*INDUROT Research Institute, GICONSIME Research Group, Department of Construction and Manufacturing Engineering, University of Oviedo, Spain*

**Corresponding author.*

10.1 Introduction and background

This chapter examines the latest developments in highway drainage focused on techniques to transition road infrastructure to better resilience and climate change adaptation. The primary focus is therefore to determine the status of Sustainable Drainage Systems (SuDS) design and implementation for highways, revealing future challenges and gaps in the field, compared with conventional approaches.

Highway drainage has been driven traditionally by engineers based on two major lines of interest: safety and pavement longevity [65]. Safety to the road user is commonly related to the rapid removal of pavement surface runoff to receiving watercourses or into conventional drainage systems [18]. Longevity of the pavement is linked to the removal of runoff before it impacts underlying structures, and in [13], Rowland's concept of the "drainage iceberg" was outlined whereby water quality was emphasised as the third key area, often not given full consideration by designers and practitioners.

Highway drainage includes surface runoff originating from road boundaries, but also surface and subsurface runoff from the surrounding natural catchment affecting the highway [23,46].

Climate change is changing the way in which engineers consider highway design and maintenance. Significant changes in extreme rainfall intensities as well as other climatic effects have increased the vulnerability of road infrastructure [50]. In addition, [72] identified other related vulnerabilities in highway drainage, such as stream crossings, drainage design parameters and maintenance regimes. The authors also included the four pillars of SuDS (water quality, water quantity, amenity and biodiversity; [81]) in their review through public engagement, to encourage their use in future designs under climate change scenarios. Following on from this reasoning, arid countries such as Qatar have developed research based on the design storm in order to take account of climate change [40]. European research of climate change effects on the road network of member

Sustainable Water Engineering. DOI: 10.1016/C2017-0-04301-X

states focused on drainage systems and pavement performance [5]. The life cycle and performance of such pavements is key, and is usually far shorter than the time span over which climate change would have an impact. In a study of temperature and precipitation patterns, McCurdy and Travis [44] applied such trends to the maintenance of the required levels of service for highway infrastructure, highlighting the need to simulate the performance of highway stormwater systems with adaptation measures to account for future climate change.

Internationally, there appears to be a lack in terms of innovation and resilience to adapt to future change [1,3]. There is also a need for proper investment and maintenance to raise levels of security in the road sector. The AICCP [1] report compared road infrastructure at a global scale, using the following factors: highway capacity, economic considerations, benefits, adaptability to future changes, issues around operation and maintenance, security factors and potential for innovation. Concerning issues were revealed in this report (see Table 10.1A–B) whereby countries such as Turkey, Mexico and Poland failed to achieve 5 out of a potential of 10, and were considered to be "precarious" whilst the UK, Italy and Japan scored over 5, but were still only considered to be "mediocre". France and Spain scored above 6 ("sufficient"), and Germany came top, scoring over 7 points ("good"). This report stresses the need to increase public investment to enhance highway performance and especially adaptation to present and future challenges.

Table 10.1: A.) Highway classification and scores. B.) Scores for individual countries based on classification in A). Adapted from [1].

1A) Classification of highway development based on overall criteria			
Scores	**Description**	**Classification**	
0–2.9	Critical	F	
3–4.9	Precarious	FX	
5–5.9	Mediocre	E	
6–6.9	Sufficient	D	
7–7.9	Good	C	
8–8.9	Very Good	B	
9–9.9	Excellent	A	
1B) Country	**Classification**	**Scores**	
Germany	Good	C	7.1
France	Sufficient	D	6.6
Spain	Sufficient	D	6.1
Ireland	Sufficient	D	6.1
UK	Mediocre	E	5.6
Japan	Mediocre	E	5.6
USA	Mediocre	E	5.3
Italy	Mediocre	E	5.3
Poland	Precarious	FX	4.9
Mexico	Precarious	FX	3.7
Turkey	Precarious	FX	3.3

10.2 Conventional systems: Why is sustainability needed in highway drainage?

Sustainable development in the United Kingdom is defined by four main factors in order to achieve higher quality life for future generations [17]:

- Social progress, which recognises the needs of everyone
- Effective protection of the environment
- Prudent use of natural resources
- Maintenance of high and stable levels of growth and employment

These are enshrined in the UN's Sustainable Development Goals (SDG) with appropriate adjustments made for individual countries, "to ensure no-one is left behind". SDG6 is the "Water" SDG, however, its main concern is with the provision of safe water and sanitation, and the drainage of excess stormwater from highways barely considered. In its report on progress, UN [75] identified pollution from roads amongst important sources of contamination, which included urban areas in general, industrial emissions, agricultural activities, oil spillages and gas, and mining industries in countries in the Middle East, Sahel and India.

In order to identify where highways drainage can achieve sustainability, consideration must be taken of the conventional approach. The following sections therefore examine highway construction, drainage approaches and costings.

10.2.1 Conventional highway drainage systems

Highway drainage systems are designed to collect surface and subsurface run-off water to decrease the risk of flooding and protect underlying structural layers; in general, therefore, they have the following objectives:

- Removal of surface water from the carriageway as quickly as possible to provide safety and minimum nuisance to the travelling public
- Provision of effective sub-surface drainage to maximise longevity of the pavement and its associated earthworks
- Minimisation of the impact of runoff on the receiving environment in terms of flood risk and water quality [83]

Drainage is achieved via surface or sub-surface approaches or a combination of both, depending on the permeability of the road during service life, and uses separate or combined piped systems. The collected runoff is then directed into a surface watercourse or sewer system but the runoff is not allowed to be conveyed to groundwater or other natural reservoirs. It cannot be directed underground if there is a risk of groundwater flooding, thus infiltration rates must be able to manage flows. If the discharge is contaminated, which may reduce water quality, it similarly cannot be allowed to infiltrate. The main types of conventional drainage system for highways will be briefly discussed to give context to their potential to be replaced by more sustainable approaches. More details of their structures and specific applications are given in [64].

10.2.1.1 Kerbs and gullies

These are widely used in urban areas and on rural embankments, as part of the hard drainage infrastructure. Kerbs direct surface water to the gully piping system as a structural element; they are not recommended for high-speed highways due to the risk to vehicles, however, combined kerb and drainage channel systems are now used routinely when refurbishing highways. A highway gully is a pit to convey runoff into the sewerage system; located at the edge of the pavement surface, they can be inset into the verge to prevent hazards to vehicles if on high speed roads.

10.2.1.2 Surface water channels

Surface water channels have various advantages over kerb and gully systems:

1. Their accessibility and visibility, which provides quick detection and maintenance opportunities when blockages occur.
2. Continuity of highway construction when using surface channels, which prevents ingress of surface water into pavement structural layers through cracks in the pavement that could arise during construction of the gullies.
3. Conveyance of larger volumes of runoff due to their higher capacities than kerbs and gullies, which can also increase the distance between water outlets and as a result, project costs are less. Surface water channels are typically concrete sections which can be constructed relatively quickly and inexpensively. They can pose less risk to high-speed vehicles as they are constructed at the highway edge, away from the travelling surface.

10.2.1.3 Drainage channel blocks

These consist of a drainage channel alongside the highway, which collects surface water using small channels crossing the verge. The three main problems associated with drainage channel blocks which need to be taken into consideration by decision makers include [64]:

1. Any settlement of adjacent unpaved surfaces can reduce their effectiveness.
2. They may be prone to rapid build-up of silt and debris in flat areas.
3. Grass cutting operations by mechanical plant can be hindered when adjacent to the channel.

10.2.1.4 Combined kerb and drainage (CKD)

CKD consists of precast concrete channel units 400 to 500 mm long with holes that convey water to an interior cavity. They are strong enough to withstand loadings from traffic and construction and have been in use for over 30 years, providing a cost-effective and flexible approach to highway water management. CKD are highly efficient in managing low to medium and high flows in both urban and rural contexts. Their design includes space for compaction of paving materials adjacent to the concrete unit and resistance to freeze-thaw, which reduces maintenance costs and increases durability. Since units are precast, they are easy to install, and due to their shape, deliver maximum drainage efficiency. Their design also makes retrofit near pedestrian crossings possible, although they are most efficient in flat areas with a minimum longitudinal gradient. Periodic inspection is necessary to remove silt traps, and if blockages occur, rodding and jetting is required.

10.2.1.5 Linear drainage channels (LDC)

LDCs are some of the most popular drainage systems in the industry. They are relatively cheap and easy to install due to minimal pipework requirements. LDCs drain runoff along the entire length of the pavement, whereas CKD collect it at particular points. LDC consist of a surface grating and the channel which comes in a range of sizes and materials including concrete, polymer concrete, plastic and galvanized steel, selected according to project requirements. Plastic channels are generally only used in residential contexts; however, polymer concrete is applicable for a variety of uses.

10.2.1.6 Combined channel and pipe drains

This combined system is constructed at the edge of the road and consists of a channel at the surface with an underlying pipe, which runs at the same gradient as the road. The channel has the same function as a surface water channel; internal pipes are the same as those typically used for pavements but the channels are generally deeper and not as wide, to eliminate the necessity of outfalls. This leads to a significantly larger flow capacity, an advantage in projects with limited verge space. This system generally does not need a separate pipe to carry excess surface water, leading to significantly lower construction costs than conventional systems.

10.2.2 Sustainability and current highway drainage practices

The majority of the approaches described above are based on hard infrastructure, mostly concrete and the use of virgin aggregates in the substructure. In order to achieve more sustainability in highway drainage, recycled materials should be considered as alternatives to virgin aggregates, however structural strength can be an issue, although the use of geosynthetics can address this [27]. This approach leads to a decrease in the demand for primary aggregates, subsequent pressure on landfills and, as a result, environmental impact is reduced. These factors must be considered when evaluating whole life costs of highway drainage construction and maintenance, both as upfront and future costs to achieve better sustainability.

Another important factor in sustainable development for highway drainage and construction is the selection of suitable drainage systems and testing programs for construction materials. The Manual of Contract Documents for Highway Works (MCHW: https://www.

standardsforhighways.co.uk/ha/standards/mchw/index.htm) and the Design Manual for Roads and Bridges (DMRB: https://www.standardsforhighways.co.uk/dmrb/) are two documents produced by the UK Highways Agency which are valuable in providing information on the selection of drainage systems and guidance on recycled materials for construction. In particular, DMRB CG 501 Revision 2 (2020) recommends that, where site conditions are suitable, vegetated systems should be given priority. In DRMB CD 532 Revision 0 (2020) the design of these systems applied to motorways and trunk roads for the conveyance, storage and treatment of road runoff is given, and CD 531 Revision 0 (2020) describes the benefits, design and structure of reservoir pavements. The sustainability and benefits of these systems are recognised, and their use encouraged in these guides.

UK highway regulations state that "*Prior to undertaking drainage design, an assessment of the flood and pollution risk from highway runoff should be undertaken in accordance with HD45 (DMRB11.3.10)*" [28]. The range of standard drainage options presented above, would be unlikely to be seen as the most attractive designs for water quality, but perhaps would be economically sustainable, particularly in the short term

The following sections describe the contaminants associated with road runoff and the potential for their treatment using Sustainable Drainage Systems or SuDS.

10.3 Sources of contamination due to road runoff

Road runoff is commonly considered to be an important contamination route into the environment and as a result, can be a major cause of water quality deterioration due to drainage from road surfaces. Flows from the road collects chemical and pathogenic compounds as well as suspended solids, which may carry contaminants adsorbed to their surfaces that will ultimately find its way into receiving water bodies, resulting in negative impacts on water quality. Increasing urban populations have resulted in significant increases in impervious areas, with less opportunity for water to infiltrate [8]. If, as forecast, urbanisation is set to double or triple between 2000 and 2030 [15,68], then this "sealing" and associated contamination is likely to worsen. For example, it has been reported from some developed cities, impervious surface areas have increased to 70 percent leading to significant runoff contamination [6]. Moreover, increasing global concern over water scarcity has started to convince authorities to consider harvesting and remediating stormwater for use rather than disposal. Therefore, characterisation of emerging contaminants in storm water runoff and identifying their main sources is of importance. The following sections outline different types and sources of contamination in road runoff.

10.3.1 Classification of runoff contaminants

The EU water framework directive [11] published a list of priority pollutants and environmental quality standards (EQS) in the field of water policy. Chemical contamination in urban runoff can be classified into two main categories: inorganic elements, such as toxic metals, and organic compounds.

10.3.1.1 Toxic metals

Water quality in many parts of the world has declined due to the increase in toxic metals associated with emissions from vehicles and impervious surfaces [53]. Concentrations of these toxic metals in road dust have been reported to be higher than standard regulations [4,30]. EQS for dissolved metals in the aquatic environment are shown in Table 10.2. Comparing these EQS against typical heavy metal concentrations in road runoff shows the potential seriousness of the problem, whereby the highest levels of all metals exceed recommendations by orders of magnitude.

Table 10.2: Environmental quality standard for priority metals (EU water framework directive) (μg/L).

Element (μg/L)	AA-EQS inland surface waters[b]	AA-EQS other surface waters[b]	MAC-EQS inland and other surface waters[b]	Typical highway runoff[c]
Cd	≤0.08–0.25[a]	0.2	≤0.45–1.5[a]	0–40
Pb	1.2	1.3	14	73–1780
Ni	4	8.6	34	0–53.3
	EU runoff specific thresholds[a,b]			
	24 h	6 h		
Zn*	60–385	120–770	56–929	
Cu	21	42	22–7033	

EQS, environmental quality standard; AA, annual average value; MAC, maximum allowable concentration; CAS, chemical abstracts service.
[a] range of values dependent on water hardness.
[b] [43].
[c] [10].

10.3.1.2 Organic compounds

Whilst organic compounds are found naturally in water, human activities have increased their concentrations to concerning levels, with [35] linking it to traffic-related activities. For example, poly aromatic hydrocarbons (PAH), pesticides, polychlorinated biphenyls and polybrominated diphenyl ethers are amongst the most toxic; previous studies have reported their existence in road dust and runoff [86,87,89]. The EU water framework directive has produced standard values for priority organic compounds to guarantee water quality (see [43]).

The following sections consider the sources of these contaminants in highway runoff, including drainage surfaces, anthropogenic activities and atmospheric deposition.

10.4 Sources of runoff contamination

10.4.1 Pavement surfaces and construction materials

The identification of sources of contamination deposited on pavement surfaces is important, particularly in identifying which contaminants are from car exhausts, tyres, engine lubricates and atmospheric deposition in order to possibly mitigate their effects. These sources can result in high concentrations of toxic metals, PAHs and nutrients being deposited onto road surfaces. However, materials used in the construction of pavement layers, such as asphalt and concrete components can themselves be damaging to receiving systems, including bitumen and its substitutes.

Although the European Commission has provided methods of testing and regulations for implementation of new construction materials, there is little known about the organic and inorganic compounds leaching from surface materials used in pavements. In addition, materials used in the construction of surrounding buildings and associated urban furniture has the potential to contribute to urban runoff quality reduction. High concentrations of PAHs [84], nonylphenol ethoxylates (NPEOs) [49] heavy metals [49,62] and pesticides [7] leached from pavement surfaces have been reported in recent years. In the last decade, the asphalt sector has increased reclaimed asphalt pavement (RAP) production rates due to lower costs and extended pavement life but as a result, more rejuvenating agents have been introduced to asphalt surfaces. RAP is a combination of natural aggregates and bituminous asphalt, from which various studies have reported leaching of heavy metals and PAHs which then enter runoff [37,45]. Application of rejuvenators, such as tall oils, organic oils or recycled waste oils can be cost effective in restoring surfaces that have degraded whilst in service, but do pose a potential risk to runoff quality [42].

10.4.2 Vehicles

Vehicles are considered to be one of the main sources of contamination in urban runoff, due to exhaust emissions, leaking engine sump oil, tyre wear, brake pad residues and vehicle body paint (see Fig. 10.1).

Important pollutants associated with vehicles include PAHs, which are present in high concentrations derived from exhausts, fuels and lubricating oils. Metals such as cadmium, copper and zinc are present in tyres and brake pads. Chlorine, sodium and calcium salts are also reported in street dust, as they are widely used as de-icing agents during snowy conditions [70]. Müller et. al. [48] reviewed different sources of vehicular contamination including attrition of the road surface due to the passage of motor vehicles, these are summarised in Table 10.3.

Microplastics, of particle size <5 mm, are considered one of the main concerning emerging pollutants, altering water quality and degrading aquatic life [63]. Liu et al. [38], identified

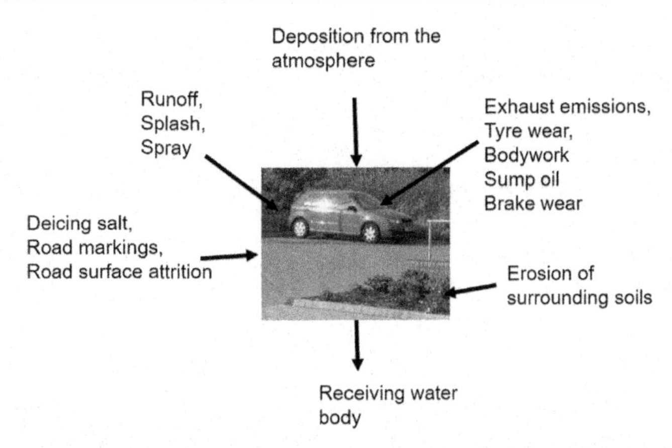

Fig. 10.1: Sources of road contamination associated with vehicles (after [76]).

Table 10.3: Vehicular sources of contamination to roads (after [48]).

Contaminant	Source
Hydrocarbons	Vehicle use
PAHs, NOx, Ni, BTEX	Exhaust gases and particulates
Rh, Pd, Pt	Catalytic converters
TSS, Cd, Cu, Zn, PAHs, microplastics	Tyre wear
W	Wear of tyre studs
TSS, Cd, Cu, Ni, Pb, Sb, Zn, PAHs	Brake wear
Cr, Ni	Engine and vehicle body wear
Pb	Bodywork paint
Pb, Fe (steel), Zn	Wheel balancing weights
Pb, Cd, Cr, Zn, phthalates, NPs, NPEOs	Vehicle washing
TSS, PAHs, microplastics	Road abrasion

Where BTEX, benzene series; NPEOs, nonylphenol ethoxylates; NPs, nonylphenols; TSS, total suspended solids.

highways as a contributor to microplastics pollution, having found them, and associated pollutants, in 7 Danish stormwater ponds. They have the potential to convey contaminants such as PAHs and heavy metals to receiving waterbodies [58], exposing aquatic life to both chemical and physical hazards.

Vehicle tyres have been implicated in the production of microplastics due to their attrition when running along a road surface producing rubber particles ranging from 10 nm to 100 μm due to shear force and resulting heat [32]. Synthetic rubbers are used in tyre manufacture (polymers derived from petroleum) as are additives sulphur, zinc oxide, carbon black and silica [54,73]. Microplastics also affects runoff water quality by transporting environmental contaminants by adsorption, but are also composed of PAHs and toxic metals, which are likely to leach into runoff.

Other particles derived from vehicle use are nanoparticles released from brake pads which are manufactured using barium, antimony, chromium, silicates and aluminium as well as different fibres and organic compounds [32] which are eventually transported in urban runoff.

10.4.3 Pathways: Atmospheric deposition

Atmospheric deposition is a substantial pathway for the transport of contaminants, leading to water quality deterioration, including PAHs [2], heavy metals [39,55] and nutrients [29]. Vehicular– and industry–associated heavy metals settle out from the overlying air to be deposited onto surfaces and thence into runoff [55]. Contaminants transported in this way can originate outside the urban area, which has made pollution monitoring challenging, as it is difficult to distinguish between vehicle and atmospheric derived pollutants.

10.5 Sustainable highway drainage

An approach, which has been widely applied across the world in highway drainage since the last part of the 20th century, is that of Sustainable Drainage Systems (SuDS). Despite SuDS being widely known, there is common confusion due to the different terminology used for the sustainable management of stormwater, depending on where they are located. Fletcher et al. [21] reviewed the most utilised terminology for SuDS over the years internationally, covering low impact development (LID), stormwater best management practices (BMP), water sensitive urban design (WSUD) and stormwater control measures (SCM) amongst others. This chapter will use the term "SuDS" to express the technique which encourages infiltration, storage of excess stormwater volume and slows downstream conveyance. The addition of blue and green infrastructure to the urban environment is described in Chapter 11, and thus the fundamentals of this approach will not be covered here.

In terms of its specific application to the highway environment, there are 2 main sources of detailed information: the "SuDS for Roads" publication [57] and also a chapter in the SuDS Manual [81] dedicated to the design and installation of SuDS associated with highways. The latter particularly recommends the use of Pervious Paving Systems (PPS) [9], swales, detention ponds and wetlands, highway filter drains and also bioretention. In their drainage design manual [23], the Georgia Department of Transport lists filter strips, sand filters, bio-slopes, swales and enhanced dry and wet swales, dry detention basins, bioretention basins, dry and wet detention ponds, stormwater wetlands and infiltration trenches as being highly suited to drain highways. They also recommend porous pavements, or open graded friction courses, as being the most cost-effective since they have the capacity to replace traditional asphalt road surfacing.

Many of these approaches have been used to reduce pavement flooding, but also as a way to treat pollutants, improve sustainability in water management and help road systems to adapt to climate change effects [34]. There are significant examples of these practices in use, for example highway filter drains [65], bioretention areas [69,74], swales [22,71] and PPS [31], amongst others.

Since the early 2000s, manuals and guidance from authorities such as Wong et al. [80] around water sensitive design in roads and highways have been released, informing the wider community of engineers of how to implement what is perceived as a new paradigm in water management. This paradigm shift began from management as a learning approach, instead of the previous paradigm based on prediction and control [56], subsequently moving towards the consideration of socially valued ecosystem functions and services [59], as well as ensuring a more sustainable and holistic approach to water management [77]. Water sensitive urban design (WSUD) included this environmentally-oriented approach to water management in the philosophy for urban areas [47]. The following sections consider the many benefits of the SuDS approach, in particular in addressing pollutants associated with road runoff and climate change.

10.5.1 SuDS, highways and water quality

Jato-Espino et al. [33] conducted a literature review of the most utilised SuDS internationally, finding that bioretention zones, swales, raingardens, permeable pavements and ponds were used extensively. Swales and constructed wetlands have been widely used to control the surface runoff from both the carriageway and the surrounding embankments, introducing the concept of water *quality* being equally important as water *quantity* [18,80]. According to [67], vegetated filter strips along roadsides can create functional hydrologic landscapes, reducing the need for conventional systems by managing stormwater runoff at source.

Bridge deck drainage is often a neglected issue; drained to watercourses without any previous pollutant treatment, it can result in pollution events downstream. Experiences reported in North Carolina, USA, by Winston et al. [79] showed the efficiency of several stormwater control measures such as permeable friction course, wet retention ponds, bioretention cells, vegetated filter strips, constructed stormwater wetlands, and grassed swales, in treating bridge deck runoff pollutants.

Sustainable water management techniques in the mining industry such as retention ponds have been used in road construction in order to treat pollutants from acidic rock runoff as reported by [26] in Tennessee, USA. This experience showed the importance of wider BMP techniques for application in civil engineering infrastructures like roads.

10.5.2 Treatment of highway runoff using SuDS

As described above, standard drainage infrastructure for highways does not have any significant treatment capacity to improve water quality, unless it has been specifically added to the design. The imperative for driver safety, comes from the requirement to drain the carriageway efficiently to prevent dangerous driving conditions. The longevity of the highway is a secondary concern and usually satisfied by the prevention of water ingress into the road foundation, and prevent the mobilisation of fines into the sub-base. In terms of water quality in the UK, research has demonstrated that stormwater control measures can and should provide pollution mitigation [28].

Despite the advances in design options available, much highway runoff in the UK is discharged untreated, with mainly particulate material removed and dissolved contaminants such as metals released into the environment. The UK [16], announced a £28.8 billion National Roads Fund (NRF) for 2020–25 which could lead to a consideration of more sustainable options for highway drainage during renovation and retrofit. However, this does not necessarily guarantee SuDS as the most attractive option, as shown in Section 10.2.2 where filter drain operation and maintenance is considered.

Very often, the treatment of highway runoff depends on the amount of space available. For example, where highways run through privately owned land, the amount of space in which to install drainage devices can limit the range of options to linear structures such as combined kerb and drainage, with little attention given to stormwater treatment. Ponds and constructed wetlands are very effective in retention and treatment of pollution, but represent a large land take [28,81].

10.5.2.1 Pervious pavement systems (PPS) for highway use

As shown by the Highways Agency [28] PPS use for low speed roads or parking infrastructure can also be applied in a highway setting, with the important proviso that there must be careful consideration of structural risk and a bespoke design process. The guidance also states that PPS is more suited to trunk roads and dual carriageways than motorways or interstate highways.

The performance of water quality improvement by PPS has been rigorously documented in a large number of studies in non-highway environments, including sediment removal and the treatment of organics and hydrocarbons by microbiological action [51,52,60,61]. To the authors' knowledge, there have been no studies specifically addressing the operation of in-use highway PPS, so it is not possible to verify that the principles of source control, infiltration, contaminant retention and bioremediation would operate on high speed roads. It is, however, reasonable to expect the same enhanced water quality treatment in comparison with conventional highway construction. The following section documents common infiltration techniques used in association with highways and their performance to reduce water volume and treat pollutants.

10.5.2.2 Swales and filter drains

Swales and highway filter drains (HFD) only require a small area adjacent to the road (typically < 3 m) and offer some treatment of pollutants. Such devices adhere to SuDS principles as they infiltrate and treat highway runoff [57]. Swales are vegetated and provide treatment due to the interaction between contaminants, vegetation and soils, where sediments and metals (e.g. lead, copper, zinc and cadmium) have been recorded as significantly removed from stormwater [71]. Swales have been used in Australia as pre-treatment systems for PPS, removing 50–75 percent of total suspended sediment within the first 10 m of the swale, thereby prolonging the service life of the pavement [36].

Filter drains remove particles, free product oil and probably dissolved hydrocarbon fractions. Where flow slows and moves through complex media, pollutant removal in HFDs increases, relative to systems where water is merely conveyed in pipes or channels. This pollutant removal occurs due to a combination of deposition, filtration through sediments and plant root networks, absorption and adsorption. Paradoxically, as the HFD surface fills with sediment, causing a decrease in drainage rate, the amount of interaction between contaminants and substrate increases, improving treatment.

As the primary function of HFDs is to remove water from the carriageway, UK maintenance regimes include inspection of drainage condition, removal of sediment by separation of fines and replacement of the cleaned aggregate [19]. Although plants, both living and dead organic matter, play an important role in removal and retention of contaminated sediments in HFDs, they can also impede water moving from the highway into the drain and therefore must be removed. There is some evidence of microbiological treatment of organics in filter drains by the action of naturally occurring soil and waterborne bacteria, fungi, protists and larger metazoans [12]. When investigating the retention and treatment of sediments and used lubricating oils in laboratory simulated HFDs, it was observed that CO_2 was significantly raised relative to non-contaminated controls. The concentration of CO_2 inside the models was 5000 ppm, at least 10 times that of the ambient air, demonstrating significant levels of aerobic treatment. This has not been fully explored, but many of the environmental factors that produce bioremediating biofilms in permeable pavements [14,41] are also encountered in HFDs. These beneficial properties for biological treatment include a highly oxygenated atmosphere that facilitates aerobic degradation, regularly wetted substrates and 620–2,300 mg/kg of nitrogen, phosphorous and potassium [13]. There is a possibility that significant concentrations of soluble pollutants, such as metals may be adsorbed onto organics in HFDs, despite sediments and aggregates being the only media interacting with runoff. It is likely that HFDs become colonised by aquatic insects, some of which may have some biodiversity value or even protected status. However, there have been instances of "nuisance insects" in some drainage systems e.g. *Psychodidae*, the drain fly, generally found in sewage treatment systems [25].

10.5.2.3 New developments in UK highway drainage

A significant change in drainage infrastructure has occurred in the UK in recent years, where upgrading to smart motorways has included removal of HFDs and their replacement with precast concrete channels. Fig. 10.2 shows a combined kerb and drainage system used in a

Fig. 10.2: A concrete combined kerb and channel system.

car park of a motorway service area, but this is similar to those which can be used alongside highways.

These channels are not specifically designed for water quality improvement, thus this development could be seen as a backward step in the treatment of highway runoff, when compared to the performance of HFDs. Channels often include a silt trap which can remove some particulate matter but not necessarily fine sediment such as that < 63 μm, colloids and certainly not dissolved matter. However, as Fig. 10.2 shows, they can be incorporated into SuDS by directing runoff into PPS or vegetative system. Extra design options to improve runoff water quality include connection to a separator or interceptor, but these are equally or even more dependent on maintenance than HFDs, particularly from silt. Runoff velocities in the channel make it unlikely that they would achieve any significant treatment. Despite this, HFDs will probably continue to feature in UK drainage, as they run parallel to approximately 7,000 km (4,350 miles) of motorways and main roads in the UK [66].

10.6 Highways and climate change

Neumann et al. [50] stress the need for more research in order to assess the effects of climate-induced extreme events affecting highway infrastructure and that design can account for this in terms of safety and the longevity of highway structures. It is accepted that precipitation may vary increasingly every year due to more extreme rainfall events worldwide [78,85]. Therefore, the design storm that translates into the size of the drainage system, is likely to have increased over the last few years according to studies such as Yilmaz and Perera [82] in Australia.

In its infrastructure report, ASCE [3] recommends that PPS should be used more for highways and that more research is needed to develop new materials and techniques, reduce stormwater runoff and increase infrastructure resilience to climate change. AICCP-IC [1] also highlights the importance of adapting roads to new technologies, real time monitoring and future requirements such as automatic driving systems.

10.7 Conclusions

There are considerable challenges in the future to take account of population growth and the risks associated with climate change, such that designing and managing of highway drainage infrastructure is likely to become more complex. However, this chapter has illustrated that there is potential for combining sustainable, ecological with engineering approaches in the highway environment in a paradigm shift away from predominantly grey infrastructure. Whilst not specifically applied to the highway environment, nonetheless, [59] assert "engineers are actively exploring how to incorporate these into infrastructure operations". This in spite of conflicting paradigms in terms of engineered highway drainage infrastructure and those that seek to mimic natural systems such as SuDS, which [59] assert, "have typically been antagonistic". Indeed, [20], would promote the use of green infrastructure combined with "the best engineering practices of today" in designing cities of the future, rather than allowing the *ad hoc* development of the past that has led to problems which future generation will have to manage.

Sustainable highway drainage can provide multiple benefits that pipes, impervious running surfaces and concrete cannot. Incorporating vegetation, storage and slow conveyance provides treatment, improving water quality as well as reducing volume and standing water on the highway, thus improving safety for the road user [57]. However, design for the specific site is key, to ensure optimum performance and longevity of the pavement.

To further promote sustainable credentials for the highway, recycled, reused and repurposed materials from the construction industry, road refurbishment and the rail network can also be used, bearing in mind that drainage flows through these materials. Thom and Dawson [88] suggest that this approach is beneficial for the environment, has long term advantages, and is also economically sound, but that it does need careful consideration. They suggest that key to the delivery of these options are guidelines on the use of secondary materials in pavement construction, including design, operation and maintenance.

Whilst this chapter considered their use in terms of drainage, the multiple benefits of a more sustainable approach clearly has advantages beyond simply managing surface of runoff as quickly as possible. Roads and highways connect communities, delivering people and goods; case studies such as the "Sustainable Streets" initiative reported by [24] show how these approaches can be incorporated into existing roads, including highways, to take account of traffic movement, community activities and the enhancement of roadside ecology in a sustainable, efficient and attractive way.

References

[1] AICCP-IC, Las Obras y Servicios Públicos a Examen, España, Informe, 2019, Madrid, 2019.

[2] S. Al Ali, X. Debade, G. Chebbo, B. Béchet, C. Bonhomme, Contribution of atmospheric dry deposition to stormwater loads for PAHs and trace metals in a small and highly trafficked urban road catchment, Environ. Sci. Pollut. Res. 24 (2017) 26497–26512.

[3] ASCE, Infrastructure report card: a comprehensive assessment of america's infrastructure, 2017, 76–80. Accessed at https://www.infrastructurereportcard.org/wp-content/uploads/2016/10/2017-Infrastructure-Report-Card.pdf.

[4] C. Barlow, L.I. Bendell, C. Duckham, D. Faugeroux, V. Koo, Three-dimensional profiling reveals trace metal depositional patterns in sediments of urban aquatic environments: a case study in Vancouver, British Columbia, Canada, Water. Air. Soil Pollut. 225 (2014) 1856, doi:10.1007/s11270-013-1856-y.

[5] K.F. Bizjak, A. Dawson, I. Hoff, L. Makkonen, J.S. Ylhäisi, A. Carrera, The impact of climate change on the European road network, Proc. Inst. Civ. Eng. Transp. 167 (2014) 281–295.

[6] K.L. Blansett, J.M. Hamlett, Runoff ratios for five small urban karst watersheds under different levels of imperviousness, in: World Environmental and Water Resources Congress 2013: Showcasing the Future - Proceedings of the 2013 Congress, 2013, 3196–3205.

[7] M. Burkhardt, S. Zuleeg, R. Vonbank, P. Schmid, S. Hean, X. Lamani, K. Bester, M. Boller, Leaching of additives from construction materials to urban storm water runoff, Water Sci. Technol. 63 (2011) 1974–1982.

[8] D. Butler, J. Davies, Urban Drainage, 3rd ed, CRC Press, 2010.

[9] S.M. Charlesworth, C. Lashford, F. Mbanaso, Hard SUDS Infrastructure. Review of Current Knowledge, Foundation for Water Research, 2014, Accessed at http://www.fwr.org/environw/hardsuds.pdf.

[10] J. Clary, M. Leisenring, P. Hobson, International Stormwater Best Management Practices (BMP) Database Pollutant Category Summary: Metals, 2011, Accessed at www.bmpdatabase.org.

[11] Council of the European Union, E.P., Directive 2013/39/EU of the European Parliament and of the Council of 12 August 2013 amending Directives 2000/60/EC and 2008/105/EC as regards priority substances in the field of water policy Text with EEA relevance, 2013.

[12] S.J. Coupe, A.P. Newman, S. Fontaneda, L.A., Hydrocarbon biodegradation in hard infrastructure, in: S.M. Charlesworth, C. Booth (Eds.), (2017) Sustainable Surface Water Management; AHandbook for SuDS, Wiley Blackwell publishing, 2016a.

[13] S.J. Coupe, L.A. Sañudo-Fontaneda, A.-M. McLaughlin, S.M. Charlesworth, E.G. Rowlands, The Retention and In-Situ Treatment of Contaminated Sediments in Laboratory Highway Filter Drain Models, in: Water Efficiency Conference 2016, Coventry, 2016a. Accessed at https://www.watefnetwork.co.uk/files/default/resources/Conference2016/Session_One/45-COUPE.pdf b.

[14] S.J. Coupe, H.G. Smith, A.P. Newman, T. Puehmeier, Biodegradation and microbial diversity within permeable pavements, Eur. J. Protistol. 39 (2003) 495–498.

[15] Department of Economics and Social Affairs, 2014 Revision of the World Urbanization Prospects, 2014, New York. Accessed at https://www.un.org/en/development/desa/publications/2014-revision-world-urbanization-prospects.html.

[16] Department of Transport, Roads Funding: Information Pack, 2018. Accessed at https://assets.publishing.service.gov.uk/government/uploads/system/uploads/attachment_data/file/757950/roads-funding-information-pack.pdf.

[17] DETR, Guidance on the Interpretation of Major Accident to the Environment for the Purposes of the COMAH Regulations, 1999, ISBN 011753501X, The Stationery Office, London.

[18] J.B. Ellis, Design considerations for the use of vegetative controls for the treatment of highway discharges, Proceedings of IUGG 99 Symposium HS5, Impacts of Urban Growth on Surface Water and Groundwater Quality, 259, IAHS Publ. Birmingham, UK, 1999, pp. 357–363.

[19] J.B. Ellis, E.G. Rowlands, Highway filter drain waste arisings: a challenge for urban source control management? Water Sci. Technol. 56 (10) (2007) 125–131.

[20] B.M. Fekete, J.J. Bogárdi, Role of engineering in sustainable water management, Earth Perspectives 2 (2) (2015) 1–9, doi:10.1186/s40322-014-0027-7.

[21] T.D. Fletcher, W. Shuster, W.F. Hunt, R. Ashley, D. Butler, S. Arthur, S. Trowsdale, S. Barraud, A. Semadeni-Davies, J.L. Bertrand-Krajewski, P.S. Mikkelsen, G. Rivard, M. Uhl, D. Dagenais, M. Viklander, The evolution and application of terminology surrounding urban drainage, Urban Water J. 12 (7) (2015) 525–542.

[22] S. Gavrić, G. Leonhardt, J. Marsalek, M. Viklander, Processes improving urban stormwater quality in grass swales and filter strips: a review of research findings, Sci. Total Environ. 669 (2019) 431–447.

[23] GDOT, Design Manual: Drainage, 2019. Accessed at http://www.dot.ga.gov/PartnerSmart/DesignManuals/Drainage/Drainage%20Manual.pdf.

[24] E. Greenberg, Sustainable streets: an emerging practice, ITE J. May 2008 (2008) 30–39.

[25] T.B. Griffith, J.L. Gillett-Kaufman, Featured Creatures: Drain fly, 2020. Accessed at http://entnemdept.ufl.edu/creatures/URBAN/FLIES/drain_fly.html.

[26] J.J. Gusek, V. Bateman, J. Ozment, L. Oliver, D. Kathman, J. Waples, T. Rutkowski, H. Moore, W. Bowden, A. Reither, Mitigating impacts from acid-producing rock in Tennessee road construction projects, in: Tailings and Mine Waste'10 - Proceedings of the 14th International Conference on Tailings and Mine Waste, 2011, 171–185.

[27] J. Han, J.K. Thakur, Sustainable roadway construction using recycled aggregates with geosynthetics, Sustain. Cities Soc. 14 (2014) 342–350.

[28] Highways Agency, Reservoir Pavements for Drainage Attenuation, 2018. Accessed at www.standardsforhighways.co.uk.

[29] S.E. Hobbie, J.C. Finlay, B.D. Janke, D.A. Nidzgorski, D.B. Millet, L.A. Baker, Contrasting nitrogen and phosphorus budgets in urban watersheds and implications for managing urban water pollution, Proc. Natl Acad. Sci. U.S.A. 114 (2017) 4177–4182.

[30] H.-M. Hwang, M.J. Fiala, D. Park, T.L. Wade, Review of pollutants in urban road dust and stormwater runoff: part 1. Heavy metals released from vehicles, Int. J. Urban Sci. 20 (2016) 334–360.

[31] H.M. Imran, S. Akib, M.R. Karim, Permeable pavement and stormwater management systems: a review, Environ. Technol. (United Kingdom) 34 (2013) 2649–2656.

[32] P. Jan Kole, A.J. Löhr, V. Belleghem, R. F.G.A.J., A.M.J., Wear and tear of tyres: a stealthy source of microplastics in the environment, Int. J. Environ. Res. Public Health 14 (10) (2017) 1–31, doi:10.3390/ijerph14101265.

[33] D. Jato-Espino, L.A. Sañudo-Fontaneda, V.C. Andrés-Valeri, Life-cycle assessment of construction materials: analysis of environmental impacts and recommendations of eco-efficient management practices, Handb. Environ. Mater. Manag. (2019) 1525–1550, doi:10.1007/978-3-319-58538-3_76-1.

[34] V.M. Jayasooriya, A.W.M. Ng, Development of a framework for the valuation of ecosystem services of green infrastructure, in: Proceedings - 20th International Congress on Modelling and Simulation, MODSIM 2013, 2020 3155–3161.

[35] A. Jonsson, U. Fridén, K. Thuresson, L. Sörme, Substance flow analyses of organic pollutants in Stockholm, Water Air Soil Pollut. Focus 8 (2008) 433–443.

[36] M.A. Kachchu Mohamed, T. Lucke, F. Boogaard, Preliminary investigation into the pollution reduction performance of swales used in a stormwater treatment train, Water Sci. Technol. 69 (2014) 1014–1020.

[37] M. Legret, L. Odie, D. Demare, A. Jullien, Leaching of heavy metals and polycyclic aromatic hydrocarbons from reclaimed asphalt pavement, Water Res. 39 (2005) 3675–3685.

[38] F. Liu, K. Borg Olesen, A. Reimer Borregaard, J. Vollertsen, Microplastics in urban and highway stormwater retention ponds, Sci. Total Environ. 671 (2019) 992–1000.

[39] A. Liu, N. Hong, P. Zhu, Y. Guan, Understanding benzene series (BTEX) pollutant load characteristics in the urban environment, Sci. Total Environ. 619–620 (2018) 938–945.

[40] A.A. Mamoon, N.E. Joergensen, A. Rahman, H. Qasem, Design rainfall in Qatar: sensitivity to climate change scenarios, Nat. Hazards 81 (2016) 1797–1810.

[41] F.U. Mbanaso, S.J. Coupe, S.M. Charlesworth, E.O. Nnadi, Laboratory-based experiments to investigate the impact of glyphosate-containing herbicide on pollution attenuation and biodegradation in a model pervious paving system, Chemosphere 90 (2013) 737–746.

[42] G. Mazzoni, E. Bocci, F. Canestrari, Influence of rejuvenators on bitumen ageing in hot recycled asphalt mixtures, J. Traffic Transp. Eng. 5 (3) (2018) 157–168.
[43] A.-M. McLaughlin, S.M. Charlesworth, S.J. Coupe, UK and EU water policy as an instrument of urban pollution, in: S.M. Charlesworth, C. Booth (Eds.), Urban Pollution, Science and Management, Wiley Blackwell Publishing, 2018.
[44] A.D. McCurdy, W.R. Travis, Simulated climate adaptation in stormwater systems: evaluating the efficiency of adaptation strategies, Environ. Syst. Decis. 37 (2017) 214–229.
[45] Y. Mehta, A. Ali, B. Yan, A.E. McElroy, H. Yin, Environmental Impacts of Reclaimed Asphalt Pavement (RAP), FHWA-NJ-2017-008, 2017.
[46] Ministerio de Obras Públicas y Urbanismo, G. de E., Boletín Of. del Estado 2016, Comunidad Autónoma, de Cataluña (2010) 18882–19023, https://www.boe.es/eli/es-ct/dlg/2010/08/03/1/con.
[47] P.J. Morison, R.R. Brown, C. Cocklin, Transitioning to a waterways city: municipal context, capacity and commitment, Water Sci. Technol. 62 (2010) 162–171.
[48] A. Müller, H. Österlund, J. Marsalek, M. Viklander, The pollution conveyed by urban runoff: a review of sources, Sci. Total Environ. 709 (2010) 136125, doi.10.1016/j.scitotenv.2019.136125.
[49] A. Müller, H. Österlund, K. Nordqvist, J. Marsalek, M. Viklander, Building surface materials as sources of micropollutants in building runoff: a pilot study, Sci. Total Environ. 680 (2019) 190–197.
[50] J.E. Neumann, J. Price, P. Chinowsky, L. Wright, L. Ludwig, R. Streeter, R. Jones, J.B. Smith, W. Perkins, L. Jantarasami, L. Jantarasami, J. Martinich, Climate change risks to US infrastructure: impacts on roads, bridges, coastal development, and urban drainage, Clim. Change 131 (2015) 97–109.
[51] A.P. Newman, C.J. Pratt, S.J. Coupe, N. Cresswell, Oil bio-degradation in permeable pavements by microbial communities, Water Sci. Technol. 45 (7) (2002) 51–56.
[52] A.P. Newman, L. Platje, T. Puehmeier, J.A. Morgan, J. Henderson, C.J. Pratt, Recent studies on oil degrading porous pavement structures, in: Proc. Standing Conf. on Stormwater Source Control, 2002. Accessed at https://www.researchgate.net/publication/254560382_Recent_Studies_on_Oil_Degrading_Porous_Pavement_Structures.
[53] V. Novotny, A. Bartošová, N. O'Reilly, T. Ehlinger, Unlocking the relationship of biotic integrity of impaired waters to anthropogenic stresses, Water Res. 39 (2005) 184–198.
[54] T.A. Okel, J.A. Rueby, Silica morphology and functionality: addressing winter tire performance, Rubber World 253 (2016) 21–33.
[55] M. Omrani, V. Ruban, G. Ruban, K. Lamprea, Assessment of atmospheric trace metal deposition in urban environments using direct and indirect measurement methodology and contributions from wet and dry depositions, Atmos. Environ. 168 (2017) 101–111.
[56] C. Pahl-Wostl, Transitions towards adaptive management of water facing climate and global change, Water Resour. Manag. 21 (2007) 49–62.
[57] C. Pittner, G. Allerton, SUDS For Roads, 2001. Accessed at http://www.scotsnet.org.uk/assets/SudsforRoads.pdf.
[58] L. Pittura, C.G. Avio, M.E. Giuliani, G. d'Errico, S.H. Keiter, B. Cormier, S. Gorbi, F. Regoli, Microplastics as vehicles of environmental PAHs to marine organisms: combined chemical and physical hazards to the Mediterranean mussels, Mytilus galloprovincialis, Front. Mar. Sci. 5 (2018) 1–15.
[59] N.L. Poff, C.M. Brown, T.E. Grantham, J.H. Matthews, M.A. Palmer, C.M. Spence, R.L. Wilby, M. Haasnoot, G.F. Mendoza, K.C. Dominique, K.C. Dominique, A. Baeza, Sustainable water management under future uncertainty with eco-engineering decision scaling, Nat. Clim. Chang. 6 (2016) 25–34.
[60] C.J. Pratt, A.P. Newman, P.C. Bond, Mineral oil bio-degradation within a permeable pavement: long term observations, Water Sci. Technol. 39 (2) (1999) 103–109.
[61] C.J. Pratt, A Review of Published Material on the Performance of Various SUDS Components, 2004. Accessed at https://www.susdrain.org/files/resources/evidence/suds_lit_review_04.pdf.
[62] S. Raffo, I. Vassura, C. Chiavari, C. Martini, M.C. Bignozzi, F. Passarini, E. Bernardi, Weathering steel as a potential source for metal contamination: metal dissolution during 3-year of field exposure in an urban coastal site, Environ. Pollut. 213 (2016) 571–584.

[63] L.C. de Sá, M. Oliveira, F. Ribeiro, T.L. Rocha, M.N. Futter, Studies of the effects of microplastics on aquatic organisms: what do we know and where should we focus our efforts in the future, Sci. Total Environ. 645 (2018) 1029–1039.

[64] S.V. Santhalingam, SCI Lecture Paper Series - Highway Drainage Systems, Highways Agency, 1999. Accessed at www.soci.org.

[65] L.A. Sañudo-Fontaneda, S.J. Coupe, S.M. Charlesworth, E.G. Rowlands, Exploring the effects of geotextiles in the performance of highway filter drains for sustainable and resilient highway drainage, Geotext. Geomembranes 46 (2018) 559–565.

[66] L.A. Sañudo Fontaneda, E. Blanco-Fernández, S.J. Coupe, J. Carpio, A.P. Newman, D. Castro-Fresno, Use of geosynthetics for sustainable drainage, in: S.M. Charlesworth, C. Booth (Eds.), 2017 Sustainable Surface Water Management; A Handbook for SuDS, Wiley Blackwell Publishing, 2017.

[67] P.L.S. Schooler, An alternate approach to size vegetative filter strips as elements of a highway LID stormwater management strategy, Low Impact Development 2010: Redefining Water in the City - Proceedings of the 2010 International Low Impact Development Conference, (2010) 1557–1570.

[68] K.C. Seto, B. Güneralp, L.R. Hutyra, Global forecasts of urban expansion to 2030 and direct impacts on biodiversity and carbon pools, Proc. Natl. Acad. Sci. U.S.A. 109 (2012) 16083–16088.

[69] P. Shrestha, S.E. Hurley, B.C. Wemple, Effects of different soil media, vegetation, and hydrologic treatments on nutrient and sediment removal in roadside bioretention systems, Ecol. Eng. 112 (2018) 116–131.

[70] M. Skorbiłowicz, E. Skorbiłowicz, Content of calcium, magnesium, sodium and potassium in the street dust from the area of Białystok (Poland), J. Ecol. Eng. 20 (2019) 125–131.

[71] J.H. Stagge, A.P. Davis, E. Jamil, H. Kim, Performance of grass swales for improving water quality from highway runoff, Water Res. 46 (2012) 6731–6742.

[72] R.L. Strauch, C.L. Raymond, R.M. Rochefort, A.F. Hamlet, C. Lauver, Adapting transportation to climate change on federal lands in Washington State, USA, Clim. Change 130 (2) (2015) 185–199.

[73] D. Susa, J. Haydary, Sulphur distribution in the products of waste tire pyrolysis, Chem. Pap. 67 (2013) 1521–1526.

[74] S.A. Trowsdale, R. Simcock, Urban stormwater treatment using bioretention, J. Hydrol. 397 (2011) 167–174.

[75] United Nations, Sustainable Development Goal 6: Synthesis Report 2018 on Water and Sanitation, United Nations, New York, NY, USA, 2018, 2018.

[76] C. Vogelsang, A.L. Lusher, M.E. Dadkhah, I. Sundvor, M. Umar, S.B. Ranneklev, D. Eidsvoll, S. Meland, Microplastics in Road Dust – Characteristics, Pathways and Measures. Report no. 7361-2019, Norwegian Environment Agency, 2018. Accessed at https://www.miljodirektoratet.no/globalassets/publikasjoner/M959/M959.pdf.

[77] N. Voulvoulis, K.D. Arpon, T. Giakoumis, The EU water framework directive: from great expectations to problems with implementation, Sci. Total Environ. 575 (2017) 358–366.

[78] S. Westra, L.V. Alexander, F.W. Zwiers, Global increasing trends in annual maximum daily precipitation, J. Clim. 26 (2013) 3904–3918.

[79] R.J. Winston, M.S. Lauffer, K. Narayanaswamy, A.H. McDaniel, B.S. Lipscomb, A.J. Nice, W.F. Hunt, Comparing bridge deck runoff and stormwater control measure quality in North Carolina, J. Environ. Eng. (United States) 141 (1) (2015) 04014045.

[80] T. Wong, P. Breen, S.D. Lloyd, Water Sensitive Road Design – Design Options for Improving Stormwater Quality of Road Runoff, CRC for Catchment Hydrology Technical. Report 00/01, 2000, 74 p. Accessed at https://pdfs.semanticscholar.org/6e28/3fe7e86bc561ab40cd8a435dadddbcc6e921.pdf?_ga=2.126430029.1343888460.1586588768-1714881280.1585918099.

[81] B. Woods Ballard, S. Wilson, H. Udale-Clarke, S. Illman, T. Scott, R. Ashley, R. Kellagher, The SuDS Manual, CIRIA, London, 2015.

[82] A.G. Yilmaz, B.J.C. Perera, Changes in intensity-frequency-duration relationship of heavy rainfalls at a station in Melbourne, Proceedings - 20th International Congress on Modelling and Simulation, MODSIM 2013, 2020, 2834–2840.

[83] S. Young, Drainage Design. Chapter 34: ICE Manual of Highway Design and Management, Institution of Civil Engineers, 2011. Accessed at https://www.icevirtuallibrary.com/doi/pdf/10.1680/mohd.41110.0285.
[84] Y. Zhang, M. van de Ven, A. Molenaar, S. Wu, Preventive maintenance of porous asphalt concrete using surface treatment technology, Mater. Des. 99 (2016) 262–272.
[85] Q. Zhou, A review of sustainable urban drainage systems considering the climate change and urbanization impacts, Water 6 (4) (2014) 976–992.
[86] M.B. Jensen, P.E. Holm, J. Laursen, H.C. Hansen, Contaminant aspects of blackish surface deposits on highway roadsides, Water Air Soil Pollut. 175 (2006) 305–321.
[87] T. Kose, T. Yamamoto, A. Anegawa, S. Mohri, Y. Ono, Source analysis for polycyclic aromatic hydrocarbon in road dust and urban runoff using marker compounds, Desalination 226 (2008) 151–159.
[88] N. Thom, A. Dawson, Sustainable road design: Promoting recycling and non-conventional materials: Sustainability 11 (21) (2019).
[89] S. Zgheib, S. Moilleron, R. Chebbo, G. Priority pollutants in urban stormwater: part 1 - case of separate storm sewers. Water Res. 46 (20) 6683–6692.

CHAPTER 11

Sustainable drainage, green and blue infrastructure in urban areas

Susanne M. Charlesworth[a,*], Frank Warwick[b]

[a]*Centre for Agroecology, Water and Resilience, Coventry University, Ryton Gardens, Wolston Lane, Coventry CV8 3LG, UK* [b]*School of Energy, Construction and Environment, Faculty of Engineering, Environment and Computing, Coventry University, Priory Street, Coventry CV1 5FB, UK*
**Corresponding author.*

11.1 Introduction

Floods are the most common disaster across Europe [30]. Fig. 11.1 shows that between 1980 and 2010, the number of floods categorised as "severe" have revealed an increasing trend, with 2002 (77 severe floods) and 2010 (71 severe floods) the highest in the 30-year period.

With predictions of the impacts of climate change suggesting more incidents of increasing severity, and associated pollution and reductions in biodiversity, an approach which is flexible and of multiple benefit is required to address these challenges. Pipes, concrete infrastructure and impermeable surfaces are very efficient at quickly conveying water away, but provide very few other benefits. Further, their actions often exacerbate flooding by collecting stormwater across the catchment in large volumes. Interest in the use of green and blue infrastructure (GBI), working with nature (WwN) and nature based solutions (NBS) is therefore growing. These approaches use vegetation, water storage, infiltration and slow conveyance to attenuate the storm peak, improve water quality, and increase both biodiversity provision, and amenity opportunities for people. However, in spite of such approaches being well known, they are often ignored in the management of water and associated policy, and thus conventional hard infrastructure dominates. In fact, the United Nations [39] report on progress with Sustainable Development Goal 6 (clean water and sanitation) suggested that spend on NBS is very much less than 5 percent of the total invested in water resources management infrastructure [39].

One approach that does utilise GBI is that of sustainable drainage systems (SuDS), which, as shown by Charlesworth [12] not only manages surface water, but also has benefits in terms of mitigating the effects of climate change, and potentially to adapt to those changes. This approach is accepted in many countries globally, for example in the USA, it is referred to as low impact development (LID) [21] or best management practices (BMP) [40,41]; in Australia and New Zealand it is water sensitive urban design or WSUD [16,34].

Sustainable Water Engineering. DOI: 10.1016/C2017-0-04301-X

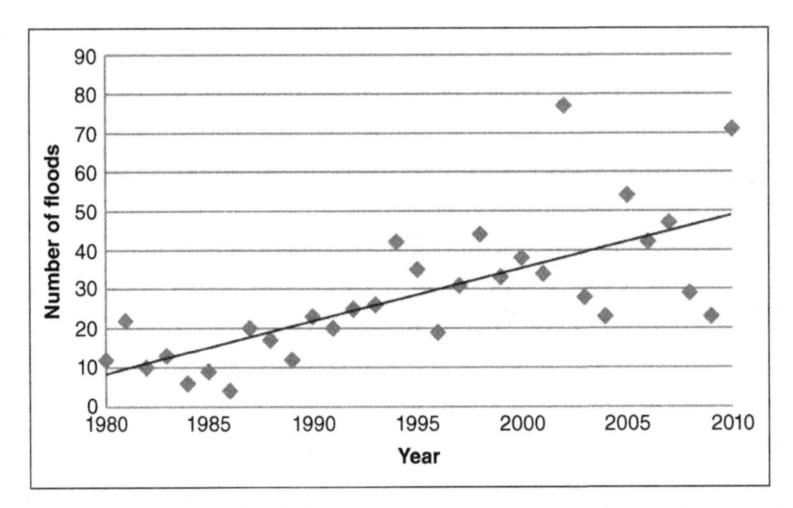

Fig. 11.1: Numbers of severe floods in Europe, 1980–2010, high severity class (data from https://www.eea.europa.eu/data-and-maps/daviz/number-of-severe-floods-in#tab-chart_3).

There are many publications describing the multiple functions of SuDS (e.g. [7,48]), but designing SuDS so that they are fit for purpose in new build, retrofit or rural areas, such as natural flood risk management (see Chapter 12), is a complex process which has to take full advantage of the various multiple benefits. This chapter begins by considering what blue and green infrastructure are, and therefore how they fit in with hard, engineered structures, both incorporated into a SuDS design, as well as conventional, piped drainage.

11.2 What is blue green infrastructure?

There are many definitions of green infrastructure, or GI, but essentially it is a set of green spaces, including water (or blue infrastructure) which is planned, designed and managed in both urban and rural areas and "conserves natural ecosystem values and functions" [13]. In the urban environment, it can include street trees, parks, private gardens, green roofs and walls, road verges, allotments, cemeteries, woodlands, rivers and wetlands. Chini et al. [13] differentiated GI from grey infrastructure which uses pipes and concrete in a centralised system to gather stormwater from across the catchment and transport it out of the city as quickly as possible. Whilst runoff across impermeable surfaces will not be quite 100 percent, in a heavy rainfall event, the traditional approach using pipes will be overwhelmed, leading to flooding downstream, increased rates of flow and higher levels in local waterbodies [23]. They state that GI is an economically and environmentally preferable method to increase the capacity of cities to cope with the increasing risk of flooding in urban areas.

The vegetation utilised in GI confers a variety of ecosystem services (ESS) on the design, or they provide a set of advantages which can contribute to quality of life, health and wider environmental benefits [44]. Examples include the provision of food and water, flood regulation (or regulating), protection of soil against erosion and the outbreak of disease, as well as less tangible benefits including amenity and aesthetics (or cultural services) (see [43]). In terms of the provision of "goods", ESS can provide food, wood and pharmaceuticals (or provisioning services), but these depend on other services based on soil formation and fertility (or supporting services). It is beyond the scope of this chapter to provide any more than this simple introduction to ESS, however, further information from Defra can be found at https://www.gov.uk/guidance/ecosystems-services; the UK National Ecosystem Assessment UKNEA [38]; and the millennium ecosystem assessment: http://www.millenniumassessment.org/en/Global.html. There are also many academic studies investigating ESS provision in SuDS, for example [20,27,32].

Blue Infrastructure can include ponds (retention or detention), wetlands (constructed i.e. engineered, or taking advantage of a naturally high groundwater table coupled with depressions in the ground); raingardens and swales also provide short term water storage, and in the case of swales, slow conveyance. For further details on individual devices, see [7,11,48].

11.3 The purpose of green and blue infrastructure in SuDS

Sustainable drainage systems have four main purposes: reduction of water volume and flow to reduce flooding and provide flood resilience, improve water quality, particularly in urban environments which are impacted by industrialisation and traffic amongst other sources of pollution, provide the opportunity for people to access opportunities for amenity including recreation and finally, provides space for biodiversity – the SuDS square [48]. Blue and green infrastructure incorporated into SuDS also provide other benefits such as climate change adaptation and mitigation (see [12]), as well as improvements in human health, opportunities for urban agriculture and addressing the urban heat island effect.

SuDS design therefore has to begin by considering what site-specific functions are required, an overview of which is given in Table 11.1A: i.e. whether source control, infiltration, detention/retention, filtration or conveyance, which are themselves based on the characteristics of the area. Since they frequently perform more than one function, only the primary role of each one is given. Table 11.1B includes examples of SuDS devices grouped according to their role, which informs the design of several devices into a management train.

Fig. 11.2 shows the SuDS management train, whereby individual devices are designed together into a cascade to maximise SuDS functionality. Water is initially managed using source controls, with water being infiltrated into the ground if this is appropriate, or it can be evaporated from vegetation or the ground surface. What water is left is then managed at the site scale, followed by regional management.

Further information regarding individual SuDS devices, including their capital and maintenance costs, land take and multiple benefits can be found in [11].

Table 11.1: (A) Physical and anthropogenic factors influencing the siting of SuDS devices. Where √ indicates the functions dependent on the physical characteristics: 'fixed' – unlikely to change over the long term, i.e. >10 years; 'variable' - may vary over shorter terms i.e. <10 years. (B) Examples of SuDS devices according to their roles (after [9]).

		Infiltration	Detention	Filtration	Conveyance	Source Control
A. Fixed factors						
Physical	Geology	√	√			
	Topography	√	√			
	Soil drainage	√	√			
	Hydrology	√	√	√	√	√
	Fluvial flood zones	√		√		
	Water table	√	√			
Anthropogenic	Groundwater source protection zones	√				
	Contamination of groundwater	√	√			
	Current and former industrial use	√				
Variable factors						
Anthropogenic	Landuse	√	√	√	√	√
	Planning constraints				√	√
B. Example SuDS devices						
Grouped by primary role		Soakaway; infiltration basin; infiltration trench	Detention basin; retention basin; pond; wetland; engineered detention	Sand filter; filter strip; filter trench; bioretention	Swale; filter strip	Green roof; trees; rainwater harvesting; permeable paving; sub-surface storage; rain garden

The following section covers the kinds of information required of the site before design can begin. This is followed by a consideration of the management train, and then the construction of the devices that make up the management train. Finally, case studies detail the processes undertaken.

11.3.1 Information required enabling SuDS design

The information required depends on the characteristics of the site, whether retrofit, new build, urban or rural. However, before design and construction of the SuDS can begin, important information is needed to ensure the proper devices are used in the correct context. These factors which influence the use of SuDS are often spatially determined, and thus their design has to take account of local landscape characteristics. The USEPA [42] website hosts detailed

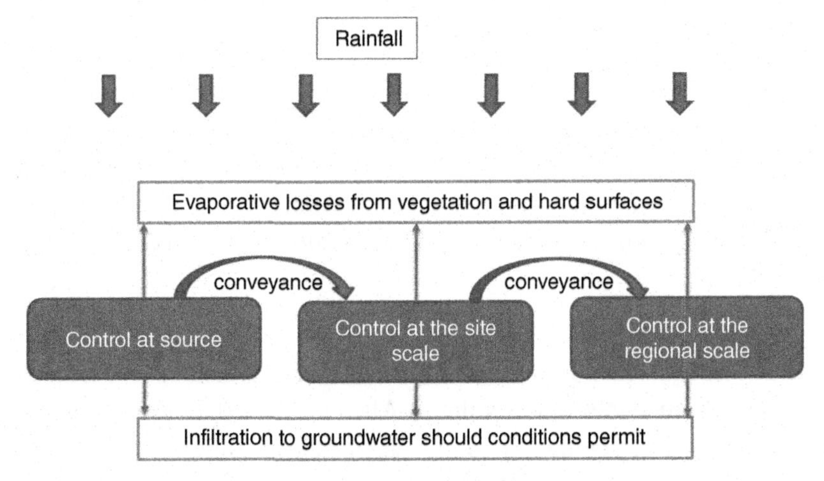

Fig. 11.2: The SuDS management train (after [26,48]).

design manuals for specific local physical and regulatory landscapes, at different scales from the whole State to river basins and cities.

Box 11.1 specifies such "baseline information" requirements, on which decision-making should be focused, and which can be used for the majority of sites globally – see case study 2.

Box 11.1 Baseline information required before design or construction of a SuDS approach [23,24,45].

1. A full topographic survey and flood risk assessment must be carried out to assess areas in the proposed development susceptible to flooding, such as low lying land, and also slopes which would promote faster flow rates. The following gives details of how to carry out such surveys: https://www.gov.uk/guidance/flood-risk-assessment-for-planning-applications#when-you-need-an-assessment
2. Location of services, which may be below or above ground, including storm sewers and associated manholes, as well as utilities
3. What street trees or other vegetation may already exist at the site, and whether they are protected in any way, and thus need preserving. This is also true of any archaeological remains.
4. The hydrology of the area including all surface and underground waterbodies which the SuDS may discharge into. Location of polluted waterbodies, and protected zones, such as nitrate vulnerable zones (for the UK see: http://apps.environment-agency.gov.uk/wiyby/141443.aspx; and [29]). The flood risk assessment in 1.) would be useful here.
5. Related to 4.), previous landuse which may include brownfield sites, land previously used for industry, old landfills etc. SuDS can be installed on polluted land, but they may need to be tanked (e.g. infiltration devices like pervious paving systems surrounded by impermeable geotextile; see [10]).
6. Investigate soil type and underlying lithology which will dictate whether water is able to infiltrate; if it cannot infiltrate, designs must focus on detention.

11.3.2 Greenfield runoff rates and calculations

Greenfield runoff is the drainage of the site before construction of any buildings or access roads. In order to ensure that a development has minimal impact on the hydrology of the site, it is recommended that post-construction, the site replicates that of its previous greenfield state as far as possible. Achieving this is set out in Woods Ballard et al. ([48], Chapter 24), and is defined as the difference in volume between the developed and greenfield scenario, or long-term storage volume.

According to the Welsh Government [47], ideally long-term storage should be discharged from the finished site at 2 L s^{-1} ha^{-1} or less, while still allowing greenfield runoff peak flow rates to be applied for the greenfield runoff volume. Also, for the 1:100-year event, all runoff from the site should be discharged either at 2 L s^{-1} ha^{-1} or the average annual peak flow rate (i.e. the mean annual flood, QBAR), whichever is the higher. Box 11.2 gives the preliminary requirements in order for the design to meet the hydraulic criteria for a SuDS drainage approach to be acceptable.

Box 11.2 Preliminary requirements to meet the hydraulic criteria for a SuDS design. After [23,48].

1. Whether for greenfield, retrofit, or redeveloped sites, an estimation is required of both greenfield peak runoff rate and volume.
2. Specifically for sites undergoing redevelopment, an estimation is needed of their previous peak runoff and volume.
3. The proposal for development/ redevelopment of the site also needs to account for future changes in climate as well as further development with time. Estimates of runoff rates and volumes are also then required of the developed site.

These estimates determine the rate of runoff and volumes allowable from the site; they may need to be throttled back to achieve these values, and this needs to be accounted for in the design. Where SuDS are concerned, this throttling can be achieved fairly simply by using some kind of weir (see Fig. 11.3), a fixed orifice, or even a short pipe, which might be useful where a driveway passes over a swale. Vortex flow controls can enable any attenuation storage, if used, to fill ([48]; Charlesworth et al., 2014 [10]). It is beyond the scope of this chapter to reproduce the many specific calculations required of engineers involved in achieving these requirements – they are given in full in Woods Ballard et al. ([48]; Chapter 24).

Fig. 11.3: The use of a weir plate to throttle flow in a stone-lined swale to reduce erosion, North Hamilton, Leicestershire, UK

11.4 Integrating SuDS into the city

It was suggested in Charlesworth [12] that design in the city was like a "Bull's Eye" in which "patches" of devices could be integrated into the city centre, such as PPS, green roofs etc. and that pocket parks and other small areas of green space could be used to install, for example, rain gardens (Fig. 11.4) where muddy grass verges outside houses due to car parking can be converted into an engineered SuDS device.

Further out of the city centre and into the suburbs, there may be opportunities to design a train of interventions, whose purpose is to manage rainfall as close to its source as possible, as shown in Section 11.3, Fig. 11.2. However, interest in SuDS to date has typically focussed on individual devices such as green roofs or pervious paving systems [33,36], with little attention being paid to combining devices into a management train. In their statutory guidance on the use of SuDS, the Welsh Government [47] prioritise the collection of excess surface water

Fig. 11.4: Raingardens constructed to convert muddy areas outside houses (A) to a device which provides flood resilience, biodiversity benefits, stormwater treatment and visual amenity, Nottingham, UK (kind permission, John Brewington, EA).

Box 11.3 Hierarchy levels to prioritise surface water runoff destination [47].

1. Collect runoff at source for other uses
2. Infiltrate into the ground
3. Discharge to suitable surface water body
4. Discharge to a surface water sewer, a highway drain or other drainage system
5. Discharge to a combined sewer

at source so that it can be used as a resource, or secondly that it is dealt with essentially using infiltration (see Box 11.3, hierarchy level 2). However, the site may not be suitable for infiltration, so hierarchy level 2 should have a caveat, in that it would require further investigation to determine its suitability for infiltration and, if it is not, then either tanked infiltration devices, or detention/ retention devices should be used. Dearden and Price [15] produced large scale geological maps of the UK which were based on the potential for infiltration SuDS, but these did not take into account ground conditions such as soil type and possible land contamination which may preclude infiltration.

Designing SuDS into new build housing or industrial estates is relatively easy but retrofitting into existing built landscapes is less so. Section 11.5 therefore addresses the potential and associated issues of retrofitting devices to existing urban areas. Section 11.6 provides some case studies of the design and implementation of SuDS management trains in the UK and in a refugee camp in the Kurdistan Region of Iraq.

11.5 New build/retrofit

Most urban areas comprise already existing buildings, although there is usually an on-going strategy of either refurbishing older buildings, or of their removal and replacement with new construction. There are therefore opportunities to install SuDS solutions at the same time. Stovin and Swan [37] applied the term SuDS retrofit to the situation where a conventional drainage system could either be replaced altogether by SuDS, or it could be improved. Thus, the volume of stormwater would be reduced or removed, leading to less water having to be treated at the treatment plant, reducing costs and potentially reducing combined sewer overflows by reducing the volume of runoff discharged to combined sewer systems [1].

The roles and function of SuDS in a retrofit context are no different to those already described for new developments. The siting and implementation of retrofit SuDS is often controlled by existing buildings or infrastructure, making it more difficult and potentially more expensive to install than in new developments. There may also be constraints due to lack of space in a high-density area, and public acceptability, restricting the use of, for example, ponds and wetlands. Smaller locations may therefore be more appropriate, such as car parks, single streets and roofs (Grant et al., 2017 [49]). Complicating factors include land cover and who owns the land, very often small-scale, individual landowners, whether the sewers are combined or not, if there are locations under redevelopment, and if there are sites for which there may be issues when considering SuDS such as abandoned industrial areas, or contaminated land. However, Grant et al., 2017 state that any arguments against the use of SuDS based on problems associated with specific sites "may be overstated", as they consider that the alternatives available are numerous, so that it should always be possible to incorporate some kind of SuDS device regardless.

However, it is perfectly possible to retrofit green roofs, pervious paving systems and raingardens, with cities such as Portland (Oregon, USA), Malmö (Sweden) and Tokyo (Japan) engaging with the approach by actively retrofitting SuDS interventions [12,17,22,35]. It is often better, and more successful, to engage with a single, large landowner, such as a Housing Association, a commercial organisation such as a supermarket chain, or even a local authority. In a retrofit SuDS scheme in Llanelli, Wales, UK, Ellis et al. [19] mainly engaged with the local water company, who in turn informed the local community of the reasons for the works, leading to a successful retrofit implementation, achieving "considerable flow reductions", but also providing more pleasant spaces for people to live.

11.6 Case studies of the design and implementation of SuDS management trains

The following sections detail the decision-making processes involved in incorporating a SuDS management train across a whole city, and then focuses on an area slated for redevelopment. The second case study is an end-of-pipe solution to problems associated with surface water flooding and greywater management in a humanitarian context in the Kurdistan Region of Iraq.

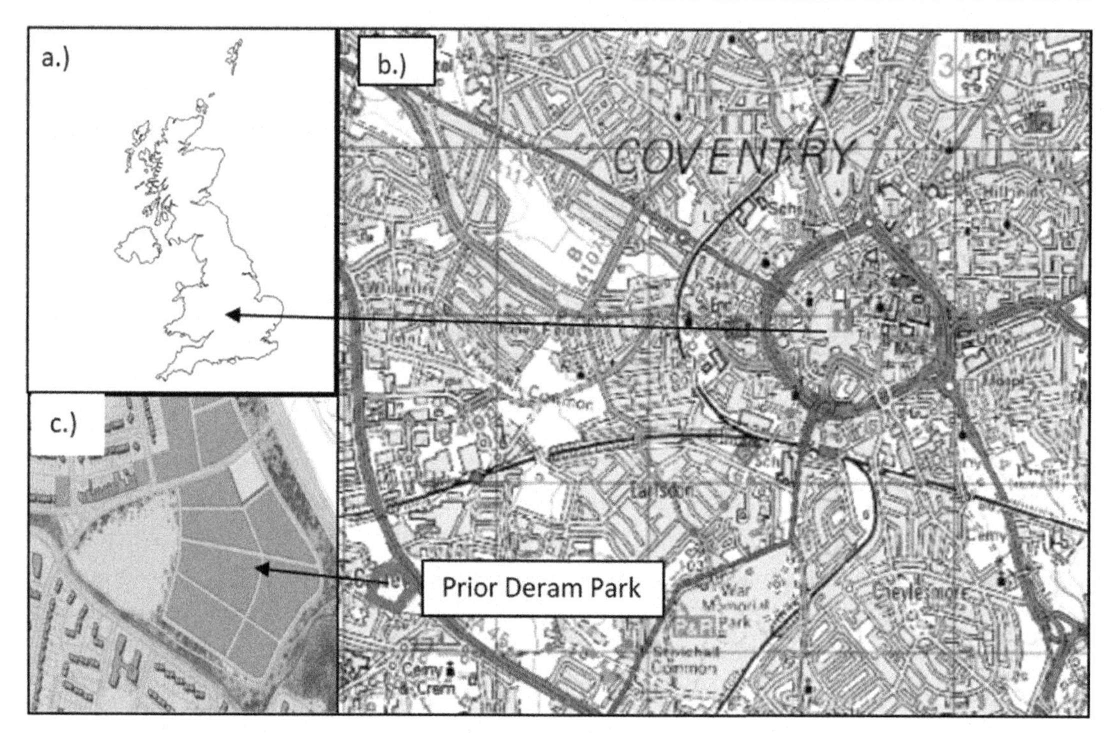

Fig. 11.5: Location of (A) the city of Coventry in the English Midlands. (B) Prior Deram Park, Canley, Coventry. (C) proposed redevelopment of Prior Deram Park, Canley Redevelopment Zone (CRZ).

11.6.1 SuDS and scale: The city of Coventry, West Midlands

This section shows how the information detailed in previous sections can be used in the design and installation of SuDS across a whole city. This case study is based in the city of Coventry, located in the West Midlands of England, UK (see: Fig. 11.5 and [11]). It is a desk-based study, illustrating the use of geographical information systems (GIS) to produce maps on which to base a decision support tool (DST) to aid in the design and suitable location of SuDS for local authorities.

As outlined above, a variety of physical attributes of the area were required in order to decide which type of device was most suitable (see Table 11.1A). These were determined as "fixed" if they did not change appreciably over the long term (e.g. lithology), or "variable" if their characteristics changed over shorter timescales (e.g. land use). The spatial distribution of each of these factors was established using data from a variety of sources, whose choice included the level of detail offered, whether it was readily available and, as the DST was focused on local authorities, at zero or low cost. An example of how the information gathered was used in the decision-making process is shown in Fig. 11.6, based on soil type, where impermeable

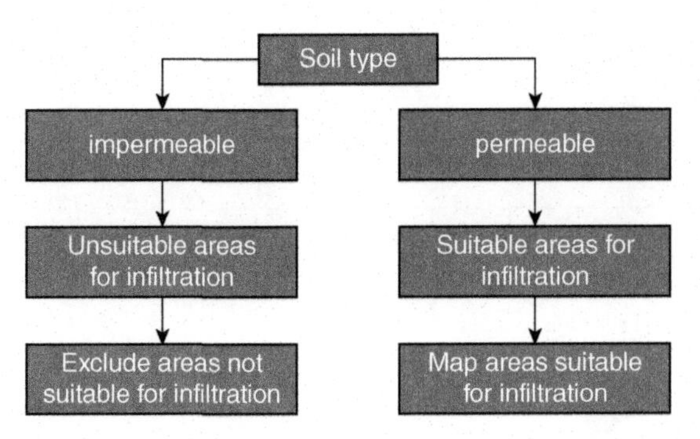

Fig. 11.6: Determining areas suitable and unsuitable for infiltration based on soil type.

soils, such as those containing clay would preclude infiltration devices like pervious pavement unless tanked, whereas soils containing a sandy matrix would readily allow water to infiltrate.

It is therefore possible to base early decisions on soil maps; in the UK, these are available either from the British Geological Survey (BGS) [4] (https://www.bgs.ac.uk/data/services/soilwms.html) or more specifically, and targeted at SuDS: http://www.bgs.ac.uk/products/hydrogeology/infiltrationSuds.html. The BGS website also hosts case studies, e.g. the use of the infiltration SuDS map from the Isle of Mull to enable an initial determination of ground conditions to identify areas for trial pits investigating whether soakaways could manage the excess surface water causing flooding downstream, see: http://nora.nerc.ac.uk/id/eprint/20840/1/Suds%20case%20study_corporate_Dec12v2.pdf. The UK Soil Observatory (UKSO) (http://www.ukso.org/) also hosts information on soil characteristics across the UK. However, infiltration testing should still take place with trial pits dug in areas where infiltration is proposed. The process is explained in Woods Ballard et al. ([48], Chapter 25) with the calculations based on Bettess [3].

The same process as shown in Fig. 11.6 was carried out for the other factors shown in Table 11.1, and a combination of the resultant layers in the GIS database enabled the determination of suitable sites for infiltration, detention and retention across Coventry. An estimate of percentage landcover yielded nearly 45 percent of greenspace, which, added to the 15 percent of landcover identified as gardens where SuDS could be more easily incorporated, contrasted with the already built 14 percent coverage or 25 percent of the area which represented transport infrastructure (Fig. 11.7).

A set of maps was developed to address new construction, on both new 'greenfield' sites and previously developed land, for five categories of SuDS (Table 11.1): infiltration, detention, conveyance, filtration and source control. The resulting maps were scalable and were able to be viewed at different resolutions from the whole city scale down to small, local sites undergoing development or regeneration. Examples of these maps are given in Fig. 11.8 and show that detention and retention is possible across the whole city, apart from where waterbodies

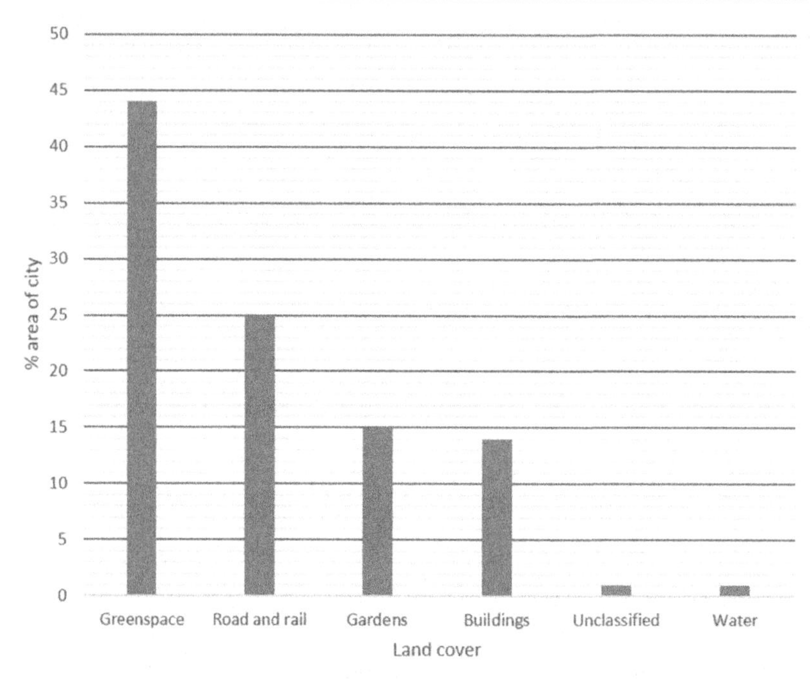

Fig. 11.7: Estimated percentage landcover types, city of Coventry.

already exist. However, infiltration was very limited and mostly confined to the western suburbs due to the city being underlain by clay.

By enabling the maps to be scalable, it was possible to focus in on either new build sites, or those undergoing regeneration. An example of an area of regeneration is that of the Canley Regeneration Zone (CRZ), the location shown in Fig. 11.9 to the south west of the city. The large block of greenspace to the north and west of the city provides opportunities to integrate interventions which would not be possible in the already built-up suburbs further south and east.

Fig. 11.9 shows that there are pockets of regeneration occurring throughout the city – this is common practice in cities worldwide which offers the opportunity to implement SuDS piecemeal as the area is redeveloped.

In the specific case of Coventry, due to the city being underlain by clay, infiltration is not possible across most of the CRZ, however the use of source control, using devices listed in Table 11.1 is possible over the whole area apart from where waterbodies already exist. Detention, whether engineered or vegetated is possible, and designs should reflect the specific context in which they are to be established, which these maps demonstrate (Fig. 11.10).

These maps were developed to enable local authorities to better engage with the SuDS approach, by supporting their decision-making processes in terms of what devices were suitable and where. This decision support tool can be used in any city which has access to the required information and, as was shown by [9], even smaller scales can be achieved.

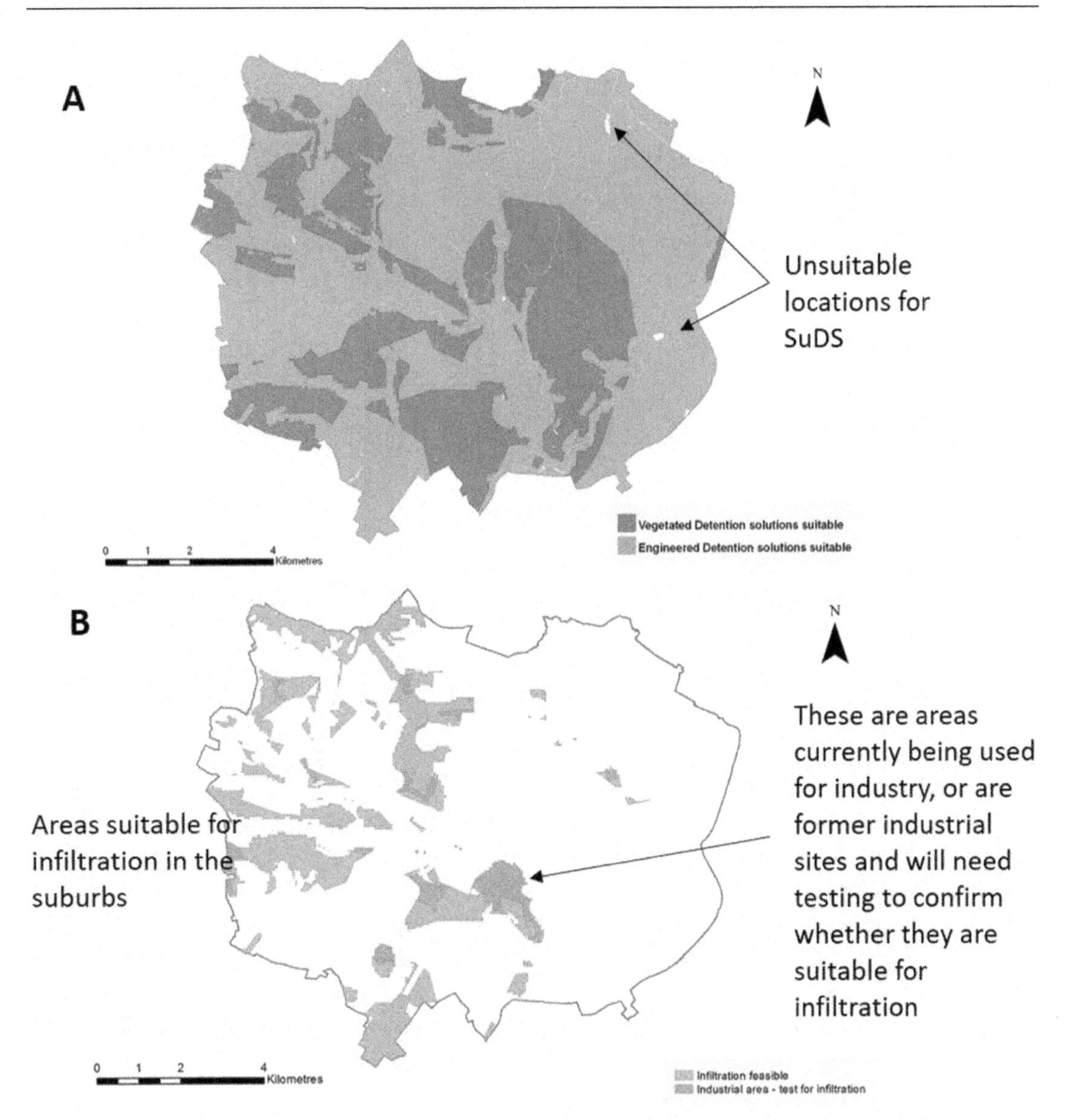

Fig. 11.8: Suitable locations SuDS devices: (A) new developments suitable for detention and retention; (B) Feasibility of infiltration.

11.6.2 Case study 2: SuDS in challenging environments

Installing SuDS in western, temperate cities is not without its problems, as shown in case study 1, but doing so in a humanitarian context requires care and sensitivity when dealing with vulnerable people [5,6].

In the humanitarian context, potable water and human waste disposal is usually prioritised in water sanitation and hygiene (WASH) strategies. However, excess surface water and greywater

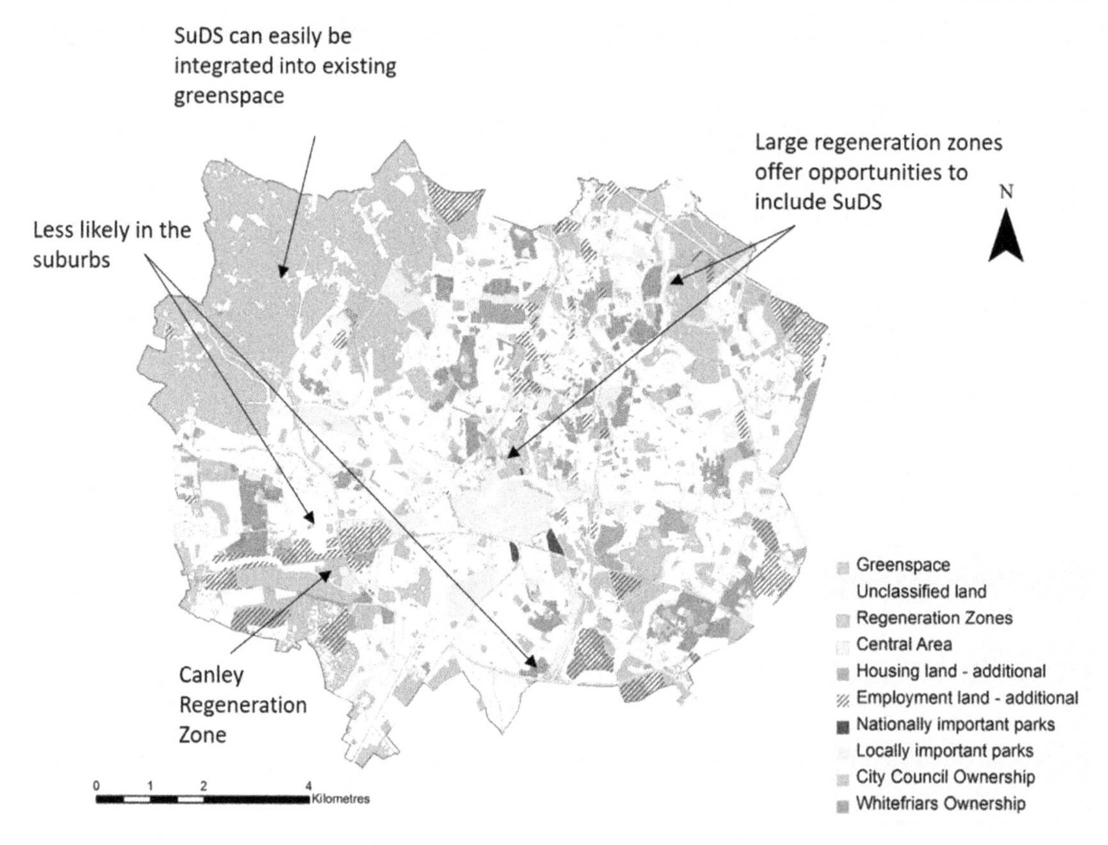

Fig. 11.9: Feasibility of SuDS implementation in new build and sites undergoing regeneration.

(which is wastewater from personal washing, food preparation etc. but not human waste which is disposed of in a toilet) management in general is not [5,6]. This is true for both refugee and internally displaced person (IDP) camps where the result is disposal of greywater on the street outside, or between shelters. These streams and puddles are potentially contaminated and eventually pools form at camp boundaries, or they pass to downstream communities (Fig. 11.11).

These pools attract children to play in them, but also disease vectors such as mosquitoes. Thus this is an example of where the SuDS approaches of managing excess surface water and management of greywater would work together.

Funded by the Humanitarian Innovation Fund (HIF), a SuDS demonstration site was designed and constructed at Gawilan Refugee Camp, Ninewah Governate, in the Kurdistan Region of Iraq.

Fig. 11.12 illustrates the preliminary work carried out to support the new drainage, which included the collection of information and data on the physical characteristics of the site, observation using site walkovers and drone overflies, and also community engagement to ensure residents knew what the aims of the SuDS management train were and to collect their

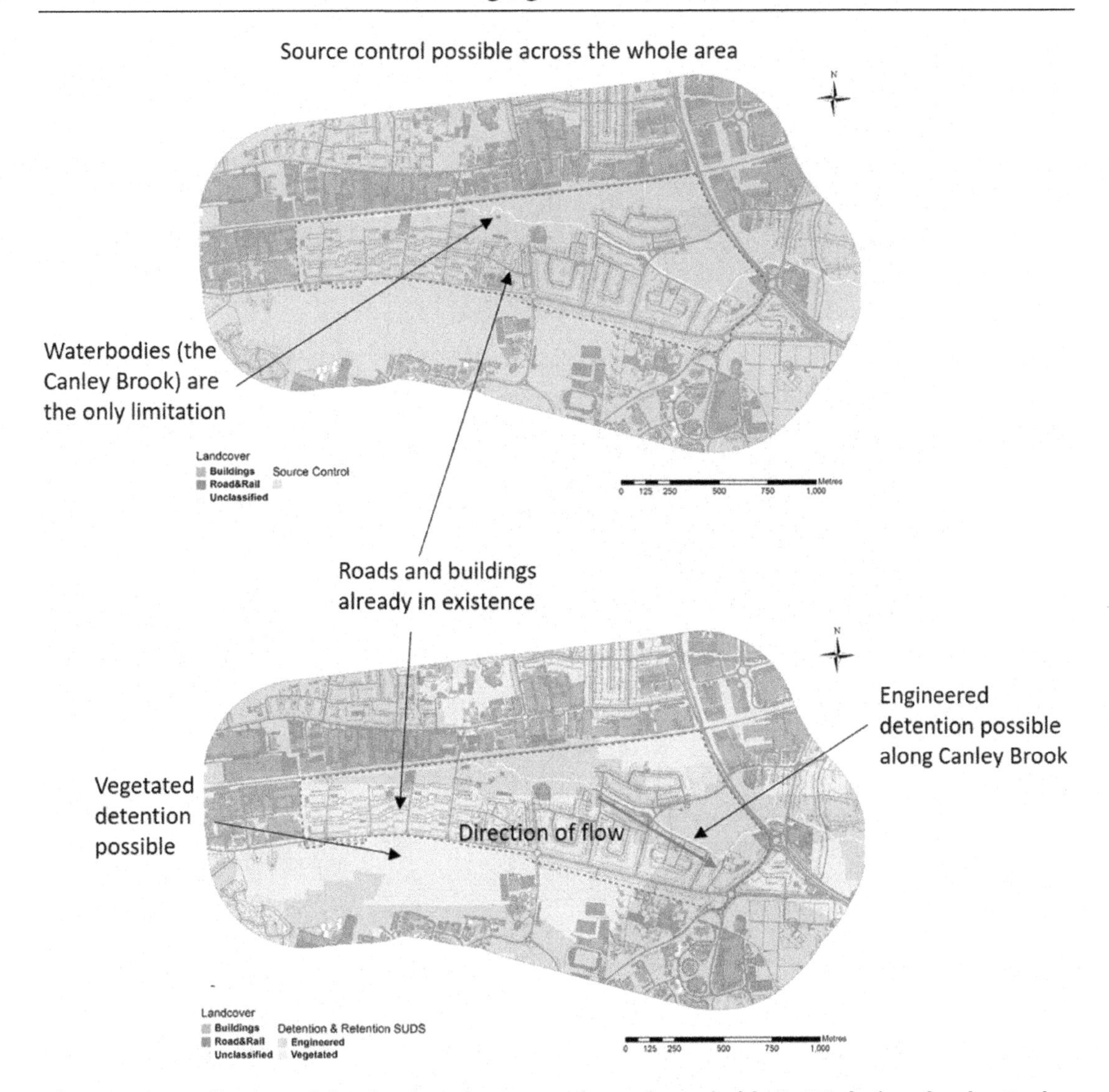

Fig. 11.10: Application of the developed maps to determine suitable SuDS devices for the Canley Regeneration Zone, Coventry.

"flood memories". The latter also enabled the refugees to have a say in how the site was designed, enabling them to include aspects of amenity (a sports pitch, an area for meetings, urban gardening etc.).

Fig. 11.13 shows a pit dug to investigate soil type which revealed that the site was made up of highly disturbed layers, possibly waste from levelling the site prior to constructing the camp. Water quality analysis before the SuDS system was installed revealed little concern with respect to metals, but BOD_5 and COD were above relevant guidelines (further details in [5]),

Fig. 11.11: Paths and accumulation of wastewater from Gawilan refugee camp, Kurdistan Region of Iraq.

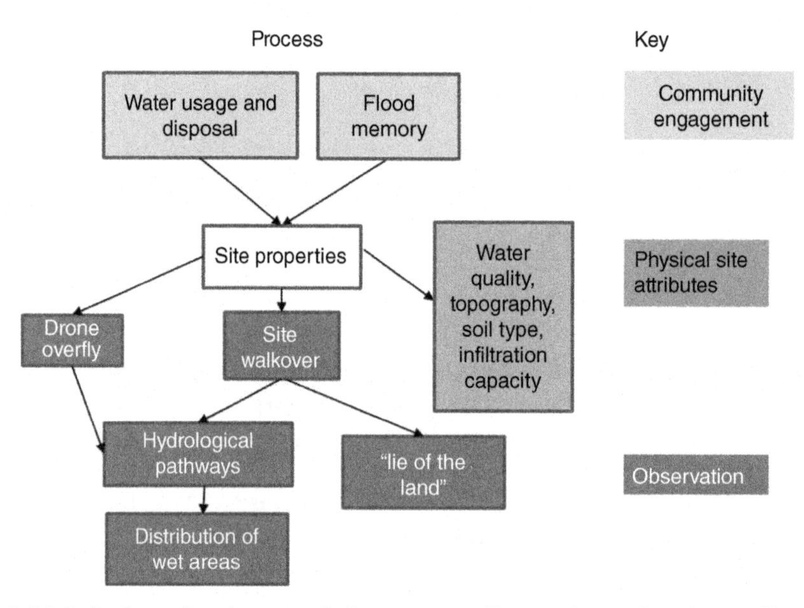

Fig. 11.12: Initial site investigations carried out at Gawilan to determine the attributes required in order to design a SuDS management train.

Fig. 11.13: Trial soil pit dug to determine soil type and depth.

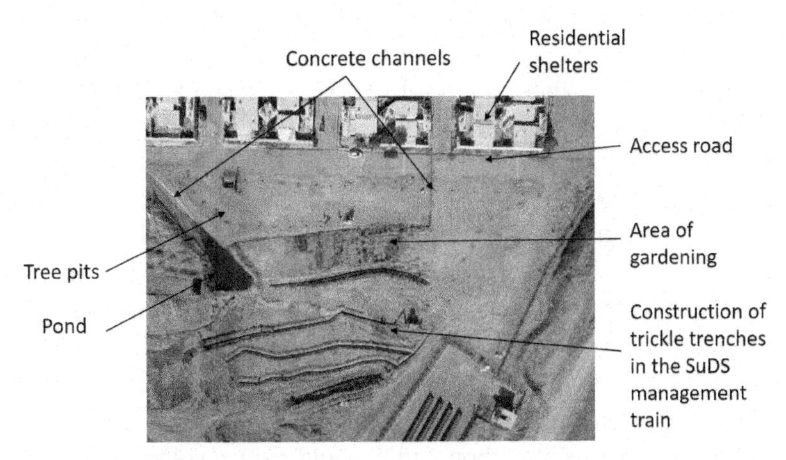

Fig. 11.14: Construction of the SuDS management train, seen from above via a drone overfly.

turbidity was high, and bacterial counts were concerning, reflecting the ingress of blackwater (i.e. from human waste) into the greywater stream.

Based on the information collected, the management train was constructed using tree pits, mainly olive, to make productive use of the water, ponds and wetland areas to allow particulates to settle, and trickle trenches to slow the flow of water and redistribute it around the site. The drone overfly in Fig. 11.14 shows the layout of the site as construction progressed.

The successful introduction of the SuDS approach into this challenging environment has illustrated the possibility of its use, by engaging with the community, and if designed correctly. The future for this site is in monitoring its effectiveness in retaining and managing excess water on-site with the ambition that as little as possible will leave the site, as well as training the community in its maintenance to safeguard its longevity.

11.7 Are SuDS actually "sustainable?"

This book is about sustainable water engineering and management, and SuDS badges itself as being sustainable in the long term, but how true is this? From an engineering perspective, the choice of materials to construct the devices can drive their eventual sustainability, with important considerations around structural strength and potential environmental pollution.

The unsustainability of SuDS is argued by Heal et al. ([50]) to include the following considerations:

1. Where in-service failures of SuDS devices have occurred, this may be due to incorrect siting, for example, of an infiltration device where water cannot percolate into the ground, e.g. where soils are clayey.
2. The fact that SuDS do require regular maintenance may add to the argument that they are unsustainable because of the need for regular inspections and interventions during operation and maintenance activities (e.g. trimming back vegetation, grass mowing, removing trash, replacing blocked media etc.; see [2]). However, so does any form of drainage, e.g. the regular cleaning of gully pots, the regular removal of contaminated sediment from highway filter drains (see [31]).
3. In their estimates of the costs of the winter floods during 2015–16 in the UK, the EA [18] came to a figure of £1.6 billion, with more than £200 billion worth of assets at risk in Britain overall. Ofwat has estimated that to upgrade the 309,000 km storm sewer network in the UK would cost approximately £174 billion at 2007–08 prices. It would take centuries to complete and close every road for a minimum of 3 months, incur heavy carbon costs and cause many spills of contaminated water. It would cause yet more flooding, lead to larger volumes requiring treatment at the water treatment plant incurring more costs both financially and in terms of carbon. SuDS could reduce volumes requiring treatment and hence extend the useful life of sewer assets. It is therefore arguable that the country cannot afford not to embrace a sustainable approach to drainage.
4. The fate of pollutants and whether they accumulate in SuDS devices to hazardous degrees as well as the management of devices such as PPS at their "end-of-life" (see [8,28]). Some metals, for instance Cu and Zn, can be utilised as a micro-nutrient by the biofilm which develops on the aggregate and geotextile forming part of the structure of block paver PPS (see [10]). Oil and grease can also be metabolised by the biota. The very few studies of potential accumulation of toxic elements and compounds in PPS e.g. Mbanaso et al. [28] and Charlesworth et al. [8] have not found cause for concern.
5. As explained in Heal et al. (2004) whilst recycled aggregate and plastics have been used to increase the sustainability of SuDS, "they have never been viewed as completely sustainable", but the design manual for Scotland and Northern Ireland states that SuDS are "designed to drain surface water in a more sustainable fashion than some conventional techniques" [14].

However, Illman et al. [25] took a wider view in their report on the lead local flood authorities achievement of SuDS, in which SuDS was considered to be sustainable when they were designed for source control (see Fig. 11.2) and when as a result, water was managed or made use of, close to where it fell and also by making use of GI. This approach was also considered most able to deliver the broadest scope in terms of multiple benefits.

11.8 Conclusions

Flooding, climate change, contamination and environmental degradation are all having negative consequences on the natural world. As stated in Charlesworth and Booth [7], sustainable drainage is a technique which can go some way to addressing these issues, however, they do acknowledge that there is not a single answer to these multiple and complex problems.

SuDS offers multiple benefits unlike the role of, for example, a pipe in a conventional system, which simply conveys water from one place to another. This chapter has shown that it is possible to utilise SuDS in new build and retrofit situations, and also in challenging environments, as long as it is designed, constructed and maintained correctly. In order to do this, information about the site is needed to ensure that the devices used can function properly and efficiently according to the SuDS Square [48]. It has also emphasised that community engagement is important; if people living alongside a SuDS device or management train understand its role, or better still, are involved in its design, it is far more likely to be accepted and will function in the long term.

References

[1] A. Alves, A. Sanchez, Z. Vojinovic, S. Seyoum, M. Babel, D. Brdjanovic, Evolutionary and holistic assessment of green-grey infrastructure for CSO reduction, Water 8 (9) (2016) 402, doi:10.3390/w8090402.

[2] N. Berwick, Sustainable drainage systems: operation and maintenance, in: S.M. Charlesworth, C. Booth (Eds.), (2017) Sustainable Surface Water Management: A Handbook for SuDS, Wiley Blackwell Publishing, 2017.

[3] R. Bettess, Infiltration Drainage – Manual of Good Practice, R156, CIRIA, London, UK, 1996. Available from https://www.ciria.org/.

[4] BGS (British Geological Society), Infiltration SuDS Maps, BGS, London, UK, 2015.

[5] S.M. Charlesworth, M. McTough, A. Adam-Bradford, The design, construction and maintenance of a SuDS management train to address surface water flows by engaging the community: Gawilan refugee camp, Ninewah Governate, Kurdistan Region of Iraq, J. Refug. Stud. Fez082 (2019) 1–17, doi:10.1093/jrs/fez082.

[6] S.M. Charlesworth, K. Winter, A. Adam-Bradford, M. Mezue, M. McTough, F. Warwick, M. Blackett, Sustainable drainage in challenging environments, New Water Policy Pract. J. 4 (1) (2018) 31–41, Special Issue: Water Governance Leadership.

[7] S.M. Charlesworth, C. Booth (Eds.), Sustainable Surface Water Management: A Handbook for SuDS, Wiley Blackwell Publishing, 2017.

[8] S.M. Charlesworth, J. Beddow, E. Nnadi, The fate of pollutants in porous asphalt pavements, laboratory experiments to investigate their potential to impact environmental health, Int. J. Environ. Res. Public Health 14 (2017) 666.

[9] S.M. Charlesworth, F. Warwick, C. Lashford, Decision-making and sustainable drainage: design and scale, Sustainability 8 (2016) 782, doi:10.3390/su8080782.

[10] S.M. Charlesworth, C. Lashford, F. Mbanaso, Hard SuDS infrastructure, Rev. Curr. Knowl. Found. Water Res. (2014). Available from http://www.fwr.org/environw/hardsuds.pdf.
[11] S.M. Charlesworth, S. Perales-Momparler, C. Lashford, F. Warwick, The sustainable management of surface water at the building scale: preliminary results of case studies in the UK and Spain, J. Water Supply Res. Technol. – AQUA 62 (8) (2013) 534–544.
[12] S. Charlesworth, A review of the adaptation and mitigation of global climate change using sustainable drainage in cities, J. Water Clim. Chang. 1 (3) (2010) 165–180.
[13] C.M. Chini, J.F. Canning, K.L. Schreiber, J.M. Peschel, A.S. Stillwell, The green experiment: cities, green stormwater infrastructure, and sustainability, Sustainability 9 (2017) 105, doi:10.3390/su9010105.
[14] CIRIA, Sustainable Urban Drainage Systems – Design Manual for Scotland and Northern Ireland, CIRIA Report C521, CIRIA, London, 2000.
[15] R.A. Dearden, S.J. Price, A proposed decision-making framework for a national infiltration SuDS map, Manage. Environ. Qual. Int. J. 23 (5) (2012) 478–485, doi:10.1108/14777831211255060.
[16] J. Donofrio, Y. Kuhn, K. McWalter, M. Winsor, Water-sensitive urban design: an emerging model in sustainable design and comprehensive water-cycle management, Environ. Pract. 11 (3) (2009) 179–189.
[17] DTI, Sustainable Drainage Systems: A Mission to the USA, Global Watch Mission Report, 2006, pp. 148. Available from http://www.oti.globalwatchonline.com/online_pdfs/36635MR.pdf.
[18] EA (Environment Agency), Estimating the Economic Costs of the 2015 to 2016 Winter Floods, Environment Agency, 2018. Available from https://assets.publishing.service.gov.uk/government/uploads/system/uploads/attachment_data/file/672087/Estimating_the_economic_costs_of_the_winter_floods_2015_to_2016.pdf.
[19] C. Ellis, R. Cripps, M. Russ, S. Broom, Transforming water management in Llanelli, UK, Proc. Inst. Civil Eng. Civil Eng. 169 (1) (2016) 25–33, doi:10.1680/jcien.15.00027.
[20] M. Everard, R.J. McInnes, H. Gouda, Progress with integration of ecosystem services in sustainable drainage systems, in: S.M. Charlesworth, C. Booth (Eds.), Sustainable Surface Water Management: A Handbook for SuDS, Wiley Blackwell Publishing, 2017.
[21] B. Ferguson, Low impact development in the USA, in: S.M. Charlesworth, C. Booth (Eds.), Sustainable Surface Water Management: A Handbook for SuDS, Wiley Blackwell Publishing, 2017.
[22] S. Haghighatafshar, J. la Cour Jansen, H. Aspegren, V. Lidström, A. Mattsson, K. Jönsson, Storm-water management in Malmö and Copenhagen with regard to climate change scenarios, VATTEN – J. Water Manag. Res. 70 (2014) 159–168.
[23] J. Hubert, T. Edwards, A.B. Jahromi, Comparative study of sustainable drainage systems, Eng. Sustain. Proc. Inst. Civil Eng. 166 (3) (2013) 138–149, doi:10.1680/ensu.11.00029.
[24] S. Illman, S. Wilson, Guidance on the Construction of SuDS, CIRIA, 2017, pp. C768. Available from https://www.ciria.org/ItemDetail?iProductcode=C768&Category=BOOK.
[25] S. Illman, B. Brown, B. Buntine, K. Bailey, A. Chisholm, J. Harris, P. Shaffer, Lead Local Flood Authorities: Achieving Sustainable Drainage, 2019. Available at https://www.ciwem.org/assets/pdf/Policy/Policy%20Area%20Documents/11689_LI_SuDS-Report_v4a-Web.pdf.
[26] A. Kirby, SuDs – innovation or tried and tested practice? Proc. Inst. Civil. Eng. 158 (2) (2005) 115–122.
[27] C. Mak, M. Scholz, P. James, Sustainable drainage system site assessment method using urban ecosystem services, Urban Ecosyst. (2017) 20 (2017) 293–307.
[28] F.U. Mbanaso, S.M. Charlesworth, S.J. Coupe, A.P. Newman, E.O. Nnadi, Reuse of materials from a sustainable drainage system device: health, safety and environment assessment for an end-of-life pervious pavement structure, Sci. Total Environ. 650 (2019) 1759–1770.
[29] A.-M. McLaughlin, S.M. Charlesworth, S.J. Coupe, UK and EU water policy as an instrument of urban pollution, in: S.M. Charlesworth, C. Booth (Eds.), Urban Pollution, Science and Management, Wiley Blackwell Publishing, 2019.
[30] B. Menne, V. Murray, (Eds.), Floods in the WHO European Region: Health Effects and their Prevention, World Health Organisation, Europe, 2013. Available from http://www.euro.who.int/__data/assets/pdf_file/0020/189020/e96853.pdf?ua=1.

[31] L.A. Sañudo-Fontaneda, S.J. Coupe, S.M. Charlesworth, E.G. Rowlands, Exploring the effects of geotextiles in the performance of highway filter drains for sustainable and resilient road drainage, Geotext. Geomembr. 46 (2018) 559–585.

[32] M. Scholz, V.C. Uzomah, S.A.A.A.N. Almuktar, J. Radet-Taligot, Selecting sustainable drainage structures based on ecosystem service variables estimated by different stakeholder groups, Water 5 (2013) 1741–1759, doi:10.3390/w5041741.

[33] M. Scholz, P. Grabowiecki, Review of permeable pavement systems, Build. Environ. 42 (2007) 3830–3836.

[34] R. Simcock, Water sensitive design in Auckland, New Zealand, in: S.M. Charlesworth, C. Booth (Eds.), (2017) Sustainable Surface Water Management: A Handbook for SuDS, Wiley Blackwell Publishing, 2017.

[35] P. Stahre, Blue-Green Fingerprints in the City of Malmö: Malmö's Way to a Sustainable Urban Drainage, VA SYD, Malmö, Sweden, 2008. Available from https://greenroof.se/wp-content/uploads/2017/04/BlueGreenFingerprintsPeterStahrewebb.pdf.

[36] V.R. Stovin, The potential of green roofs to manage urban stormwater, Water Environ. J. 24 (2010) 192–199.

[37] V. Stovin, A. Swan, Retrofit SuDS—cost estimates and decision-support tools, Proc. Inst. Civil Eng. Water Manag. 160 (4) (2007) 207–214, doi:10.1680/wama.2007.160.4.207.

[38] UKNEA, UK National Ecosystem Assessment Follow-on: Synthesis of the Key Findings, UNEP-WCMC, LWEC, UK, 2014. Available from http://uknea.unep-wcmc.org/Resources/tabid/82/Default.aspx.

[39] United Nations, Sustainable Development Goal 6 Synthesis Report 2018 on Water and Sanitation, United Nations, New York, 2018.

[40] USEPA (United States Environmental Protection Agency), Low Impact Development (LID), 2018. Available from https://www.epa.gov/nps/urban-runoff-low-impact-development.

[41] USEPA, National Menu of Stormwater Best Management Practices, 2019. Available from https://www.epa.gov/npdes/national-menu-best-management-practices-bmps-stormwater-documents.

[42] USEPA, Green Infrastructure Design and Implementation, 2019. Available from https://www.epa.gov/green-infrastructure.

[43] R. Wade, Urban pollution and ecosystem services, in: S.M. Charlesworth, C. Booth (Eds.), (2019), Urban Pollution, Science and Management, Wiley Blackwell Publishing, 2019.

[44] R. Wade, N. McLean, Multiple benefits of green infrastructure, in: C. Booth, S.M. Charlesworth (Eds.), (2014), Water Resources in the Built Environment - Management Issues and Solutions, Wiley Blackwell Publishing, 2014.

[45] F. Warwick, S. Charlesworth, M. Blackett, Geographical Information as a Decision Support Tool for Sustainable Drainage at the City Scale, NOVATECH, Lyons, France, 2013. Available at: http://documents.irevues.inist.fr/bitstream/handle/2042/51250/1A7P04-111WAR.pdf?sequence=1.

[46] Water UK, Lessons Learnt from Summer Floods 2007: Phase 2 report – Long-term Issues, Water UK's Review Group on Flooding, 2008. Available at: https://www.yumpu.com/en/document/read/34316511/flooding-report-phase-2-final-water-uk.

[47] Welsh Government, Statutory Standards for Sustainable Drainage Systems – Designing, Constructing, Operating and Maintaining Surface Water Drainage Systems, Welsh Government, 2018. Available from: https://gweddill.gov.wales/docs/desh/publications/181015-suds-statutory-standards-en.pdf.

[48] B. Woods Ballard, S. Wilson, H. Udale–Clarke, S. Illman, R. Ashley, R. Kellagher, The SuDs Manual, CIRIA, London, 2015.

[49] L. Grant, A. Chisholm, R. Benwell, A Place for SuDS? Assessing the effectiveness of delivering multifunctional sustainable drainage, Chartered Institute of Water and Environmental Management (CIWEM), London, 2017. http://www.ciwem.org/wp-content/uploads/2017/10/A-Place-for-SuDS-Online.pdf.

[50] K. Heal, N. McLean, B. D'Arcy, SUDS and Sustainability, 26th Meeting of the Standing Conference on Stormwater Source Control, Dunfermline, 2004. September 2004.

CHAPTER 12

From umbrellas to sandbags—An integration of flood risk management, engineering and social insights

Tom Lavers[a], Ian Berry[b], Colin A. Booth[c,*]

[a]School of Energy, Construction and Environment, Coventry University, Sir John Laing Building, Much Park Street, Coventry CV1 2LT, UK [b]School of Architecture and the Built Environment, University of the West of England Frenchay Campus, Coldharbour Lane, Bristol BS16 1QY, UK [c]Architecture and the Built Environment, University of the West of England, Bristol, UK

*Corresponding author.

12.1 Introduction

Flooding may be fluvial (river), pluvial (surface water), coastal or groundwater related, or caused by a combination of these hydrological processes. Moreover, it is widely considered the most destructive of all natural hazards. In the 20th century, 12 percent of all natural hazard deaths were attributed to flooding, claiming approximately 93,000 lives across the world (OECD International Disasters Database, 2019). In 1953, the extensive North Sea coastal floods caused approximately 2500 deaths across the UK, Netherlands, Germany and Belgium; since then, intermittent flash floods have also caused many fatalities, such as Lynmouth in Cornwall (over 30 deaths in 1952) and Vaison-la-Romaine in France (38 deaths in 1992). In the UK, there have been less fatalities in recent flood events, with 5 deaths in Central England during the 1998 floods, 4 deaths during the autumn 2000 floods and 13 fatalities attributed to the July 2007 floods [113,52].

Development on floodplains has increased the levels of exposure to people, property and infrastructure to floods. In many cases it is not practical, cost effective or politically feasible to relocate communities, property and economic infrastructure away from areas prone to flooding, so measures are put in place to manage flood risk by reducing the probability of inundation, and in the events of a flood, the negative associated consequences. This chapter focuses on the methods used to manage flood risk across a "mosaic" approach, from catchment-scale options to localised property level resistance and resilience measures.

Sustainable Water Engineering. DOI: 10.1016/C2017-0-04301-X

12.2 Managing flood risk: A mosaic approach

Flooding is by no mean a recent phenomenon, there have been many momentous flood events. In the UK, for instance, a catalogue of historical flood events has been compiled by the British Hydrological Society in the Chronology of British Hydrological Events (www.cbhe.hydrology.org.uk). Previously, when buildings had stone flagged floors and undecorated walls, with no electricity and few possessions, the losses would not have been the same compared to modern timber framed buildings, with electricity and with high-value items. The July 2007 floods resulted in an average insurance loss of £20,000 per household affected, and a total cost to the UK economy of £3.2 billion [114].

The current challenge is to manage and, where feasible, reduce this risk, but complete flood prevention is now considered an improbable ambition. Compelling as the aim of modern, integrated catchment-based flood management certainly is, it brings considerable complications. The risk-based approach involves analysing the likely impacts of flooding under an extensive range of conditions and the effect of a varied range and type of mitigation measures. As the systems being managed expand in scope, hydrological scale and timeframes, so too does the number of potential uncertainties [41]. There are many possible components to a mosaic of traditional "hard" measures, like flood walls, levees and large storage areas, and the newly emerging "soft" flood risk management measure, such as natural flood risk management (NFRM) and sustainable drainage systems (SuDS). Communicating risks and building the consensus that is necessary to engage effectively with flood risk management stakeholders requires an aptitude for facilitation, communication and arbitration [115,37,39,55].

The concept of "risk" has long been recognised as a central condition when considering where it is appropriate to deliver flood defences and long-term management. The Waverley Report, produced in response to the 1953 North Sea floods that devastated the UKs east coast recommended that flood defence standards should be proportionate to the area of protection, (i.e. large conurbations ought to be protected from more severe storms than farmland and sparsely populated villages) [48,103]. However, the engineering principles of designing flood defence infrastructure were, and remains, often opportunistic and not risk based [96]. Take designing a river flood wall, for instance, firstly engineers establish the appropriate standard for the protection (SoP) (e.g. the 1 percent annual exceedance probability or 1-in-100-year flood height, also known as river stage); secondly, an estimate of the hydraulic loading capacity is made from the river stage for a specified event; thirdly, there is a process in which primary physical characteristics (such as bank crest levels) are assessed to consider how it could withstand that specified event, and lastly; engineers incorporate safety factors, such as a freeboard allowance, based on individual circumstances.

As flood forecast modelling and early warning systems were progressively introduced and refined since the 1950s, the decision-making process was adhering to essentially deterministic principles, based on comparing water level forecasts with levels that would trigger the need for, and the dissemination of, a warning to a designated flood warning area. Over the last thirty years the limitations of such an approach in delivering efficient and sustainable flood risk management have become clear. Because ad hoc methods for decision-making have evolved in different ways in the various domains of flood risk management (flood warning, flood defence design, land use planning, urban drainage, etc.), they inhibit the catchment-based integrated approach on which flood risk management is now predicated. This is motivated by the recognition that there is no single completely effective scheme to all sources and types of flood risk [90]. Instead, a mosaic of flood risk management measures—be they "hard" structural measures, or "soft" measures (such as tree planting, land use planning and flood warning systems) are amassed in order to reduce risk in an efficient and sustainable way.

The composition of the flood risk management mosaic is matched to the functioning and needs of individual localities at risk and should be adapted as more understanding of the sources and pathways of flooding is acquired and as systems, both natural and social, change. However, there are institutional implications here. Implementing this approach involves the collective action of a range of different and often disparate stakeholders [41]. This places an increasing emphasis upon communication and mechanisms to reach unanimity. In the mosaic approach, risk estimates and assessments of changes in risk provide a vital common reference for comparing and choosing between alternatives that might contribute to flood risk reduction [116].

The principles of considering a risk-based approach have become well established in flood risk assessment [19]. However, it is worth reviewing how the risk-based approach addresses some of the main challenges of assessing flood risk in a systems approach [96]. In these systems of managing risk, there is a great deal of uncertainty and variability. Rainfall is naturally variable in space and time; flood forecasts cannot be beyond a few days and, therefore, forward planning cannot be too early in advance, and often worse-case scenarios are considered in flood warning areas. Extreme events that may never have been observed in practice (such as the 1 in 100 plus any climate change allowance) must be accounted for in design and risk assessment. Extrapolating loads to these extremes are uncertain, particularly when based on limited historical data and relative unexperienced climate change influences. Load and response combinations are critical; the severity of flooding is usually a consequence of a combination of conditions. So, for example, overtopping or breach of a river wall is usually a consequence of a combination of high waves and surge water levels, rather than either of these two effects in isolation. In complex river basins the timing of rainfall and runoff response at different locations in the catchment determines the severity of the downstream flood peak. The severity of any resultant flooding will typically be governed by the number of defences breached or overtopped, as well as the vulnerability of the assets and preparedness of the people within the floodplain. Therefore, analysis of loads and system response is based on an understanding of the probability of combinations of arbitrary loading conditions and the system's responses, including the social dimension.

Improved understanding of system behaviour has illustrated the importance of increasingly large combinations of variables. Spatial interactions are important; river and coastal systems show a great deal of interactivity. It is well recognized that construction of flood defences or upstream urbanisation may increase the water levels downstream. Similarly, construction of coastal structures to trap sediment and improve the resistance of coasts to erosion and breaching in one area may deplete beaches down drift [22,31,] and exacerbate erosion or flooding there, leading to economic damage or environmental harm. These interactions can be represented in system models, but engineering understanding of the relevant processes, particularly sedimentary processes over long timescales, is limited. Even where we have a detailed understanding of the physical processes, there may be fundamental limits to our ability to predict behaviour due to the chaotic nature of some of the relevant processes and loading. Complex and uncertain responses must be accommodated; models of catchment processes are known to be highly uncertain due to the complexity of the processes involved and the scarcity of observed data to effectively calibrate at appropriate scales [9]. The response of river and coastal defences to loading is highly uncertain. The direct and indirect impacts of flooding depend upon unpredictable or perverse human behaviours for which relevant measurements are scarce [34].

Risk has two components—the probability of an event occurring and the impact associated with that event. Intuitively it may be assumed that risks with the same numerical value have equal "significance", but this is often not the case. In some cases, the significance of a risk can be assessed by multiplying the probability by the impacts. In other cases, it is important to understand the nature of the risk, distinguishing between rare, catastrophic events and more frequent less severe events. For example, risk methods adopted to support the targeting and management of flood warnings represent risk in terms of probability and consequence, but low probability/high consequence events are treated very differently to high probability/low consequence events. The former can be catastrophic leading to substantial loss of life, whereas the latter are "nuisances".

Evolution and establishment of the risk-based approach—and perhaps what above all distinguishes it from other approaches to flood risk engineering or decision making—is that it deals with providing outcomes. Thus, it enables flood risk management options to be compared on the basis of the potential benefits that they are expected to provide to different events in a specified area. A risk-based approach, therefore, enables informed choices to be made based on comparison of the expected outcomes and costs of alternative courses of action, known as optioneering. This is distinct from, for example, a standard-based approach that focuses on the severity of the event that a particular flood defence is expected to withstand and the design to match that event.

12.3 Drivers of flood risk management

The range of drivers that may influence flood risk management was surveyed in the UK Foresight Future Flooding project [117]. The drivers identified as being of relevance are adapted in Table 12.1. The Foresight study went on to rank drivers of change in terms of their potential for increasing flood risk in the future, in the context of four different socioeconomic and climate change scenarios. Whilst the ranking was based largely upon expert judgement and a broad scale of quantified risk analysis, it did provide some indications of the relative importance of different drivers of change for flood managers in the future in relation risk–based approach (Source-Pathway-Receptor classification).

Table 12.1: Summary of drivers of change (adapted from Hall and Penning-Rowsell, 2010).

Driver	Drivers	SPR classification	Explanation
Catchment runoff	Precipitation	Source	Quantity, spasmodic rainfall and intensity. Rain/snow proportion.
	Urbanisation	Pathway	Land cover changes, e.g. construction of impermeable surfaces.
	Rural land management	Pathway	Influences surface and subsurface runoff. Changes include intensification of agricultural land use.
Fluvial processes	River morphology	Pathway	Changes that influence flood storage and conveyance, e.g. channelization and straightening.
	River vegetation	Pathway	Changes in river vegetation extent and type, e.g. in response to climate change or due to changed maintenance or regulatory constraints.
Societal changes	Public behaviour	Pathway	Behaviour of floodplain occupants before, during and after floods can significantly modify the floods severity.
	Social vulnerability	Receptor	Changes in social vulnerability to flooding, e.g. equity and systems of social provision.
Economic changes	Buildings and contents	Receptor	Changes in the cost of flood damage to domestic, commercial and other buildings and their contents.
	Urban vulnerability	Receptor	Changes in the number and distribution of domestic, commercial and other buildings in floodplains.
	Infrastructure	Receptor	Systems of communication (physical and telecommunication), energy distribution.
	Agriculture	Receptor	Changes in the intensity and seasonality of agriculture, including removal of agricultural land from production and hence changes in vulnerability to flood damage.

Most recently, working with natural processes (WwNP) and natural flood risk management (NFRM) has been advocated as a low-cost, sustainable form of long-term flood risk management, addressing the sources and pathways of flood risk [15,39]. There have been multiple policy drivers to encourage flood risk management that seeks to WwNP. Working with natural processes is considered the protection, restoration and emulation of the natural regulatory functions of catchments, rivers, floodplains and coasts [35,118]. With the implications of climate change and the drive for building resilience, it is recognised that society cannot rely on ever higher defensive engineering practices (that were much of the focus during the 1980–90s). More sustainable “soft-engineered” approaches must be considered. This is reflected in a host of policies and government responses to events outlined in Fig. 12.1, which states that the concept of sustainable development must be firmly rooted in all flood risk management and coastal erosion decisions and operations, seeking to meet multiple objectives and policy goals with every activity.

Reversed chronological order (Present → 1947)	Legislative Drivers	Strategies and Agendas	Flood Events
Present	Flood and Water Management Act (2010)	Synergies Project (2014)	November, December and January 2015-16 Northen UK, Northern Ireland and Wales
	Flood Directive (2007/60/EC)	Woodland for Water (2012)	December 2013-14 Extensive across England, Wales and Northern Ireland
	European Eel Regulation (2007/60/EC)	UK Flood and Coastal Erosion Risk Management R&D Strategy (2012)	April 2012 Western England and Wales
	Natural Environment and Rural Communities Act (2006)	Natural Environment White Paper (2012)	November 2009 Cumbria
	Water Framework Directive (2000/60/EC)	Flood and Coastal Erosion Risk Management Strategy for England (2011)	July 2007 Extensive across England and Wales
	Environment Act (1995)	Biodiversity 2020 (2010)	December and January 2000-01 Extensive across UK
	Habitats Directive (1992)	Making Space for Nature (2010)	April 1998 Eastern England
	Wildlife and Countryside Act (1982)	Pitt Review (2008: Recommendation 27)	April 1953 Eastern England and Scotland
1947	Birds Directive (1973/60/EC)	Making Space for Water (2004)	March 1947 Extensive across England

Fig. 12.1: International and national WwNP drivers, including legislation, strategies/agendas and notable flood events, highlighting the relationship between events, governmental and inter-governmental legislation and strategies.

12.4 Catchment based flood management

Owing to the recognition of how significantly altered land cover has become in the UK, due to rapid urban creep, agricultural intensification and wide-spread deforestation, catchment based flood management (CBFM) has sought to restore and encourage more “natural” hydrological processes. Whilst the evidence for the detrimental effects on flood risk of land cover alteration are more compelling at the small spatial scales (<20 km^2), observed impacts in larger catchments (>100 km^2) are limited [77]. Analysis of 943 catchments across the UK in the Flood Estimation Handbook, identified urban creep as the only land cover type that alters river response and the magnitude of the QMED flood (known as the mean annual flood) in UK Rivers [76,119,120]. Numerical modelling, undertaken by [121], suggests land cover changes on river flows in the Thames catchment is small compared with natural climatic variability between flood-rich and flood-poor periods (Hannaford, 2015).

Urbanisation has been shown to increase peak flood flows due to reduced capacity for infiltration, with asphalt and concrete surfaces and their associated low hydraulic roughness leading to rapid runoff; along with straightened and channelized streams, and through the use of pipes and culverts [122]. Urban surface water flooding is generally the most common from intense summer convective storms [123]. Sustainable drainage systems (SuDS), such as permeable paving, swales and stormwater retention ponds, can mitigate and, in some instances, reverse the adverse runoff effects of urbanisation [16,17,112].

Restoration of heavily modified watercourses with re-vegetated riparian corridors, plus reconnection of their floodplains can be used to convey or store urban run-off, while encouraging infiltration and improving water quality [124]. Whilst SuDS are commonly applied in the built environment, the principles of managing surface water across an interconnected management train to slow, store, disconnect and infiltrate flood flows can also be applied in a rural and coastal settings using NFRM. The ability to address agricultural sources of runoff are critical for successful CBFM, changing agricultural land management over the 19th and 20th centuries has led to a series of adverse changes of agricultural response to rainfall events, such as:

A) Field enlargement: with the advent of larger farm machinery and farm businesses shifting to macro-economic operations, there has been a wide-spread removal of hedgerows. This was also encouraged during the Second World War, to increase food self-sufficiency across the UK and much of continental Europe [2]. However, since the 12th century, agricultural land was enclosed by hedgerows and vegetated swards to consolidate individual farm holdings within a boundary. This became an "Act of Parliament" during the 18th century as part of the Enclosures Act 1750, encouraging a network of interconnected hedgerows to separate individual farmers and landowner's holdings. In some parts of the UK, 50 percent of hedgerows have been removed, equating to a rate of removal of 9500 km per year [125]. Hedgerows are recognised to provide increased runoff interception, hydraulic roughness and soil infiltration. A study in Pontbren, mid-Wales, identified infiltration losses to be sixty–seven times greater in tree shelters belts and hedgerows, compared to sheep grazed improved grassland [60,107]. Field enlargement has also led to encroachment on the river network, narrowing or removing buffers strips that provided interception and floodplain roughness in high-flows [126].
B) Soil compaction: arable and livestock farming practices can have detrimental effects on soil conditions and enhancing associated surface runoff. JBA [47] identified soil compaction due to higher livestock density increased the flood peak after a "Storm Desmond", in

the northwest of England in 2015, by 7 percent simulated in a hydraulic model. In terms of arable farms, Palmer and Smith [85] identified plots that had been planted with late harvesting crops, including potatoes, beans and maize, suffered a significant reduction in organic matter due to compaction and associated runoff.

C) Extensive land drainage: alongside the removal of hedgerows and intensification of farming practices with larger and heavier machinery, arterial land drainage and ditching became ubiquitous to increase the available farming space, including many upland blanket peat bogs [1,80]. The Peak District Derwent River restoration project, making space for water, identified greater flood peaks and shorter lag-times to reach peak flow associated with peatland degradation and removal of vegetation cover [25,127]. In terms of types of vegetation, sphagnum moss along slopes and in watercourses have been removed and burnt leading to faster overland and in-channel flows [128]. Pipe based land drains have a more complex effect on flood flows in traditional arable environments. At the site scale, in impermeable clay soils arterial piped drainage actually reduces runoff and increases storing capacity between rainfall events. However, higher flood peaks arise from extensive drainage networks, including methods like "mole-ploughing" that increases the volume and rate at which water is conveyed into a receiving watercourse via sub-surface pipes [133].

D) Crop types: arable macro-economics have not only led to field enlargement, but also changes in crop type and growing seasons, switching from spring-sown to autumn-sown cereals and vegetables. By sowing during the wettest time of year, it can lead to bare earth, less hydraulic roughness and greater overland and sediment laden "muddy" flows [129].

NFRM techniques (Table 12.2; Fig. 12.2) aim to ameliorate, if not reverse, the adverse hydrological effects of the aforementioned land management practices. These have been categorised into areas of application, including: (i) runoff management; (ii) river and floodplain management; (iii) woodland management; and (iv) coastal and estuary management. These techniques can be applied across a catchment area, from the headwaters, where flood generation processes often originate, to the lowlands and estuaries, where extensive and long-duration flooding occurs [64,65].

Pluvial and fluvial NFRM techniques aim to reduce flood risk by reducing the hydrological connectivity from the landscape and receiving watercourses to downstream communities. These must be carefully assessed in order to provide necessary upstream storage, not to detrimentally generate backwater effects to upstream residents, as well as manage the relative sub-catchment timings of flood peaks to de-synchronise flood flows and attenuate the flood peak [87,130,131]. The latter has been recognised to greatly affect the performance of a CBFM scheme, [87] found inappropriately situated measures exacerbated flooding due to the convergence of flood peaks across the Pickering Beck, Yorkshire. Coastal NFRM techniques must principally reduce flood risk by reducing wave and tidal energy in front of a line of defence in order to enhance the standards of protection (SoP) commonly provided by existing

Table 12.2: NFRM techniques (collated from [15,21,33]).

Runoff management	
Soil and land management	Conservation tillage, winter cover crops, crop rotation, stocking density, vegetation cover and buffer strips
Headwater drainage	Track drainage and grip/gully blocking
Runoff pathway management	Bunds, ponds, swales and sediment traps
River and floodplain management	
River restoration	Re-meandering, deculvert and two-staged channels
Floodplain/wetland restoration	Embankment removal and restoring wetlands
Leaky barriers	Leaky debris dams, coarse woody debris and beaver dams
Offline/online storage areas	Washlands, offline pond and online pond
Woodland management	
Catchment woodland	Hill top woodland and large-scale woodland cover
Cross-slope woodland	Woodland belt and shelter belt
Hedgerows	Hedges and cross-slope interceptors
Wet woodland	Woodland water retention area and leaky deflectors
Floodplain woodland	Floodplain zone woodland and floodplain roughening
Riparian woodland	Riparian zone woodland and bank crest roughening
Coast and estuary management	
Saltmarshes and mudflats	Saltmarsh and mudflat restoration and management
Sand dunes	Sand dune management and restoration
Beach nourishment	Beach management and replenishment

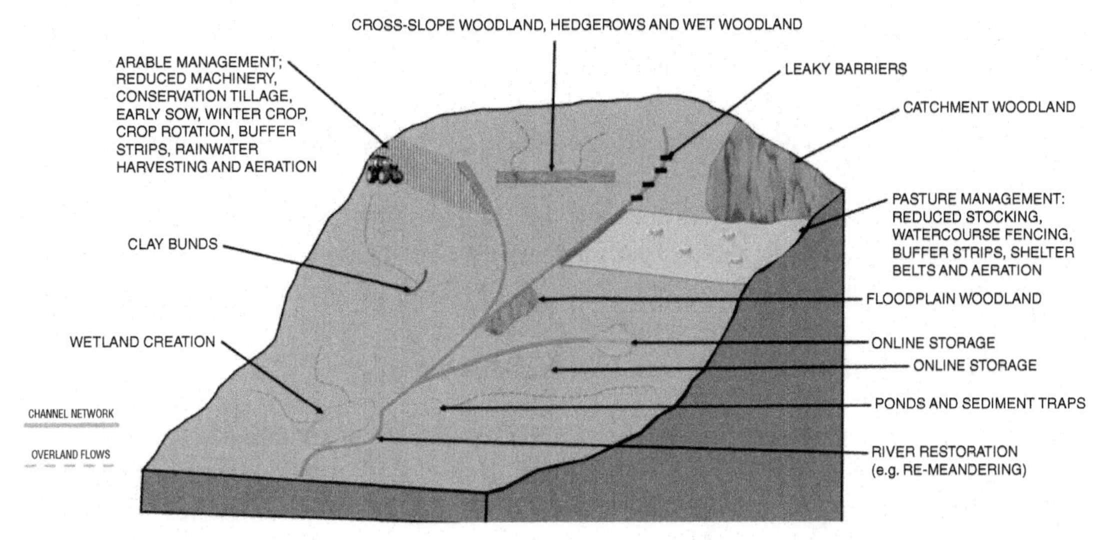

Fig. 12.2: NFRM techniques—source to sea (Burgess-Gamble et al., [15]).

hard engineered schemes [22]. Large-scale realignment and managed retreats, such as the Medmerry scheme, West Sussex, that reduces the risk to 348 properties and businesses must ensure any further storm surge intrusion does not adversely generate heightened or additional river flooding problems upstream [132].

In terms of flood risk benefits, a series of recent evidence reviews identified the risk associated with minor floods in small catchments (<20 km^2) can be reduced by CBFM that utilise SuDS and NFRM [21,77]. However, evidence does not suggest CBFM will have a positive flood ameliorating effect on the larger flood events and large catchment scales (>100 km^2). This is likely due to large fluvial floods being commonly attributed to saturation excess overland flows when prolonged rainfall events lead to a reduced capacity for any further infiltration [97]. It is important to recognise that from an engineering perspective, any flood risk management schemes effectiveness diminishes proportionate with the size of the rainfall event, and it is possible that any scheme can be overwhelmed, whether hard or soft. Where multiple interventions have taken place, it can be difficult to disentangle the effects of an individual technique, the effect of which depends upon catchment characteristics (in particular size, shape, topography, geology, soils, and both hydrological and sediment connectivity) and the extent and location of the intervention within the catchment. With the existing state of scientific understanding, it is not possible to state unequivocally whether the lack of demonstrable effect at large hydrological scales is because noticeable flood mitigation could not be achieved in a large catchment, or because a sufficiently large-scale set of techniques have not yet been implemented. However, some key lessons can be learnt from the CBFM schemes assessed, such as:

A) Techniques that increase the infiltration and retention capacity of soils, including tree planting, shelter belts and soil aeration, are considered most effective in small flood events at small hydrological scales. Once soils become saturated, the infiltration and retention of water in the soils macropores becomes negligible;
B) Techniques that generate additional upstream storage, including bunds, ponds and larger offline storage areas can be effective in reducing flood risk, depending on volumes of storage generated and how those distributed storage volumes effect sub-catchment interactivity; and
C) Techniques that enlarge the cross-sectional area of rivers, floodplains and coastal area by reconnecting wetlands, setting back embankments and restoring channel morphology have greater levels of confidence of reducing flood risk by attenuating the flood peak.

12.5 But what about businesses and homeowners?

There is an obvious and understandable focus at the societal level on the "hard" and "soft" elements of structural based flood risk management. However, the underlying principle in the UK is that a property owner must take responsibility for protecting themselves and their property from the impacts of flooding.

12.5.1 National strategy toward property level flood resilience

The UK Government's current strategy for flood risk management (FRM), propounded initially in the publication "Making Space for Water" [25], envisages property owners shouldering greater responsibility for protecting either their own homes or businesses. Despite this viewpoint, only 25 percent of those in the UK whose property has been flooded have taken steps to protect assets [43,101]. And of those who have not been flooded but are at risk, only 6 percent are known to have installed property level flood measures [43,101] that will either hold back the ingress of flood waters into a property (resistance) or facilitate the rescue, recovery and ultimate repair of a property (resilience) after it has been inundated. The characteristics and considerations of flood resistance measures and flood resilience measures and some options available to property owners are summarised below:

Property level flood resistance: resistance measures aim to defend a property from the adverse effects of floodwaters either by sealing apertures in the external envelope of a building, such as doors, windows, vents and drains, or through the deployment of barriers in the curtilage of a property (Table 12.3). The latter is akin to those measures that may be deployed by water authorities and the emergency services to protect wider areas, such as housing estates and critical infrastructure. Measures that seal the envelop of a building can be either temporary - that is they are put in place when a flood risk increases, such as barriers across doors—or built-in, such as water-tight doors or automatic sealing air-vents. Crucially, the suitability and, hence, effectiveness of resistance measures depends on both property specific characteristics, such as the age of the property and the material it is built with, and the nature of anticipated flooding - depth and speed of conveyance. At the property level, temporary resistance measures tend to be more cost-effective than permanent measures, particularly against shallow flooding and flood water levels that reduce quickly [3].

Property level flood resilience: resilience relates to the ability of a material or structure to cope with the effects of flooding and, hence, the associated ease of recovery and repair of a property. Resilience can be improved through choice of materials and the form of construction and fitting-out of the property (Table 12.3). For example choosing cement or lime-based render rather than gypsum based plasterboard and raising electrical sockets above anticipated flood levels are amongst the measures available. Again, the choice of material and

Table 12.3: A list of commonly adopted property flood resistance and resilience measures.

Resistance	Resilience
Free-standing barriers	Raised electrical circuits
Door barriers/guards	Solid impermeable flooring
Air brick/vent covers (manual and automated)	Cement/lime-based wall covering.
Waterproof external doors	Replace doors, windows and frames with water-resistant alternatives
Low-level bunds	Removable doors
Raised thresholds	Central heating/hot water boiler
Boundary Walls	Resilient kitchens
External Wall Treatment	Water-resistant cabinets
Basement or cellar tanking	Repositioned white goods
Drainage—one-way valves	

construction methods are influenced by property specific characteristics, as well as the nature of anticipated flooding.

Crucially, flood resistance and resilience measures, covered by the acronym PLFRA (property level flood risk adaptation), are not exclusive and, hence, both types could be deployed by property owners at the same time. Nevertheless, most people continue to perceive the responsibility for flood protection as belonging to Government [105], rather than the individual or the local community.

The UK National Policy Planning Framework (NPPF), which sets out the UK Government's planning policies for England and how these should be applied, aims to ensure flood risk is considered at all stages in the planning process, by avoiding inappropriate development in areas at risk of flooding [70]. Irrespective of this laudable position, there is no policy in the NPPF that relates to installation of either resistance or resilience measures on existing properties [70]. The only impetus to incorporate PLFRA measures is limited to a property owner's decision to minimise the tangible effects of flooding on their property. The corollary of a decision to install PLFRA is likely to be a reduction in the intangible effects of flooding.

The economic impact of flooding in Tewkesbury and Hull in 2007, which affected 2,500 homes and damaged critical infrastructure, such as elements of the electricity and water supply networks, plus roads and bridges, was estimated at £3.5Bn, while the financial cost of the intangible effects of the flooding, partly characterised as impacts on mental health and days of schooling lost, was estimated to be £250M [35]. The latter figure is likely to be lower than the final cost because of delayed manifestation of some effects and the inability to make clear causal links between flooding and some health conditions [4,61]. The Committee on Climate Change Report on the effectiveness of property level measures [18] and others now make strong cases for installation of PLFRA [5–8,13,27,62,66].

12.5.2 Drivers to property level flood resilience

The motivations of property owners to install flood protection have been extensively examined [40,56,92]. Before instigating works to protect property, it is quite reasonable to expect individuals to transition through the stages of "Awareness", "Perception" and "Ownership" that can be encapsulated under the heading of "Desire", followed by the stages of "developing knowledge", "sourcing capital" and then the "belief that action will achieve the desired affect". The last three stages can be summed up under the heading of "ability". Lamond and Proverbs [56] propounded this view of a growing awareness of the need to act to protect property, whereas both [92] and [40] view motivation to install PLFRA measures as being based on extended versions of the protection motivation theory (PMT). The PMT was originally described by Rogers [94] and focuses on the relationship between risk perception and demographics as a means of forecasting take-up of adaptation. Although there is a link between risk perception and adaptation, there is not a strong correlation between demographics and willingness to incorporate protection

[59]. Analysis by simple comparison of the distribution of PLFRA installed versus demographics and perception of risk fails to identify any clear linkage between most factors, although Lawrence et al. [58] found that previous experience of flooding did increase preparedness of households for flooding and established greater willingness to make household-level changes to reduce the impact of flooding. Joseph [51] does identify some demographic factors, such as age, income and educational level, as determinants of willingness to install PLFRA.

Regardless of these viewpoints on motivation to fit property protection, a large proportion of researchers conclude that to enhance and ultimately improve resilience, at both a community and individual property level, there is a need for more information about the nature and frequency of potential hazards, as well as detail on the efficacy and efficiency of protection measures [40,56,92,104].

Despite there being a body of academic research that quantifies both the tangible and intangible benefits of property level protection measures [27,50,51,66], there remains reluctance among property owners to incorporate protection measures. In some cases, it is simply through fear of labelling their homes and communities as "flood zones" [26]. Moreover, worryingly it appears doubtful that climate change, which is likely to be manifest by more frequent and more severe episodes of flooding, will drive adaptation and take-up of PLFRA [11,12,42,44,91].

As well as understanding the nature of hazards and benefits of fitting protection measures, some literature suggests there would be greater willingness to incorporate measures if adaptation was incentivised [11,38,40,46,54,56,93]. Hence, research is now examining how financial mechanisms might promote installation at both property level and possibly community level instead [,46,89,98]. Although a large number of researchers extol development of policies that incentivise the installation of flood protection, others add caution to the debate, believing there is risk of social injustice where some property owners in high flood risk areas have low income [49].

12.5.3 Flood risk insurance

Part of the socio-technical response to flooding in the UK, includes provision of voluntary private flood risk insurance, as well as improved flood warning systems. Although Central Government continues to build structural flood defences - "hard and soft" - that protect whole communities from flooding, it has also set about promoting improvements to the provision of access to property flood insurance [29], as well as promoting much greater community resilience.

Until 2016, the insurance based response to flooding, particularly for domestic property, was governed by a Statement of Principles - an agreement between the UK Government and the Association of British Insurers—whereby the UK Government agreed to underwrite insurance losses. The Statement of Principles was replaced by Flood Re in 2016. Through insurance companies, Flood Re effectively "underwrites" potential losses and, therefore, supports the provision of affordable flood risk insurance aimed at low-income households. Provision currently excludes insurance for those in the rental sector (private and social)—tenants and landlords. Originally, it was intended that over the period that Flood Re was operating (Flood Re was set-up initially for a limited duration before it was replaced by reflective pricing of insurance) the company would promote greater take-up of PLFRA measures [75]. That responsibility was put on hold so that Flood Re could be launched on time [63], but has not been re-introduced since.

At this moment the insurance sector's emphasis is driven by financial considerations rather than adaption to the effect of climate change [82]. In fact, O'Hare et al. [82] go further suggesting that insurance is maladaptive in that "it promotes resilience through the diffusion of responsibility, yet abates the incentive to act" (pp. 13; 2015). In February 2015, the Chairman of the Committee on Climate Change Adaptation, Lord Krebs, wrote to the CEO of Flood Re about the value for money of the Flood Re scheme. He also took the opportunity to remind the CEO of the need to use the scheme for promoting greater resilience in those homes at high risk of flooding [53]. Horn and McShane [45] suggests Flood Re will succeed in increasing adaptation to mitigate the effect of climate change only if it can participate in debates on land–use planning, building regulations and water management.

12.5.4 Relationship between growing rental sector and flood insurance

In the UK, it is anticipated that the size of the rental sector will grow in the coming years [23]. Without access to affordable property flood risk insurance and the potential absence of a feeling of "placement", which might be due to "churn" (frequency of moving home), it is possible that those in private rented housing, and to some degree those in social rented housing, will remain exposed to the impacts of flooding. Tenants could buy contents insurance but they would remain beholden to their landlord to buy property insurance at a commercial rate that reflects the risk of flooding. Hence, there is risk that some landlords will balance out the high cost of flood insurance premiums versus the cost of repairs and potential loss of income, ultimately leaving their property, and their tenants, exposed to the effects flooding.

A large proportion of these persons are likely to be categorised as being vulnerable. Some of the factors identified by sociological researchers examining social vulnerability include: socio-economic status, gender, race and ethnicity, age, rural/urban, type of residential property, renters, occupation, family structure, education, medical services and socially dependent needs populations [20,108]. Although, admittedly not everyone who could be categorised vulnerable is likely to be renting, greater exposure of the vulnerable in society to the impacts of flooding is likely to increase the burden on local communities, as well as society, as a whole.

12.5.5 Community resilience

Although there have been moves in the UK to build greater levels of community resilience, partly by involving individuals and communities in decision making, the most vulnerable in those communities, including those in rented accommodation, are likely to remain marginalised. DEFRAs Flood Pathfinder Report identified in the majority of community resilience projects attempts to take account of the needs of the most vulnerable in communities, although all admitted that more needed to be done [102]. Only one community resilience project had specifically targeted the most vulnerable initially, but those objectives set out to address issues affecting the most vulnerable were removed in the second year of the project.

Nevertheless, even if there is involvement in decision making at the local level, national policy setters still have influence over local action. For instance, in Norway, in their assessment of the interplay between governance structures at local and national level, Naess et al. [74] shows that despite local councils being an appropriate institutional level for adapting to new flood risk and having high legitimacy among those at risk, they are often stymied by agencies of central government seeking to shift and spread financial risks and heightening vulnerability.

12.6 Conclusions

Flood protection is a multifaceted and complex issue. Umbrellas and sandbags fail to provide adequate shelter form torrential downpours and flashy floodwater to safeguard individuals and properties. Forecasts suggest the issue will worsen when the impacts of climate change are further realised. The chapter has described a suite of risk management approaches being instigated across the catchment scale and at the localised property-level.

The water engineer's solution and involvement in future flood risk management requires holistic knowledge and an integrated understanding of acceptable economic and social frameworks. Minimising natural processes, so as to reduce runoff and increase infiltration, are the preferred future–facing solutions over the traditional means of building higher defences or installing greater capacity drainage pipes. Moreover, passing the baton of flood risk to businesses and homeowners to deal with themselves has not proven to be an achievable way forward, to date, because property-level flood risk adaptation has been limited and is further complicated by the flood insurance industry.

References

[1] M. Acreman, J. Holden, How wetlands affect floods, Wetlands 33 (5) (2013) 773–786.

[2] ADAS, Development of a Database of Agricultural Drainage, DEFRA ES0111, Department for Environment, Food and Rural Affairs, London, 2002.

[3] R. Ayton-Robinson, Resilient repair strategy, in: J.E. Lamond, C.A. Booth, D.G. Proverbs, F. Hammond (Eds.), Flood Hazards—Impacts and Responses for the Built Environment, CRC Press, Boca Raton, 2012, pp. 141–154.

[4] K. Azuma, K. Ikeda, N. Kagi, U. Yanagi, K. Hasegawa, H. Osawa, Effects of water-damaged homes after flooding: health status of the residents and the environmental risk factors, Int. J. Environ. Health Res. 24 (2014) 158–175.

[5] D.W. Beddoes, C.A. Booth, Property level flood protection: a new effective and affordable solution, in: D. Wrachien, D.G. Proverbs, C.A. Brebbia, S. Mambretti (Eds.), Flood Recovery Innovation and Response II, WIT Press, Southampton, 2010, pp. 271–280.

[6] D.W. Beddoes, C.A. Booth, Property level flood resistance versus resilience measures: a novel approach, Int. J. Saf. Secur. Eng. 1 (2011) 162–181.
[7] D.W. Beddoes, C.A. Booth, Holistic property-level flood protection, in: C.A. Booth, S. Charlesworth (Eds.), Water Resources in the Built Environment: Management Issues and Solutions, John Wiley and Sons Ltd, Oxford, 2014, pp. 271–280.
[8] D.W. Beddoes, C.A. Booth, J.E. Lamond, Towards complete property-level flood protection of domestic buildings in the UK, in: S. Hernández, S. Mambretti, D.G. Proverbs, J. Puertas (Eds.), Urban Water Systems and Floods II, WIT Press, Southampton, 2018, pp. 27–38.
[9] K. Beven, A manifesto for the equifinality thesis, J. Hydrol. 320 (2006) 18–36.
[10] K. Beven, R. Romanowicz, P. Young, I. Holman, H. Posthumus, J. Morris, S. Rose, P.E. O'Connell, J. Ewen, FD2120, Analysis of Historical Data sets to Look for Impacts of Land Use and Management Change on Flood Generation, Department for Environment, Food and Rural Affairs, London, 2008.
[11] E. Bichard, A. Kazmierczak, Are homeowners willing to adapt to and mitigate the effects of climate change? Clim. Change 112 (2012) 633–654.
[12] K. Blennow, J. Persson, Climate change: motivation for taking measure to adapt, Global Environ. Change 19 (2009) 100–104.
[13] C. Broadbent, Improving the flood resistance of domestic property, Struct. Surv. 22 (2004) 79–83.
[14] S.B. Broadmeadow, H. Thomas, T.R. Nisbet, Opportunity Mapping for Woodland Creation to Reduce Diffuse Water Pollution and Flood Risk in England and Wales, Forest Research, Farnham, Surrey, 2014.
[15] L. Burgess–Gamble, R. Ngai, M. Wilkinson, T. Nisbet, N. Pontee, R. Harvey, K. Kipling, S. Addy, S. Rose, S. Maslen, H. Jay, A. Nicholson, T. Page, J. Jonczyk, P. Quinn, Working with Natural Processes – Evidence Directory, Report SC150005, Environment Agency, Bristol, 2018.
[16] S. Charlesworth, E. Harker, S. Rickard, A review of sustainable drainage systems (SuDS): a soft option for hard drainage questions? Geography 88 (2) (2003) 99–107.
[17] CIRIA, The SuDS Manual, Report C753, CIRIA, London, 2015.
[18] R. Haskoning, Assessing the Economic Case for Property Level Measures in England, Committee on Climate Change, 2012. Project no.: 9X1055 [online]. Available at http://hmccc.s3.amazonaws.com/ASC/2012%20report/Royal%20Haskoning%20PLM%20Report%20Final.pdf.
[19] CUR/TAW, Probabilistic Design of Flood Defences. Centre for Civil Engineering Research and Codes (CUR), Technical Advisory Committee on Water Defences (TAW), Gouda, The Netherlands, 1990.
[20] S.L. Cutter, B.J. Boruff, W.L. Shirley, Social vulnerability to environmental hazards, Soc. Sci. Q. 84 (2003) 242–261.
[21] S.J. Dadson, J.W. Hall, A. Murgatroyd, M. Acreman, P. Bates, K. Beven, L. Heathwaite, J. Holden, I.P. Holman, S.N. Lane, E. O'Connell, E. Penning-Rowsell, N. Reynard, D. Sear, C. Thorne, R. Wilby, A restatement of the natural science evidence concerning catchment-based "natural" flood management in the UK, Proc. R. Geogr. Soc., A 473 (2017) 20160706, doi10.1098/rspa.2016.0706.
[22] R.J. Dawson, M.E. Dickson, R.J. Nicholls, J.W. Hall, M.J.A. Walkden, P.K. Stansby, M. Mokrech, J. Richards, J. Zhou, J. Milligan, A. Jordan, S. Pearson, J. Rees, P.D. Bates, S. Koukoulas, A.R. Watkinson, Integrated analysis of risks of coastal flooding and cliff erosion under scenarios of long–term change, Climate Change 95 (2009) 249–288.
[23] DCLG, English Housing Survey—Private Rented Sector 2015–16, Department of Communities and Local Government, London, 2017.
[24] DEFRA, Making Space for Water. Developing a New Government Strategy for Flood and Coastal Erosion Risk Management in England. A Consultation Exercise, Department for Environment, Food and Rural Affairs, London, 2004.
[25] DEFRA Making space for water: Taking forward a new Government strategy for flood and coastal erosion risk management in England: First Government response to the autumn 2004 Making space for water consultation exercise, Department for Environment, Food and Rural Affairs, London, 2005.
[26] DEFRA, Flood Risk and Insurance: A Roadmap to 2013 and Beyond (Final report of the flood insurance working groups), Department for Environment, Food and Rural Affairs, London, 2011.

[27] DEFRA, Best Practice in Property Level Protection Systems Advice for Local Authorities, Department for Environment, Food and Rural Affairs, London, 2014.
[28] DEFRA, Post–Installation Effectiveness of Property Level Flood Protection Final report FD2668, Department for Environment, Food and Rural Affairs, London, 2014.
[29] DEFRA, Cover Note to the Impact Assessment for Managing the Future Financial Risk of Flooding, Department for Environment, Food and Rural Affairs, London, 2016.
[30] DEFRA, Crop Rotation [online]. In Integrated Crop Management (CPA Book), published by the Crop Protection Association in 1996 and made available online by Department for Environment, Food and Rural Affairs, 2017. Available from http://adlib.everysite.co.uk/adlib/DEFRA/content.aspx?id=000IL3890W.17USY7NEWZ4R1 [Accessed 4 April 2019].
[31] M.E. Dickson, M.J. Walkden, J.W. Hall, Modelling the impacts of climatic change on an eroding coast over the 21st century, Clim. Change 81 (2007) 141–166.
[32] S.J. Dixon, Investigating the Effects of Large Wood and Forest Management on Flood Risk and Flood Hydrology, Unpublished PhD thesis, University of Southampton, UK, 2013.
[33] A. Duffy, S. Moir, N. Berwick, J. Shabashow, B. D'Arcy, R. Wade, CREW rural sustainable drainage systems, A Practical Design and Build Guide for Scotland's Farmers and Landowners, CRW2015/2.2, Aberdeen, Centre for Expertise in Water (CREW), (2016).
[34] R. Egorova, JMv Noortwijk, S.R. Holterman, Uncertainty in flood damage estimation, J. River Basin Manage. 6 (2) (2008) 139–148.
[35] Environment Agency, The Costs of the Summer 2007 Foods in England, (Report SC070039/R1), 51, 2010.
[36] Environment Agency, Working with Natural Processes to Reduce Flood Risk. RandD Framework: Science Report, Report SC130004/R2, Environment Agency, Bristol, 2014.
[37] Environment Agency, How to Model and Map Catchment Processes when Flood Risk Management Planning, Project SC120015/R1, Environment Agency, Bristol, 2016.
[38] T. Filatova, Market-based instruments for flood risk management: a review of theory, practice and perspectives for climate adaptation policy, Environ. Sci. Policy 37 (2014) 227–242.
[39] H. Forbes, K. Ball, F. McLay, Natural Flood Management Handbook, Scottish Environmental Protection Agency, Stirling, 2016.
[40] T. Grothmann, F. Reusswig, People at risk of flooding: why some residents take precautionary action while others do not, Nat. Hazards 38 (2006) 101–120.
[41] J.W. Hall, E.C. Penning–Rowsell, Setting the scene for flood risk management, in: G. Pender, H. Faulkner (Eds.), Flood Risk Science and Management, Blackwell Publishing Ltd, 2010, pp. 1–16.
[42] T. Harries, Feeling secure or being secure? Why it can seem better not to protect yourself against a natural hazard, Health Risk Soc. 10 (2008) 479–490.
[43] T. Harries, The anticipated emotional consequences of adaptive behaviour–impacts on the takeup of household flood-protection measures, Environ. Plan. A. 44 (2012) 649–668.
[44] T. Harries, E. Penning–Rowsell, Victim pressure, institutional inertia and climate change adaptation: the case of flood risk, Global Environ. Change 21 (2011) 188–197.
[45] D. Horn, M. McShane, Flooding the market, Nat. Clim. Chang. 3 (2013) 945–947.
[46] P. Hudson, W.J. WouterBotzen, W.J.W., L. Feyen, J.C.J.H. Aerts, Incentivising flood risk adaptation through risk based insurance premiums: trade-offs between affordability and risk reduction, Ecol. Econ. 125 (2016) 1–13.
[47] JBA, The Rivers Trust Life-IP Natural Course Project: Strategic Investigation of Natural Flood Management in Cumbria, JBA Consulting, Skipton, UK, 2016.
[48] C.L. Johnson, S.M. Tunstall, E.C. Penning–Rowsell, Floods as catalysts for policy change: historical lessons from England and Wales, Int. J. Water Resour. Dev. 21 (2005) 561–575.
[49] C.L. Johnson, E.C. Penning–Rowsell, D. Parker, Natural and imposed injustices: the challenges in implementing 'fair' flood risk management policy in England, Geogr. J. 173 (2007) 374–390.
[50] R.D. Joseph, Development of a Comprehensive Systematic Quantification of the Costs and Benefits (CB) of Property Level Flood Risk Adaptation Measures in England, Unpublished PhD Thesis, University of the West of England, Bristol, UK, 2014.

[51] R.D. Joseph, D.G. Proverbs, J. Lamond, Assessing the value of intangible benefits of property level flood risk adaptation (PLFRA) measures, Nat. Hazards 79 (2015) 1275–1297.
[52] I. Kelman, Climate change and the Sendai Framework for disaster risk reduction, Int. J. Disaster Risk Sci. 6 (2) (2015) 117–127.
[53] J. Krebs, Designing Flood Re to Encourage Flood Risk Reduction, (2015), Available online at: https://www.theccc.org.uk/wp-content/uploads/2015/02/2015-02-02-Lord-Krebs-to-Brendan-McCafferty-Flood-Re.pdf.
[54] H. Kreibich, A.H. Thieken, Th Petrow, M. Müller, B. Merz, Flood loss reduction of private households due to building precautionary measures – lessons learned from the Elbe flood in August 2002, Nat. Hazard. Earth Sys. Sci. 5 (2005) 117–126.
[55] S.N. Lane, Natural flood management, Wiley Interdiscip. Rev.: Water 4 (3) (2017) e1211.
[56] J.E. Lamond, D.G. Proverbs, Resilience to flooding: lessons from an international comparison, Proceedings of the ICE – Urban Design and Planning, 162, 2009, pp. 63–70.
[57] S.N. Lane, J. Morris, P.E. O'Connell, P.F. Quinn, Managing the rural landscape, in: C.R. Thorne, E.P. Evans, E.C. Penning-Rowsell (Eds.), Future Flooding and Coastal Erosion Risks, Thomas Telford, London, 2007, pp. 297–319.
[58] J. Lawrence, D. Quade, J. Becker, Integrating the effects of flood experience on risk perception with responses to changing climate risk, Nat. Hazards 74 (2014) 1773–1794.
[59] M.K. Lindell, R.W. Perry, Household adjustment to earthquake hazard: a review of research, Environ. Behav. 32 (2000) 461–501.
[60] M.R. Marshall, C.E. Ballard, Z.L. Frogbrook, I. Solloway, N. Mcintyre, B. Reynolds, H.S. Wheater, The impact of rural land management changes on soil hydraulic properties and runoff processes: results from experimental plots in upland UK, Hydrol. Processes 28 (4) (2014) 2617–2629.
[61] V. Mason, H. Andrews, D. Upton, The psychological impact of exposure to floods, Psychol. Health Med. 15 (2010) 61–73.
[62] P. May, Assessing the benefits of property level protection, Scotland's Flood Risk Manage. Conf. 2014.
[63] B. McCafferty, Flood Re: next steps to affordable and accessible flood Insurance, Flood Expo Conf. 2015.
[64] N. McIntyre, M. Marshall, Identification of rural land management signals in runoff response, Hydrol. Processes 24 (24) (2010) 3521–3534.
[65] N. McIntyre, C. Thorne, Land Use Management Effects on Flood Flows and Sediments – Guidance on Prediction, CIRIA Report C719, CIRIA, London, 2013.
[66] S. Merrett, Evaluation of the DEFRA property-level flood protection scheme: 25918 JBA Project, (2012) Available online at: http://nationalfloodforum.org.uk/wp-content/uploads/Evaluation-of-the-Defra-PL-Flood-protection-Scheme-25918.pdf.
[67] P. Metcalfe, K. Beven, J. Freer, Dynamic TOPMODEL: a new implementation in R and its sensitivity to time and space intervals, Environ. Modell. Softw. 72 (C) (2015) 155–172.
[68] P. Metcalfe, K. Beven, B. Hankin, R. Lamb, A modelling framework for evaluation of the hydrological impacts of nature-based approaches to flood risk management, with application to in-channel interventions across a 29 km 2 scale catchment in the United Kingdom: a modelling framework for nature-based flood risk management, Hydrol. Processes 31 (9) (2017) 1734–1748.
[69] V. Meyer, S. Priest, C. Kuhlicke, Economic evaluation of structural and non-structural flood risk management measures: examples from the Mulde River, Nat. Hazard 62 (2) (2012) 301–324.
[70] Ministry of Housing Communities and Local Government, National Planning Policy Framework, 2019.
[71] C. Morris, C. Potter, Recruiting the new conservationists: farmers' adoption of agri-environmental schemes in the UK, J. Rural Stud. 11 (1) (1995) 51–63.
[72] J. Morris, A. Bailey, C. Lawson, P. Leeds–Harrison, D. Alsop, R. Vivash, The economic dimensions of integrating flood management and agri-environment through washland creation: a case from Somerset England, J. Environ. Manage. 88 (2) (2008) 372–381.
[73] J. Morris, A.P. Bailey, D. Alsop, R. Vivash, C. Lawson, Integrating flood management and agri-environment through washland creation in the UK, J. Farm Manage. 12 (1) (2004) 33–48.
[74] L.O. Naess, G. Bang, S. Eriksen, J. Vevatne, Institutional adaptation to climate change: flood responses at the municipal level in Norway, Global Environ. Change 15 (2005) 125–138.

[75] National Flood Forum, Consultation on the implementing regulations for the Flood Reinsurance Scheme—National Flood Forum response, (2014).
[76] NERC, Flood Studies Report, 5 Volumes, Natural Environment Research Council, London, 1975.
[77] R. Ngai, M. Wilkinson, T. Nisbet, R. Harvey, S. Addy, L. Burgess–Gamble, S. Rose, S. Maslen, A. Nicholson, T. Page, J. Jonczyk, P. Quinn, Working with Natural Processes Evidence Directory—Literature Review, Project SC150005, Environment Agency, Bristol, 2018.
[78] A.R. Nicholson, M.E. Wilkinson, G.M. O'Donnell, P.F. Quinn, Runoff attenuation features: a sustainable flood mitigation strategy in the Belford catchment UK, Area 44 (4) (2012) 463–469.
[79] NWRM, Synthesis document no 1: introducing natural water retention measures: what are NWRM? Brussels: european commission, (2015) Available from: http://nwrm.eu/implementing–nwrm/synthesis–documents [Accessed 1 June 2019].
[80] P.E. O'Connell, K.J. Beven, J.N. Carney, R.O. Clements, J. Ewen, H. Fowler, G.L. Harris, J. Hollis, J. Morris, G.M. O'Donnell, J.C. Packman, A. Parkin, P.F. Quinn, S.C. Rose, M. Shepherd, S. Tellier, Review of Impacts of Rural Land Use and Management on Flood Generation – Impact Study Report, RandD Technical Report FD2114/TR, Department of Environment, Food and Rural Affairs, London, 2004.
[81] P.E. O'Connell, J. Ewen, G. O'Donnell, P. Quinn, Is there a link between land-use management and flooding? Hydrol. Earth Syst. Sci. 11 (1) (2007) 96–107.
[82] P. O'Hare, I. White, A. Connelly, Insurance as maladaptation: resilience and the "business as usual" paradox, Environ. Plan. C: Govern. Policy 34 (6) (2015) 1175–1193.
[83] J.C. Packman, P.F. Quinn, F.A.K. Farquharson, P.E. O'Connell, Review of Impacts of Rural Land Use and Management on Flood Generation, Report C1: Short-term Improvement to the FEH Rainfall Runoff Model: User Manual, 2004.
[84] J.C. Packman, P.F. Quinn, J. Hollis, P.E. O'Connell, Review of Impacts of Rural Land Use and Management on Flood Generation, Report C2: Short-term Improvement to the FEH rainfall–runoff model: Technical Background, 2004.
[85] R.C. Palmer, R.C. Smith, Soil structural degradation in SW England and its impact on surface-water runoff generation, J. Soil Use Manage. 29 (4) (2013) 567–575.
[86] I. Pattison, S.N. Lane, The link between land–use management and fluvial flood risk: a chaotic conception? Prog. Phys. Geog. 36 (1) (2012) 72–92.
[87] I.S. Pattison, S.N. Lane, R.J. Hardy, S.M. Reaney, The role of tributary relative timing and sequencing in controlling large floods, Water Resour. Res. 50 (7) (2014) 5444–5458.
[88] E.C. Penning-Rowsell, E.P. Evans, J.W. Hall, A.G.L Borthwick, From flood science to flood policy: the foresight future flooding project seven years on, Foresight 15 (3) (2013) 190–210.
[89] E.C. Penning–RowsellFlood insurance in the UK: A Critical Perspective2, Wiley Interdisciplinary Reviews: Water, 2015, pp. 601–608.
[90] M. Pitt, The Pitt Review: Lessons Learned from the 2007 Floods, Cabinet Office, London, 2008.
[91] J.J. Porter, S. Dessai, E.L. Tompkins, What do we know about UK household adaptation to climate change? A systematic review, Clim. Change 127 (2014) 371–379.
[92] J.K. Poussin, W.J.W. Botzen, J.C.J.H. Aerts, Factors of influence on flood damage mitigation behaviour by households, Environ. Sci. Policy 40 (2014) 69–77.
[93] J.K. Poussin, W.J.W. Botzen, J.C.J.H. Aerts, (2015) Effectiveness of flood damage mitigation measures: empirical evidence from French flood disasters, Global Environ. Change 31 (2014) 74–84.
[94] R.W. Rogers, A Protection motivation theory of fear appeals and attitude change, J. Psychol. 91 (1975) 93–114.
[95] M. Rogger, M. Agnoletti, A. Alaoui, J.C. Bathurst, G. Bodner, M. Borga, V. Chaplot, F. Gallart, G. Glatzel, J. Hall, J. Holden, L. Holko, R. Horn, A. Kiss, S. Kohnova, G. Leitinger, B. Lennartz, J. Parajka, R. Perdigao, S. Peth, L. Plavcova, J.N. Quinton, M. Robinson, J.L. Salinas, A. Santoro, J. Szolgay, S. Tron, J.J.H. Van Den Akker, A. Vigilone, G. Blöschl, Land use change impacts on floods at the catchment scale: challenges and opportunities for future research, Water Resour. Res. 53 (7) (2017) 5209–5219.
[96] P.B. Sayers, J.W. Hall, I.C. Meadowcroft, Towards risk-based flood hazard management in the UK, Civil Eng. 150 (2002) 36–42.

[97] E.M. Shaw, K.J. Beven, N.A. Chappell, R. Lamb, Hydrology in Practice, 4th edn., Spon Press (Taylor and Francis), London, 2010.
[98] S. Surminski, J.C.J.H. Aerts, W.J.W. Botzen, P. Hudson, J. Mysiak, C.D. Pérez-Blanco, Reflections on the current debate on how to link flood insurance and disaster risk reduction in the European Union, Nat. Hazard. 79 (2015) 1451–1479.
[99] The Flow Partnership 5% Future. Holding water in the landscape: a new practical deal for farmers and land managers,(2015) [online]. Available from: http://research.ncl.ac.uk/proactive/5future/ (Accessed 10 November 2019).
[100] C.R. Thorne, E.C. Lawson, C. Ozawa, S.L. Hamlin, S.A. Smith, Overcoming uncertainty and barriers to adoption of blue–green infrastructure for urban flood risk management, J. Flood Risk Manage. 11 (2015) S960–S972.
[101] N. Thurston, B. Finlinson, R. Breakspear, N. Williams, J. Shaw, J. Chatterton, Developing the Evidence Base for Flood Resistance and Resilience: Summary Report, R and D Technical Report FD2607/TR1, Department for Environment, Food and Rural Affairs, London, 2008.
[102] C. Twigger–Ross, P. Orr, K. Brooks, R. Sadauskis, H. Deeming, J. Fielding, T. Harries, R. Johnston, E. Kashefi, S. McCarthy, Y. Rees, S. Tapsell, Flood Resilience Community Pathfinder Evaluation—Final Evaluation Report FD2664, Department for Environment, Food and Rural Affairs, London, 2015.
[103] Waverley Committee, Report of the Departmental Committee on Coastal Flooding, Cmnd. 9165, HMSO, London, 1954.
[104] G. Wedawatta, B. Ingirige, D.G. Proverbs, Small businesses and flood impacts: the case of the 2009 flood event in Cockermouth, J. Flood Risk Manage. 7 (2014) 42–53.
[105] A. Werritty, D. Houston, T. Ball, A. Tavendale, A. Black, Exploring the Social Impacts of Flood Risk and Flooding in Scotland, Scottish Executive, Central Research Unit, Edinburgh, 2007.
[106] H. Wheater, E. Evans, Land use, water management and future flood risk, Land Use Policy 26 (1) (2009) S251–S264.
[107] H. Wheater, B. Reynolds, N. McIntyre, M. Marshall, B. Jackson, Z. Frogbrook, I. Solloway, O. Francis, J. Chell, Impacts of Upland Land Management on Flood Risk: Multi–Scale Modelling Methodology and Results from the Pontbren Experiment, FRMRC Research Report UR16, Flood Risk Management Research Consortium, Manchester, 2008.
[108] I. Willis, J. Fitton, A review of multivariate social vulnerability methodologies: a case study of the River Parrett catchment, UK, Nat. Hazard. Earth Syst. Sci. 16 (2016) 1387–1399.
[109] M.E. Wilkinson, K. Holstead, E. Hastings, Natural Flood Management in the Context of UK Reservoir Legislation, Centre of Expertise for Waters (CREW), Aberdeen, 2013.
[110] M.E. Wilkinson, E. Mackay, P.F. Quinn, M.I. Stutter, K.J. Beven, C.J.A. Macleod, M.G. Macklin, Y. Elkhatib, B. Percy, C. Vitolo, P.M. Haygarth, A cloud based tool for knowledge exchange on local scale flood risk, J. Environ. Manage. 161 (2015) 38–50.
[111] M.E. Wilkinson, P.F. Quinn, N.J. Barber, J. Jonczyk, A framework for managing runoff and pollution in the rural landscape using a catchment systems engineering approach, Sci. Total Environ. 468 (2014) 1245–1254.
[112] Q. Zhou, A review of sustainable urban drainage systems considering the climate change and urbanization impacts, Water (Basel) 6 (2014) 976–992.
[113] Environment Agency, Preliminary Flood Risk Assessment for England, Report. (2018). October 2018. Online at: https://assets.publishing.service.gov.uk/government/uploads/system/uploads/attachment_data/file/764784/English_PFRA_December_2018.pdf. (Accessed 15 October 2020).
[114] O. Carpenter, S. Platt, F. Mahdavian, Disaster Recovery Case Studies: UK Floods 2007, Cambridge Centre for Risk Studies at the University of Cambridge Judge Business School, 2018.
[115] H. Faulkner, D. Parker, C. Green, K. Beven, Developing a translational discourse to communicate uncertainty in flood risk between science and the practitioner, AMBIO: J. Human Environ. 36 (8) (2007) 692–704.
[116] R.J. Dawson, L. Speight, J.W. Hall, S. Djordjevic, D. Savic, J. Leandro, Attribution of flood risk in urban areas, J. Hydroinform. 10 (4) (2008) 275–288.

[117] E Evans, R Ashley, J Hall, E Penning-Rowsell, A Saul, P Sayers, C Thorne, A Watkinson, Foresight future flooding, scientific summary vol 1 Office of Science and Technology, London, 2004.
[118] B. Hankin, P. Metcalfe, I. Craigen, T. Page, N. Chappell, R. Lamb,..., D. Johnson, Strategies for testing the impact of natural flood risk management measures, Flood risk manag. (2017) 1–39.
[119] IoH (Institute of Hydrology), Flood Estimation Handbook, Institute of Hydrology, Wallingford, 1999.
[120] G. Watts, R.W. Battarbee, J.P. Bloomfield, J. Crossman, A. Daccache, I. Durance,..., T. Hess, Climate change and water in the UK–past changes and future prospects, Prog. Phys. Geogr. 39 (1) (2015) 6–28.
[121] T. Marsh, C.L. Harvey, The Thames flood series: a lack of trend in flood magnitude and a decline in maximum levels, Hydrol. Res. 43 (3) (2012) 203–214.
[122] M. Mansell, F. Rollet, Water balance and the behaviour of different paving surfaces, Water Environ. J. 20 (1) (2006) 7–10.
[123] T.P. Burt, J. Holden, Changing temperature and rainfall gradients in the British Uplands, Climate Res. 45 (2010) 57–70.
[124] R.K. Hall, D. Guiliano, S. Swanson, M.J. Philbin, J. Lin, J.L. Aron,..., D.T. Heggem, An ecological function and services approach to total maximum daily load (TMDL) prioritization, Environ. Monit. Assess. 186 (4) (2014) 2413–2433.
[125] J. Martin, Agricultural Development and Britain's Natural Heritage. In: The Development of Modern Agriculture, Palgrave Macmillan, London, 2000, pp. 167–195.
[126] S. Broadmeadow, T.R. Nisbet, The effects of riparian forest management on the freshwater environment: a literature review of best management practice, 2004.
[127] M. Pilkington, J. Walker, R. Maskill, T. Allott, M. Evans, Restoration of Blanket bogs; flood risk reduction and other ecosystem benefits, Moors for the Future Partnership, Edale, 2015.
[128] L. Wilson, J.M. Wilson, J. Holden, I. Johnstone, A. Armstrong, M. Morris, Ditch blocking, water chemistry and organic carbonflux: evidence that blanket bog restoration reduces erosion and fluvial carbon loss, Sci. Total Environ. 409 (2011) 2010–2018.
[129] L. Oudin, V. Andréassian, C. Perrin, C. Michel, N. Le Moine, Spatial proximity, physical similarity, regression and ungaged catchments: A comparison of regionalization approaches based on 913 French catchments, Water Resour. Res. 44 (3) (2008).
[130] S.J. Dixon, D.A. Sear, N.A. Odoni, T. Sykes, S.N. Lane, The effects of river restoration on catchment scale flood risk and flood hydrology, Earth. Surf. Proc. Land. 41 (7) (2016) 997–1008.
[131] P. Metcalfe, K. Beven, B. Hankin, R. Lamb, Simplified representation of runoff attenuation features within analysis of the hydrological performance of a natural flood management scheme, Hydrol. Earth Syst. Sci. 22 (2018) 2589–2605.
[132] Harvey, R. Case study 50.Medmerry Managed Realignment, 2015. Online at: https://www.therrc.co.uk/sites/default/files/projects/50_medmerry.pdf. (Accessed October 2020).
[133] M. Robinson, D.W. Rycroft, The impact of drainage on streamflow, In: W. Skaggs, J. van Schilfgaarde (Eds.), Agricultural Drainage, American Society of Agronomy, Madison, Wisconsin, 1999, 753–786.

CHAPTER 13

Energy harvesting in water supply systems

Armando Carravetta[a,*], Miguel Crespo Chacon[b], Oreste Fecarotta[a], Aonghus McNabola[b], Helena M. Ramos[c]

[a]Department of Civil, Architectural and Environmental, Engineering, Università degli Studi di Napoli "Federico II", Napoli, Italy [b]Trinity College of Dublin, Department of Civil, Structural and Environmental Engineering, Dublin, Ireland [c]Instituto Superior Técnico (IST), Departamento de Engenharia Civil, Arquitectura e Georrecursos, Lisbon, Portugal

**Corresponding author.*

13.1 Introduction

Water networks are essential civil structures for water transportation. These waterworks have an essential role in many industrial sectors. Water supply systems (WSS) are used to link water source to user demand. Major sectors of the WSS include: drinking water transmission and distribution systems; irrigation systems; and industrial transmission systems [50]. Other important water networks are used for drainage: to collect wastewater and transfer to water treatment plants, to collect and displace storm water to final treatment, or, finally, to drain excess water from soils [31].

An essential concept is embedded in the water energy food nexus and it is related to the energy, environmental and social costs of water services. The flux of any water flow, Q, is associated to a power, P, given by:

$$P = gQH$$

where H is called the total head and is related to the water pressure p as the sum of three terms:

$$H = z + \frac{p}{\gamma} + \frac{V^2}{2g}$$

with z geodetic elevation, $\frac{p}{\gamma}$ pressure head (the ratio between the pressure p and the specific weight of water, γ), $\frac{V^2}{2g}$ kinetic head of the water current, and V the mean flow velocity, related to the flow rate Q [54]. In addition, a source of water is characterized by a specific quality of the water body, expressed by the concentration of a number of chemicals and by the water temperature. Finally, any specific use - domestic, industrial or agricultural - requires a specific optimal level of pressure and of water quality. Therefore, the matching between water source and water use is not straightforward but has to be a part of a management strategy.

Sustainable Water Engineering. DOI: 10.1016/C2017-0-04301-X

An example of water resources management in the presence of multiple sources and multiple water uses is shown in Fig. 13.1. Any water source, represented by a different geodetic elevation, flow rate and water quality is used to fulfil the requirements of the different water uses.

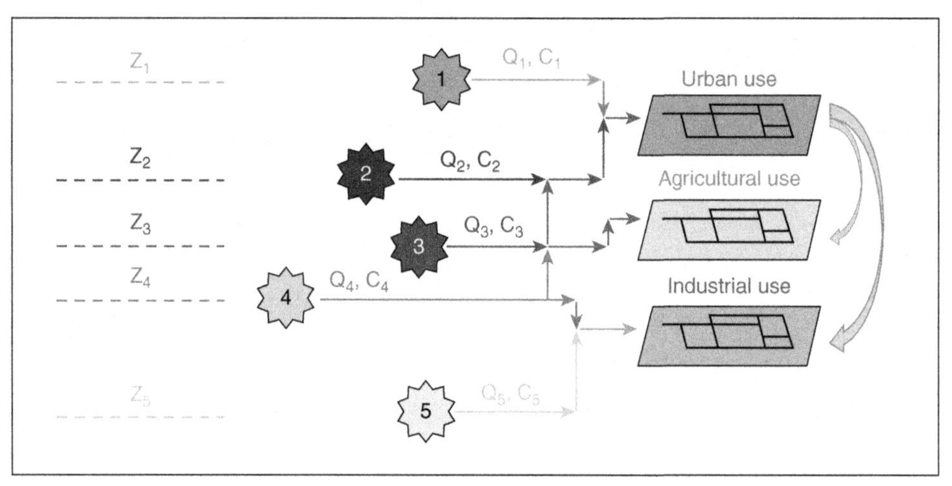

Fig. 13.1: Water resources management in presence of multiple water sources and multiple water uses.

Based on this depiction, it is clear that any water source, including drainage and waste waters, has a value, based on the flow rate, Q, the source elevation, z, and the water quality, C, and any water use has a cost based on total user demand, optimal pressure head distribution and minimum required water quality. As a matter of fact, for any end user the water has a footprint based on the sum of all direct and indirect costs connected with water extraction, treatment and transportation [21].

The relevance of the water energy food nexus comes from a perceived condition of scarcity, in terms of a lack of water and food, or in terms of a degraded environmental quality due to excessive energy use. The general tendency in the 20th century WSS design and management was in providing a technical solution to the first two issues. Nowadays, a great emphasis is given to the environmental concerns and this chapter gives a detailed description of the modern techniques for energy recovery in WSS and to the potential benefit of such techniques on WSS sustainability.

13.1.1 Energy issues in WSS

Hydraulic machines are able to increase, in a pumping station, the total head of the stream and, therefore, its power, or to decrease, in a hydropower station, the total head of the stream and its power [22,23,41]. These machines are used to transfer energy to/from the water from/to other forms of energy. These include electrical energy if the pump/turbine is connected to a motor/

generator, or mechanical energy if the pump/turbine is simply connected to a mechanical converter [3]. The efficiency is represented by the ratio between the produced and adsorbed power in a pump, or by the ratio between the adsorbed and produced power in a turbine. The largest efficiency is obtained for the design point of the machine, namely the best efficiency point (BEP). In BEP conditions the reliability of the machine is also the largest [25].

Fig. 13.2 shows a sketch view of a WSS which includes a water transmission (WT) and a water distribution system (WD). Generally, in the water transmission network, which is far from the end users, the flow rate and the pressure are fairly constant. On the contrary, the water distribution network, which delivers water directly to the end users, is affected by a large variability of flow rate and pressure due to the fluctuation in demand [12].

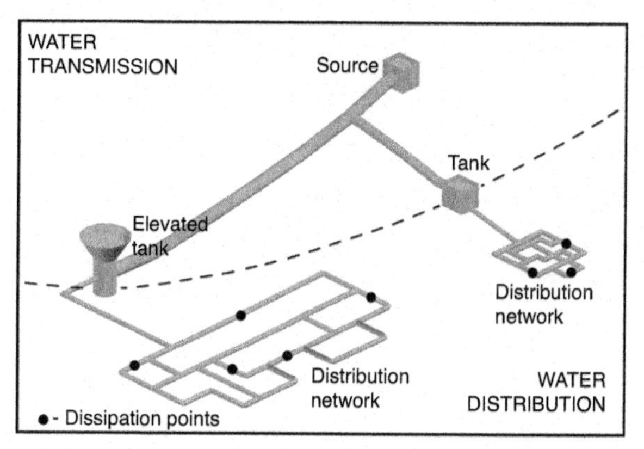

Fig. 13.2: Water transmission and water distribution in a WSS.
From: [14].

Pumping or hydro power stations in WT are frequently introduced in order to address the differences between the water source and water demand characteristics, as represented in Fig. 13.1. Considering the stability of the flow rate and head drop, traditional hydraulic machines are suitable for these plants, ensuring large system efficiencies.

In WD, the presence of a pumping station is commonly required in order to overcome the differences in the ground elevation of the different nodes of the network [27,36,43]. Thus, a water distribution network is usually divided into different water districts, with homogeneous elevation levels. Fig. 13.3 represents an example and shows the distribution of water districts for the city of Naples.

Nevertheless, the pressure head in a water district could vary by several tens of meters, and in a part of the water district an excess of energy could be present. This is due to the effect of pipe friction and the presence of ground level variations. An environmental cost has to be considered every time that a difference between water source and water demand characteristic exists. If the pressure or the flow rate of the water is higher than requested, the water stream embeds an amount of power that is not strictly necessary to the distribution [24,37].

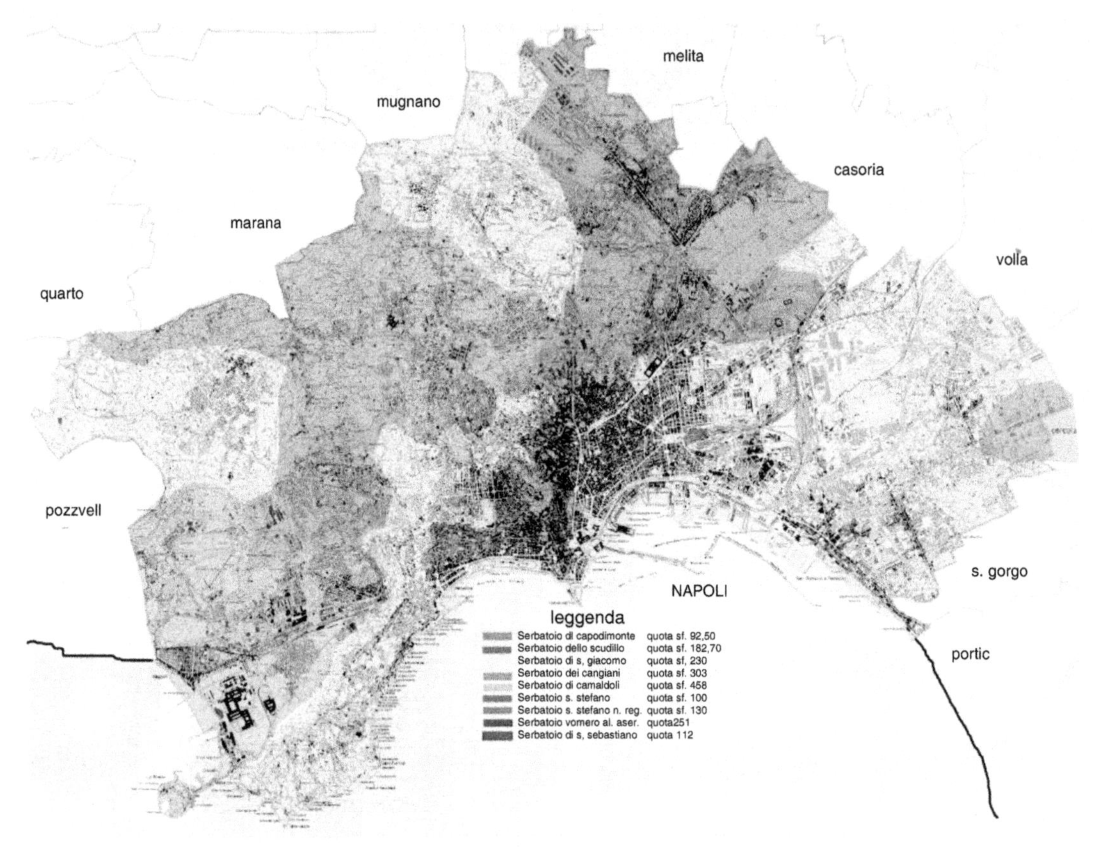

Fig. 13.3: Water districts in the WSS of Napoli, Italy.
Source: City of Naples Water Utility (ABC).

Pressure reducing valves (PRV) are commonly used to obtain a distribution of pressure head closer to the optimal pressure head distribution [32]. The pressure reducing strategy is generally considered as one of the best technical solutions for reducing water leakage. Thus they considerably reduce also the flow rate, saving considerable resource [51]. If the use of PRVs represents a suitable solution in order to reduce the use of water, they merely dissipate a part of the energy embedded in the water stream. This is a critical aspect for WSS if it is considered that 20 percent of the energy use of the EC electric motors is used for water transportation. Basically, the dissipation produced by any PRV could be recovered by a hydropower device, to both control the pressure and save a part of the energy by power production. Therefore, as well as pump eco-design, increasing the efficiency of the pumping systems and limiting the water demand of the pressure reducing strategy, the new trend in water resources management is in the use of techniques for energy harvesting within WSS [28,52].

13.2 Energy harvesting in WSS by micro and pico hydropower

The construction of hydropower stations along the water transmission system has a proven economical value, whether there is a small variability of flow rate and pressure drop, or large flow rates and available power – and these stations are widely diffused. The installed power could be as high as a few hundreds of kW. The additional cost for a grid connection will not strongly influence the payback period even in remote areas. In the water distribution networks, the viability of the installation of a conventional hydropower plant for energy harvesting in a WSS dramatically changes for a number of reasons:

1. The available power in each dissipation point is small and the connection to the grid is not assured [45].
2. The flow rate and the pressure drop present a strong daily variability reducing the efficiency of a traditional turbine [11].
3. The cost per kW of a traditional turbine is high at small power output capacities [42].
4. Every additional cost, other than the cost of the energy production device, significantly increases the payback period [12].
5. The income could be very small and is not fully attractive for WSS management [39].
6. Physical space limitations may be present which prevents the installation of turbines and control equipment [38].

All of these critical points result in a reduction in the viability of energy harvesting in WD for traditional turbines. Therefore, new strategies fully implemented for micro and pico hydropower generation in pressurized systems have been developed, based on a number of developments responding to each of the aforementioned critical points.

In particular:

1. On site and/or self-use of the energy has to be preferred where there is small available power at the dissipation point [1].
2. A regulation system has to be installed to sustain the power plant efficiency where there is a strong daily variability of flow rate and pressure drop [9].
3. This technology is generally not implemented when cost of the traditional turbine is high [15].
4. Additional costs are contained by new installation techniques, or by a proper choice of the dissipation point [53].
5. For the income to be attractive for the WSS management, a new policy has been developed in order to include the benefit of the energy harvesting in a more complete water footprint perspective [29].

These new statements are implemented in the guidance and support tool described in this chapter, which was developed as a part of the EU Interreg Atlantic Area Project: REDAWN.

The use of specific electromechanical devices - pumps working as turbines (PATs), granting low purchase and maintenance costs, is suggested in order to reduce the payback period of the hydropower plant costs. At first, a description of possible design conditions in a micro hydro power plant using PAT is given. Then, the steps to be performed for the design of a hydropower plants in WD will be described.

13.3 Micro and pico hydropower technologies for water supply systems

The economic viability of an energy recovery in WSSs is strictly connected to the potential use of energy at the dissipation point. This problem to be posed to the water manager can be categorized in four main questions:

1. Can the produced energy be possibly used at the site?
2. Is this self-consumption or use?
3. Is a connection to the electric grid possible at the site?
4. Is there capacity in the existing PRV housing unit to fit the hydropower plant components?

The answers to these questions are fundamental to evaluate the technical feasibility of the plant. In the industrial sector, a self-use of the produced energy is possible in all cases. In the urban WSSs, the self-use is often possible and a connection to the electric grid is frequent. In the irrigation WSSs, the potential for energy use of the produced energy is critical, the self-use is often possible but the connection to the electric grid can be impractical. The opportunities for energy use are also different depending on the hydropower plant size. When the power output is too large, the opportunities for on-site self-use of the produced energy becomes less, and the connection to the grid becomes the only viable solution. When the plant size is too small, with power output lower than 1–2 kW, a reasonablly small payback period becomes possible only in the presence of very low-cost technical solutions and reducing all the site costs, like piping, control systems, PAT housing, etc. Therefore, opportunities for these pico hydropower plants will come from further technological advancements.

Table 13.1 summarizes the potential energy uses by sector. A number of available technologies related to the use of PATs will be described in detail.

13.1.2 Consolidated PAT technology and design

PATs are becoming a mature technology for energy production in WSS. A large literature has been produced in the last fifteen years on the different aspects of their design. A comprehensive review can be found in the book "Pumps as Turbines: Fundamental and Applications" [4,5], together with the fundamental aspects of the design and a number of applications [6–8]. Two possible installation schemes of a PAT in a micro hydropower plant, together with the corresponding working conditions, are shown in Fig. 13.4. Both, hydraulic regulation (HR mode) or electric regulation (ER mode), are used in order to achieve a regulation of the power plant.

Table 13.1: Energy use for different sectors.

Sector	Use	Technology
Urban water	Energy supply to the grid	PAT based technology
	Water supply management	PAT based technology
	Water pumping	Turbo-pump
	Pico-hydro power	Serial/parallel regulation
	Monitoring	Ancillary technologies
	Demonstrator	All technologies
	Vehicle charging	PAT based technology
Wastewater drainage	All applications	Propeller turbines
Irrigation	Energy supply to the grid	PAT based technology
	Water pumping	PAT based technology/turbo-pump
	Water filtering	PAT based technology
	Monitoring	Ancillary technologies
Industry	Self-use of energy	PAT based technology, pico-turbines, propeller turbines, etc.
	Monitoring	Ancillary technologies
	Demonstrator	All technologies

In HR mode, where two valves (A and B) are respectively placed in series and in parallel to the PAT, two options are possible: i) whenever the network operating point is above the PAT characteristic curve, the bypass valve B (see Fig. 13.4) will be fully closed and an additional dissipation will be performed by control valve A; ii) whenever the network operating point is below the PAT characteristic curve, control valve A will be fully opened, and control valve B will be partially opened. In HR, only a part of the stream power will be converted because of the difference between the network operating conditions, and the PAT efficiency. In ER mode, the rotational speed of the impeller will be changed, by acting on the motor with a variable speed drive. In this case, the PAT working conditions will match the network operating conditions and all of the stream power will be converted, except for a component of electromechanical losses in the PAT

The best design solution can be found using the variable operating strategy (VOS), ensuring the maximum plant effectiveness, based on the daily distribution of the network operating conditions and on available PAT performance curves. When considering the hydropower plant payback period, the choice between HR and ER is determined by the balance between the larger cost of the ER equipment compared to HR, and the increment of the income from energy selling determined by the larger amount of energy converted in the ER mode compared to the HR mode.

13.1.3 Other PAT technologies

13.1.2.1 The PAT-pump turbocharger

A PAT-pump turbocharger (P&P plant) is an energy recovery device that couples the hydraulic components of two pumps [3]. The first works in direct pump mode and the second in reverse mode (PAT). This configuration can be used when it is possible to move two WSS

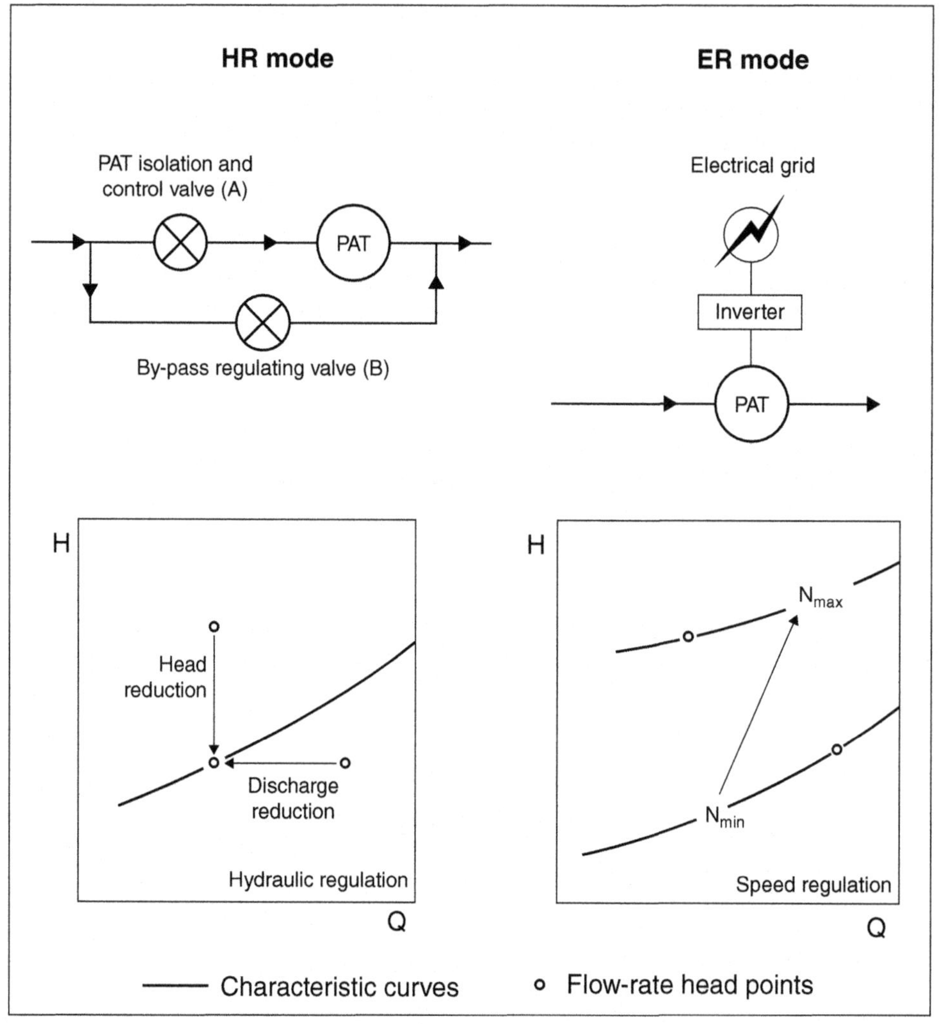

Fig. 13.4: Hydropower plants regulation modes.
From: [10].

nodes belonging to different water systems at the same point. One node is a dissipation node of a district where excess pressure has to be dissipated (district 1), while the other node is a pumping node for a district where an increase in stream pressure is required (district 2). This situation can be observed in urban networks presenting a large variation of ground levels, as in the topography represented in Fig. 13.3. In this case it may happen that a small DMA at a higher terrain elevation has to be supplied by a part of a main water stream having an excess pressure. Considering that residual excess energy is frequently present in any part of urban WSSs at the lower terrain levels, another application of P&P could be for wastewater pumping to the treatment plant.

A simplified scheme of the P&P plant is shown in Fig. 13.5. The operation of a P&P plant depends on the mutual behaviour of the PAT and the pump: the PAT acts like a motor for the pump, while the pump is the PAT load. Thus, the two devices interact, the PAT influencing the pump and vice-versa. By solving the equation representating the P&P plant for a multistage pump and a single stage PAT, the ratio between the pumping head (Hp) and the turbined head (Ht) can be calculated as a function of the ratio between the pumping discharge (Qp) and the turbined discharge (Qt). These functions are plotted in Fig. 13.6 for different numbers of stages. As shown in Fig. 13.6, as the number of stages of the pump increases, a smaller amount of discharge can be pumped at higher elevations.

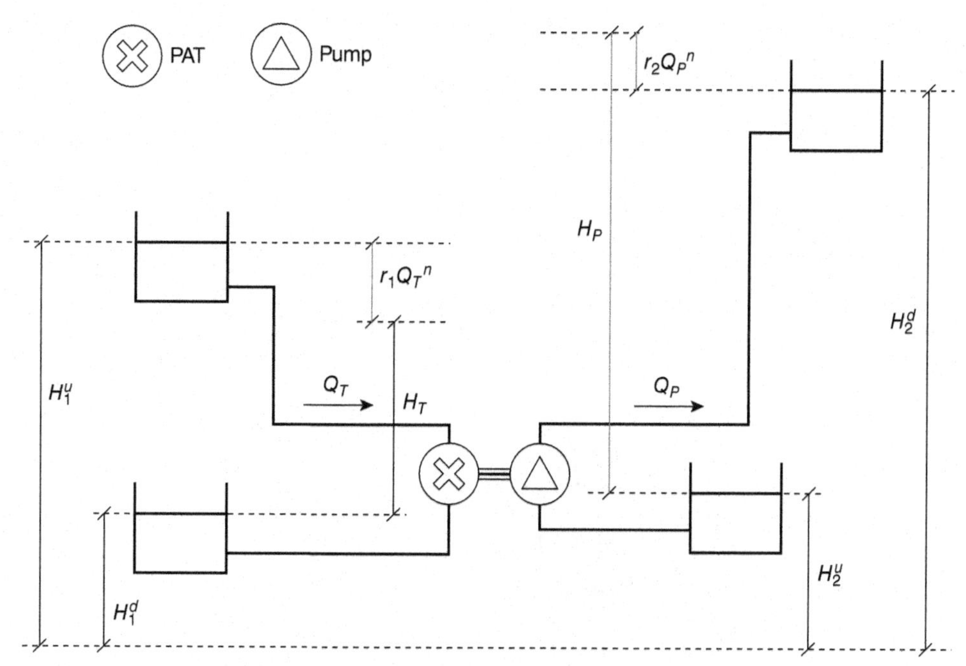

Fig. 13.5: Simplified scheme of the P&P plant.
From: [3].

13.1.2.2 Pico-hydropower solutions

The daily distribution of flow rate and head drop in the peripheral areas of a WSS is diverse. The peak coefficient in users' demand is also very high, leading to a variability of operating conditions that cannot be accommodated by the working conditions of a single PAT. In addition, the available stream power is very low and the cost of regulation valves, PAT housing, control systems, etc. has to be limited in order to obtain a viable economic solution for energy recovery. Therefore, a number of simplified technologies have been proposed based on different design approaches, to achieve low cost of the equipment.

The GreenValve [26,35] is a modification of an existing ball valve with the insertion of a vertical axis impeller in the body of the sphere (Fig. 13.7). The device has been tested under different

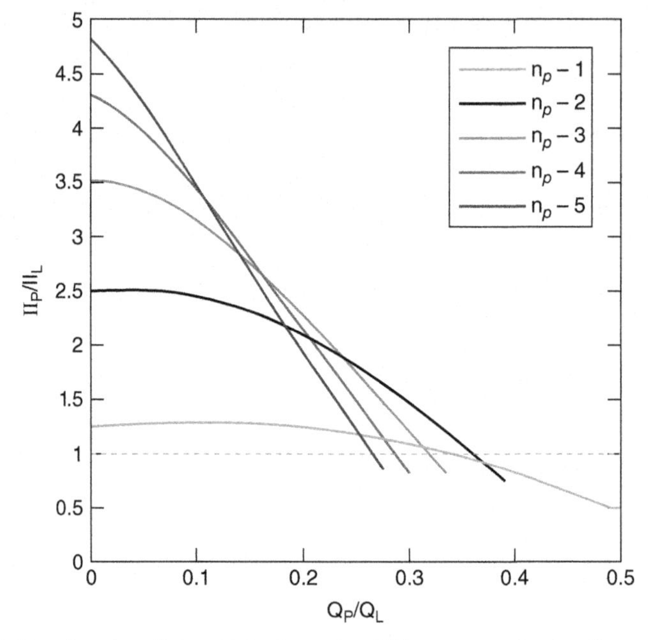

Fig. 13.6: Head ratio of the P&P plant for different numbers of pump stages.
From: [3].

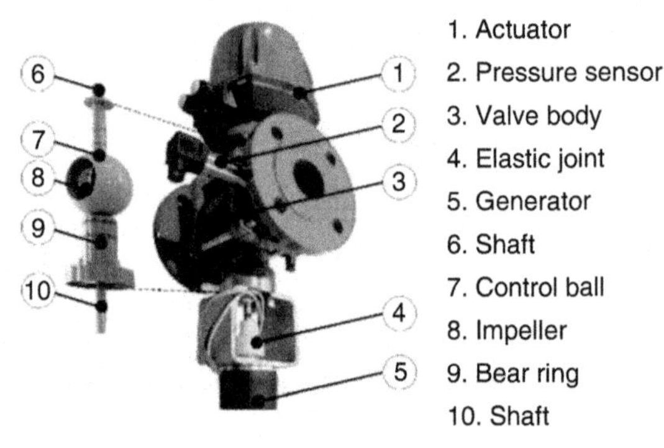

Fig. 13.7: Anatomy of the GreenValve.
From: [34].

flow conditions in order to evaluate the valve flow coefficient and the energy recovery efficiency. The results for impeller B are plotted in Fig. 13.8. An efficiency close to the maximum efficiency (assessed as 14.5 percent) was obtained for a 40 percent valve opening. The installation of such a device can be a valid choice to use the produced energy for the self-regulation of the valve.

A tubular propeller turbine [49] has also been proposed in order to contain the installation costs. The installation scheme is plotted in Fig. 13.9. The 5-blade propeller was used moving a variable speed generator. The energy converter and the performance curves were experimentally developed, and the turbine efficiency seems very promising for a small scale device (Fig. 13.10).

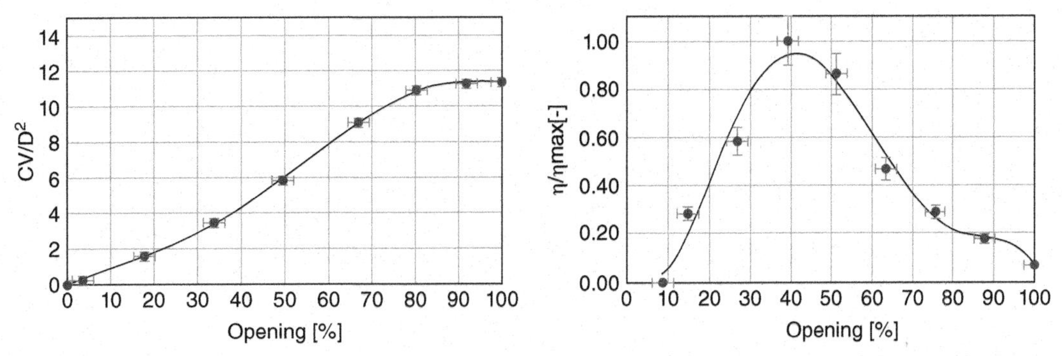

Fig. 13.8: Green valve: discharge coefficient and efficiency as functions of the opening.
From: Ferrarese and Malavasi, n.d..

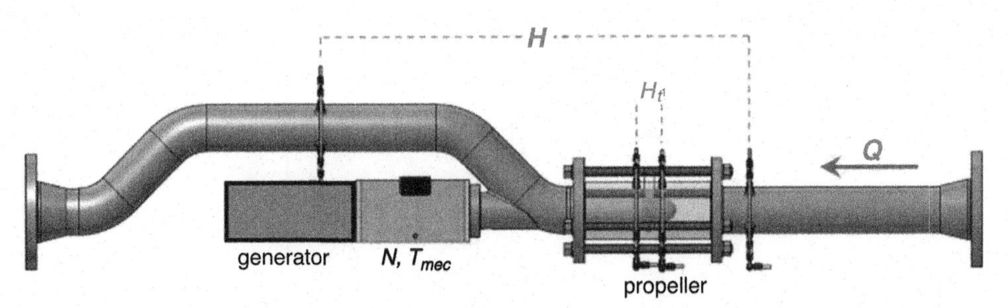

Fig. 13.9: Installation scheme of a tubular propeller turbine.
From: [47].

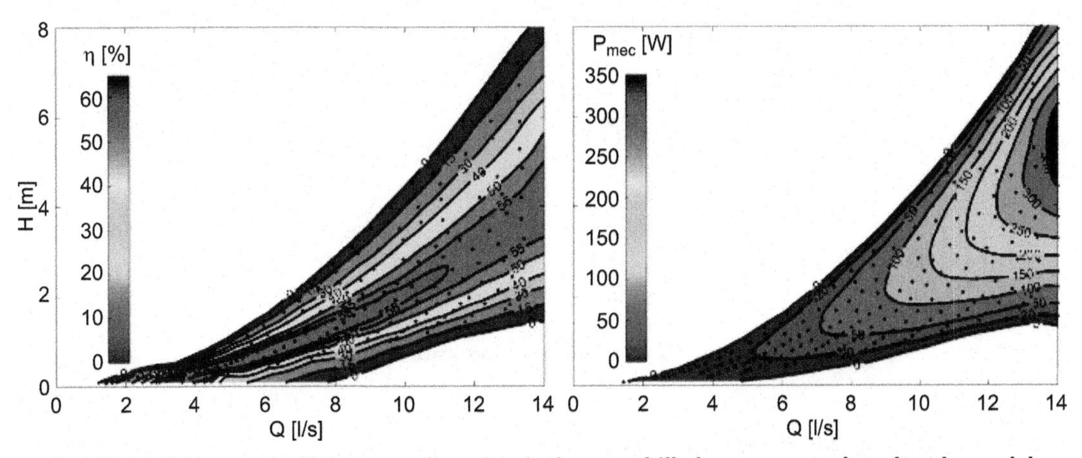

Fig. 13.10: Measured efficiency and mechanical power hill charts on a reduced scale model.
From: [47].

In addition to a high maximum efficiency of 64 percent, the variable speed control allows an electric regulation with acceptable efficiency in a wide range of WSS operating conditions.

More recently, a simplified regulation scheme to be used in pico hydropower plants has been proposed. The new low-cost plant is provided with two PATs, that can work as single, series or parallel (SSP) turbines, and three on/off valves, whose synchronous operations regulate the flows and the head loss [11]. The installation scheme of an SSP is shown in Fig. 13.11, and its working conditions are shown in Fig. 13.12. By switching the three valves on or off, three operating conditions can be obtained in order to arrive at the best network working conditions. For lower power flow rates and larger heads drops, the two PATs will work in series, in the intermediate range of flow rate and head drop, a single PAT will produce h energy, and, finally, for high flow rate and low head drops the two PATs will work in parallel. As observed in Fig. 13.12, this regulation scheme will not allow the exact matching of operating and working conditions, but highly reduces the payback period, by using low cost on/off valves and constant speed PAT operation.

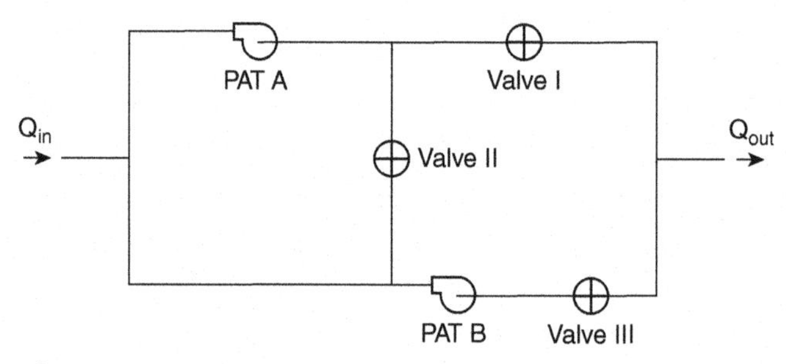

Fig. 13.11: New PAT regulation based on the SSP mode.

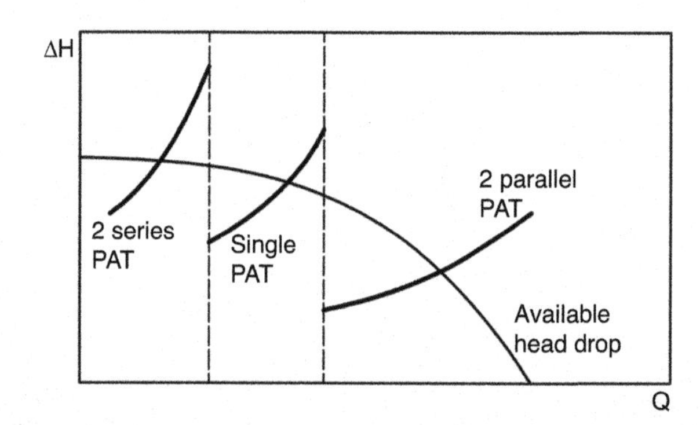

Fig. 13.12: Operating and working conditions of the SSP plant.

13.1.4 Identifying potential power plant locations

The problem of the identification of the hydropower potential in WD at a national or regional scale has been discussed in specific projects. This allows an estimation of the total potential coming from the exploitation of the energy harvesting in WSS in some countries in the EUs Atlantic Area. The preliminary results are shown in Table 13.2.

Table 13.2: Preliminary results of hydropower potential for energy harvesting in WSS for different sectors

	Theoretical potential of MHP at existing sites (kW)				
Country	Drinking Water	Waste water	Irrigation	Process industry	Total
Ireland	883	301	na	22	1206
Portugal	1805	50	26	10	1891
(Funchal WDS in Madeira Island)	(184)				
Spain	5	63	2969	na	3037
UK	34,243	na	na	na	34,243
Total	36,936	414	2995	32	40,377

Source: Resource Assessment on Micro Hydropower Potential in the Atlantic Area. REDAWN Project Work Package 4 Preliminary Report.

The exact localization and design of the hydropower plant for each potential site require detailed knowledge of the network and the territory. The final solution comes from a cross check between different aspects [2,32,48]:

i. the most effective distribution of the dissipation nodes in the network, determined by optimizing the hydraulic constraints, such as satisfaction of user demand, reduction of leakage, optimal pressure head level, etc.;
ii. lower plant payback period;
iii. best opportunities for energy use;
iv. highest social impact etc.

It is a common experience in most European countries that the administrative procedure connected with the authorization to produce energy in the network and to release energy to the grid is onerous, time consuming, and costly. In addition, in many countries, the feed in tariff does not include energy production in pressurized systems [46]. Therefore, our first concern will be a territorial analysis of the energy use looking for an opportunity for the local use of energy. In many cases, self-consumption of the produced energy is the best solution. Obviously, matching the production sites to the local energy demand by users could be more complicated. For example, in rural areas connected with irrigation networks, the opportunities for a rational use of the produced energy are few and the visibility of these plants are small [18,20]. The location problem of power plants is therefore different for fresh water WSS, irrigation, or industrial outflow. The large differences in working conditions and opportunities for energy harvesting in different sectors are considered in the next sections.

13.4 Energy recovery: applications and case studies

13.4.1 Localization for fresh water supply

Numerical methods make it possible to predict the effect of replacing one or more pressure reducing valves (PRVs) with PATs in a network [40]. The sustainable management of WSSs can be achieved by monitoring and controlling losses, which requires a deep knowledge of the network, in particular its characteristics and operating mode [33]. In recent years, significant research effort has been committed to proposing a general numerical model for the optimal location of the energy recovery node using different criteria, including leakage reduction and power plant production [16,32,48]. A complete analysis of these contributions is out of the scope of this paper, but a comprehensive review can be found in Fecarotta and McNabola [24]. At the present state of knowledge, these general models can handle only small hydraulic networks. As a result, in most of the real-life situations more simplified methods for PAT localization and sizing are implemented.

Nevertheless, the problem of the optimal location of PATs within a water distribution network is still an unresolved problem. Thus, the purpose of this case study is to analyse the energy recovery potential of the water distribution system of Funchal (Portugal) through the replacement of PRVs by PATs, ensuring both an adequate pressure management and valuable energy savings and showing the hydropower potential of a real distribution system. Only a section of the Funchal water distribution system was studied – the pilot zone selected for the study of leakage reduction carried out by the Funchal water industry – which comprises of roughly 40 percent of the entire municipality of Funchal and corresponds to the area of influence of the reservoirs of Terça, S. Martinho, Penteada, Ribeira Grande and Nazaré.

In fig. 13.13A, the pipeline and the terrain elevation distribution are reported, showing very high ground slopes in large parts of the network. The water demand distribution is reported in fig. 13.13B. In order to reduce water leakage, a pressure reducing strategy is in place, by the use of a large number of PRVs. The current pressure head distribution in the network at 13.00 H is plotted in Fig. 13.14A.

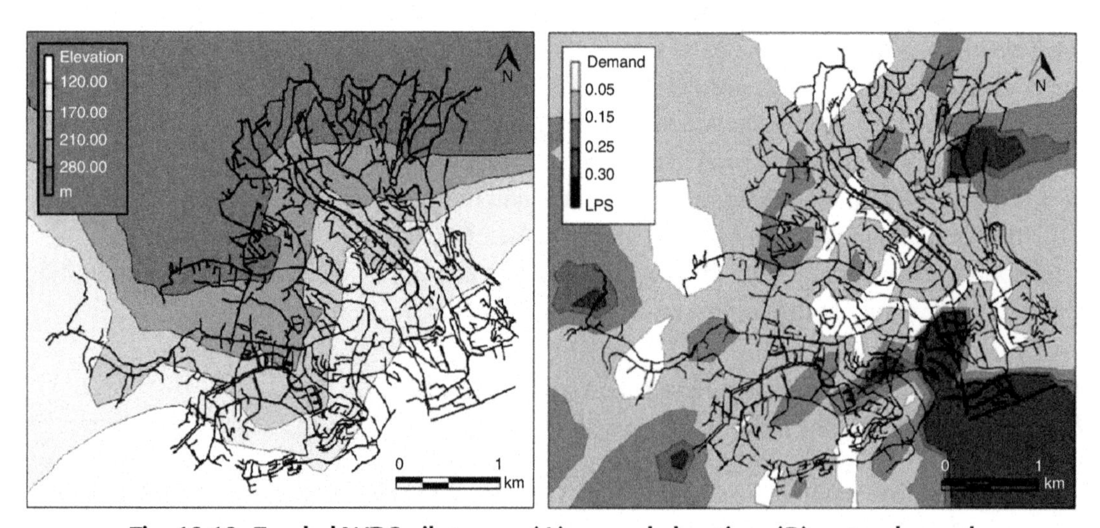

Fig. 13.13: Funchal WDS pilot zone: (A) ground elevation, (B) water demand
From: [44].

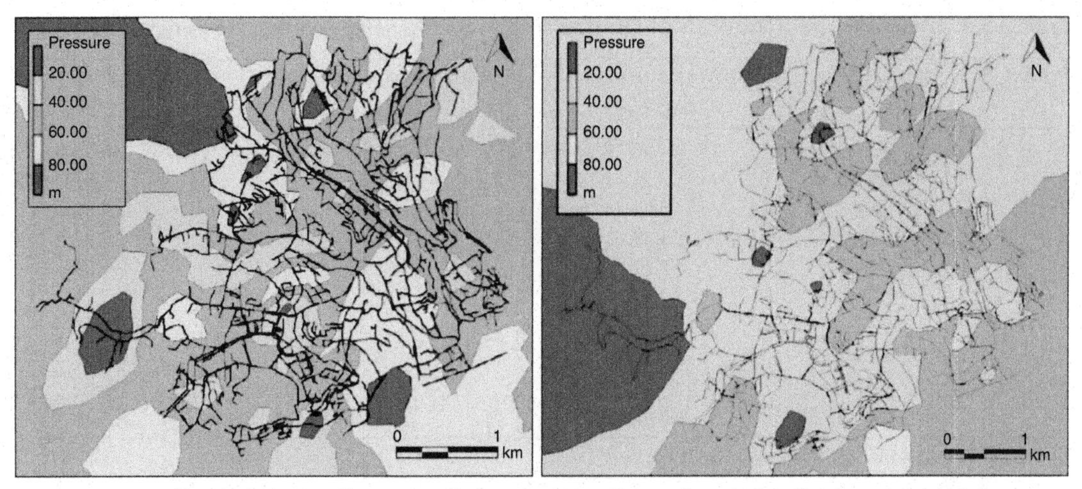

Fig. 13.14: Funchal WDS pilot zone: (A) present pressure head distribution, and (B) future pressure control effect.
From: [44].

At the time of the study, almost 70 percent of the total water entering the water network of Funchal was lost within the system, mostly as a result of inadequate pressure regulation. This poses a serious threat to the environment and presents a significant economic impact for the involved water utilities. However, the Funchal water industry estimated these losses may be reduced to 15 percent by 2033 if the correct measures are taken, particularly the creation of district metered areas (DMAs) in the water distribution network and the correct placement of PRVs. After these modifications, the installation of micro hydro power plants within the system through the use of PATs can further improve the efficiency of the system.

The first required step to evaluate the possibility of recovering energy within the network was to investigate the best possible locations for the implementation of PATs, in order to make sure these would provide the optimal conditions for energy production. Since RSS proposed the implementation of 50 different PRVs, there are 50 available locations. Despite the considerable number of possible locations, only 10 of these PRVs were selected for the implementation of PATs since that would provide sufficient data to conclude the network's potential for energy recovery. This required a selection process which consisted of a basic preassessment of each PRVs potential for the implementation of a PAT, wherein the head drop defined for the PRV was multiplied by the flow rate, and the 10 PRVs which provided the highest values ($Q \times H$) were selected for further analysis. The 10 selected PRVs are displayed in Table 13.3 (values correspond to peak discharge conditions).

Considering the predictable reduction of water losses which results from the leakage control program, the PRV head drop values display a tendency to increase and the flow rate values tend to decrease throughout the years. For this reason, to be able to perform an energy

Table 13.3: Hydraulic characteristics of the pressure reducing valves within the Funchal network.

Valve (–)	Q (L/s)	Upstream pressure head (m)	Downstream pressure head (m)	Head drop (m)
1	32.20	44.75	23.71	21.04
2	20.66	52.27	22.35	29.92
3	25.68	40.10	17.80	22.30
4	43.12	53.36	29.80	23.56
5	24.65	55.66	31.10	24.56
6	152.41	44.26	25.55	18.71
7	24.12	49.70	25.25	24.45
8	22.47	48.60	24.30	24.30
9	26.89	47.37	24.28	23.09
10	82.28	44.15	21.60	22.55

Source: RSS/IST study within REDAWN project for WP4.

recovery analysis up to 2033, a network mathematical model was required to calculate these values for each PRV and each year. This implied the creation of a specific model for each year, which was accomplished using the Funchal water network model provided by the network manager as a starting point. This provided model corresponded to the 2018 situation and included every water loss control measure previously defined, such as the implementation of newly installed PRVs and the establishment of new DMAs. After concluding the PRV selection process and the calculation of hydraulic models, specific PATs had to be selected for each PRV. The PAT has been chosen among a database of 29 machines, for which the hydraulic behaviour was known. By associating the PAT curves to the selected PRVs in the network model, the energy production could be calculated for every year from 2018 to 2033.

There are several different PAT configurations which can be applied to generate power, namely: hydraulic regulation (HR); electric regulation (ER); hydraulic and electric regulation (HER); no regulation (NR). All these configurations can provide interesting energetic results depending on the water distribution network conditions, presenting a great economic benefit for water utilities if implemented correctly. However, in the case study of the Funchal water network, only ER operating modes were tested and analysed.

In order to calculate the variation of the hydraulic characteristics with the rotational speed for each PAT for both ER modes, the affinity law of turbomachines was applied to the respective nominal rotational speed curve.

After defining the configuration for both modes of operation, the most appropriate PATs had to be selected for each PRV location. However, having in consideration that there were 29 different PAT models to choose from and apply to the 10 PRV locations, a digital algorithm has been developed to simulate every possible scenario throughout the years, enabling the development of a fast and detailed testing analysis. In fact, with the aid of this program, all the previously mentioned PATs could be tested for every PRV location and every rotational speed, not to mention the different modes of operation. The range of rotational speeds analysed by the digital algorithm correspond to the minimum and maximum rotational speed limits at which the considered PATs could operate, from 770 rpm to 3500 rpm. Summed up briefly, the nominal speed curves of each PAT have been used as reference curves and alternative rotational speed curves have been obtained through dimensional analysis, treating them as distinct PATs. Finally, the energy production results relative to each possible scenario for every year along with the total accumulated energy from 2018 to 2033 has been calculated.

To study the feasibility of the energy recovery solutions proposed in the WDS of Funchal, an economic analysis was performed for each PAT application separately, considering each mode of operation and storage method. This was essential to accurately determine the most advantageous investment and exclude unprofitable PAT applications which could undermine the economic potential of the project. To evaluate the economic benefit, two scenarios have been considered, differing for the use of energy, i.e. local use and grid connection. For each solution the adopted electricity selling prices are 0.057 €/kWh for the grid connection and 0.098 €/kWh for the local use, and the discount rate considered is 7.5 percent. Table 13.4 shows the figures of the optimal investment for the water distribution system, such as where the investment costs along with the total produced energy in 15 years, the internal return rate (IRR), the 15 years net present value (NPV), the ratio between the benefits and the costs (B/C) and the payback period.

Table 13.4: Economic analysis of the optimal hydropower investment in Madeira WSS for grid or stand-alone energy use. The total energy refers to a period of 15 years.

Energy connection	Grid	Local
Number of PATs	2	3
Total investment (€)	24,648	38,851
Total energy production (MWh)	1716	2063
IRR (percent)	23 percent	34 percent
NPV (percent)	24,855	69,203
B/C (-)	2.0	2.8
Payback period [years]	5	4

Source: RSS/IST study within REDAWN project for WP4.

13.4.2 Localization for irrigation

The identification of potential sites for hydropower generation in the irrigation WSS, is in some respects more complicated, both from an economic and a design point of view [30]. The network operation is focused on the irrigation season, and this factor will increase the payback period of the power plants. In addition, a limited number of flow control points are available in the network, constituting limiting factors in the design for the stochastic nature of users' demand. An ad hoc methodology for hydropower plant design in irrigation networks has been recently developed and applied to a real case study in Southern Spain. The monthly behaviour of the network was based on a statistical experiment known as a Bernoulli Experiment, by considering the flow rates as a discrete random variable, and defining the probability of their occurrence each month. A pre-set design hypothesis concerns the knowledge of the maximum number "n" of hydrants that are simultaneously open. The flowchart of the methodology is reported in Fig. 13.15 [18]. After a preliminary identification of the network areas with excess pressure (Step 1), the hydropower design is performed for all the nodes of the network falling in those areas, determining the open hydrant probability in terms of possible combination of open hydrants (Step 2), the monthly distribution of flow rate in the network branches (Step 3), then, the PAT was designed by a simplified VOS procedure (Step 4), and, finally the economic viability of the proposed design was determined (Step 5). The PAT with the minimum payback period was then selected.

The innovation of the methodology can be explained by looking at Fig. 13.16. The flow rate in pre-selected excess pressure point (EPP) of the network is a random variable based on the distribution of the open and closed hydrants of the network. An EPP was considered for each branch in which the existing overpressure was at least 3 m under pre-set design hypothesis aforementioned. The overpressure required was fixed to a minimum of 3 m in a Q-H space to locate the BEP conditions for a large set of PATs presented by Novara et al. [42], where the head could reach up to 3 m for certain flows. By varying this distribution, it is possible to explore, by means of a hydraulic model, the range of operating conditions in the EPP during the irrigation season. Then, different PATs were tested to define their theoretical BEP (Q_{BEP}, H_{BEP}). Hydraulic regulation was used to control the working conditions of the PAT, due to the variability in demand and the significant flow fluctuations, and since it is generally more efficient and less expensive than electric regulation [13]. The proposed scheme can be observed in Fig. 13.17. VOS is used for the PAT selection and the economic revenue of a power plant located in the EPP is evaluated (see Fig. 13.18). This network includes 162 hydrants, whose levels vary between 47 m.O.D. and 97 m.O.D, and irrigates about 920 ha. The hydraulic infrastructure is composed of pipes with diameters between 150 mm and 800 mm and a pumping station located at 86 m.O.D. with a total power installed of 810 kW. The network was designed to supply 1.2 L s^{-1} ha^{-1} on demand, which means that water is continuously available 24 h per day, for 100 percent of simultaneity (all hydrants open). Table 13.5 show the results of the design procedure reported, and the payback period of hydro power plants realized in each of the EPPs are compared for the 2017 irrigation season.

This methodology was validated comparing the predicted results obtained in sector II of the Canal del Zujar Irrigation District (CZID) located in southwestern Spain, to the actual

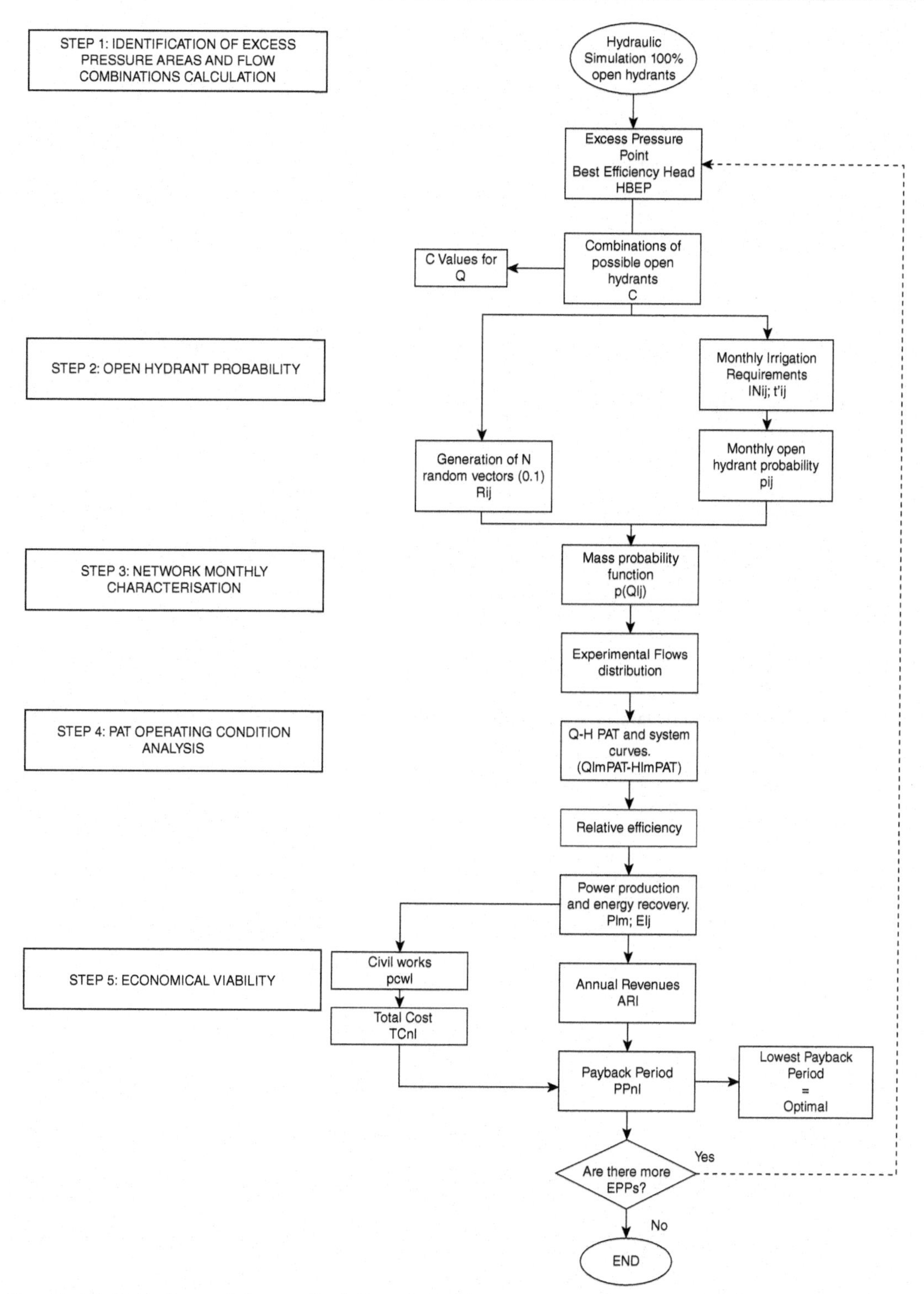

Fig. 13.15: Flowchart of the methodology for hydro power plant selection in irrigation networks. From: [18].

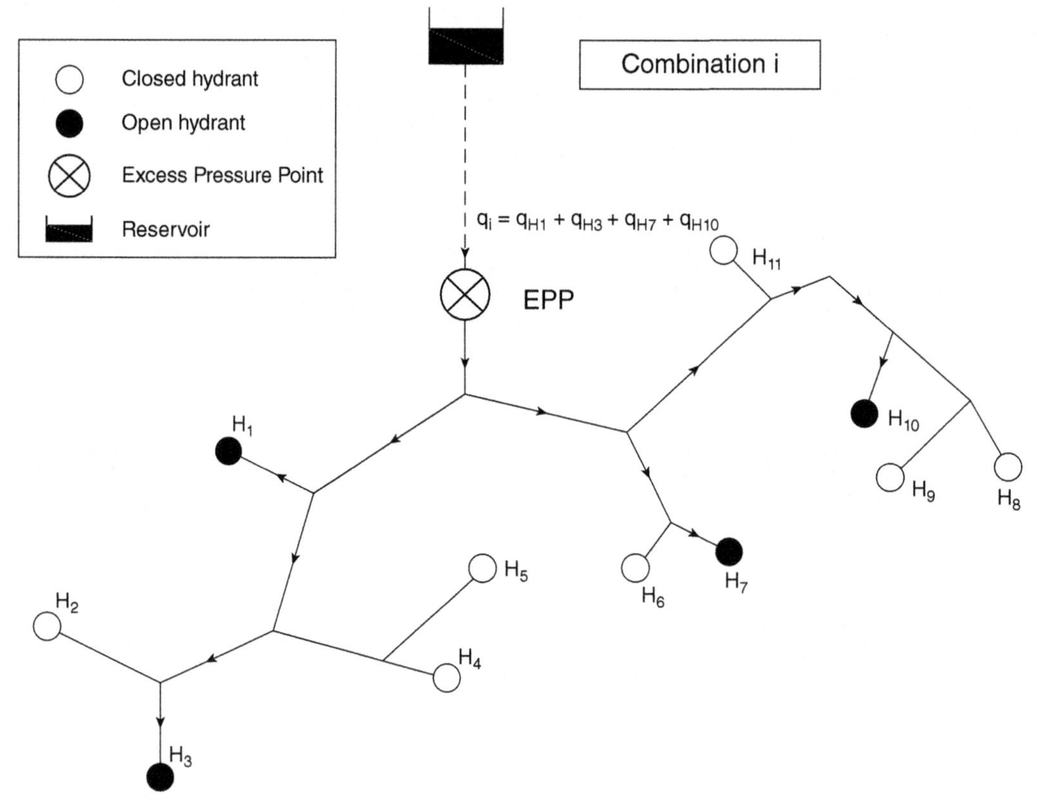

Fig. 13.16: Random combination of downstream open hydrant for a general EPP.
From: [18].

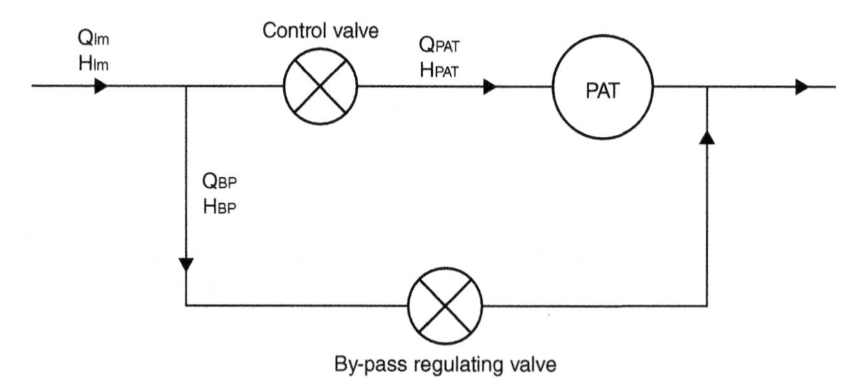

Fig. 13.17: Typical PAT installation scheme for hydraulic regulation.
From: [18].

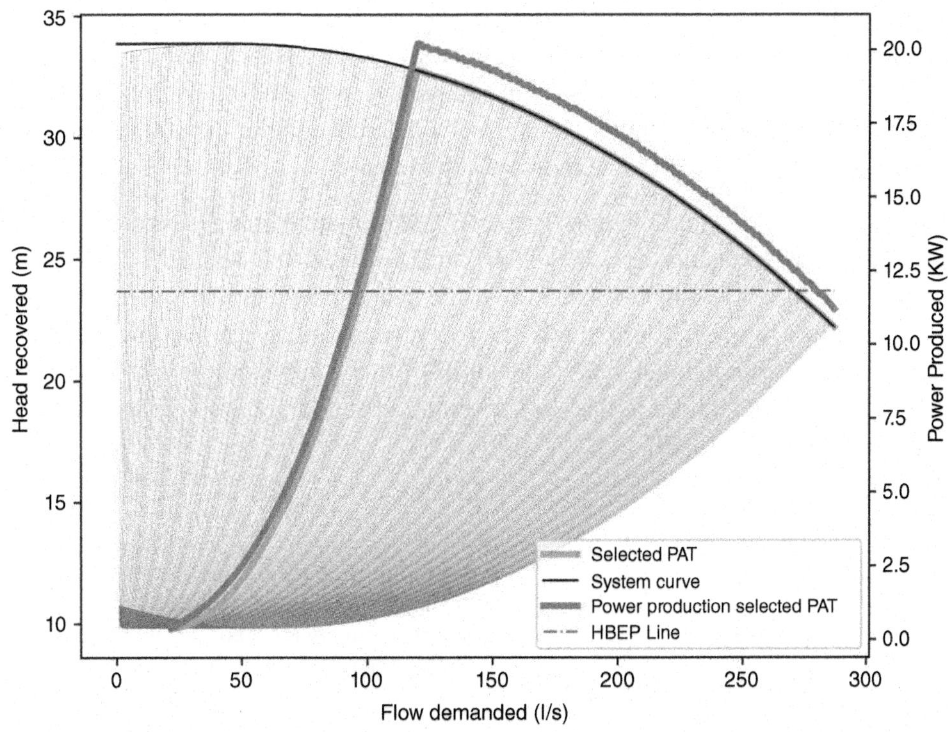

Fig. 13.18: 205 experimental PATs tested in an EPP using VOS. The selected PAT flow-head recovered curve and its power generation depending on the flow, represented by the thickest lines.
From: [17].

Table 13.5: Summary of the results of the methodology for hydro power plant selection in irrigation networks.

EPP	Optimal scenario	BEP flow (L/s)	BEP power (kW)	Polar pairs	Cost (€)	Energy (MWh)	PP (years)
1	2,743,236	88	9.1	1	16,438	40.8	3.5
2	13	39	2.9	2	12,339	6.9	15.8
3	631,784	54	5.8	2	14,207	29.5	4.2
4	30,122,847	46	4.5	2	13,352	11.1	10.6
5	1,051,433	36	2.8	2	12,278	5.6	19.4
Total	-	-	**25.1**	-	**68,614**	**93.9**	**6.4**

Source: [18].

measurements recorded by a telemetry system during the 2015 irrigation season. This network irrigates 2,691 ha, where the water is supplied by 196 hydrants. Their levels vary between 250 m.O.D. and 285 m.O.D. The pipe diameters vary between 80 mm and 1000 mm. The network was designed to supply 1.2 L s^{-1} ha^{-1} on demand under the 100 percent simultaneity hypothesis. The service pressure required at hydrant level is 35 m.

An average coefficient of determination (R^2) of 0.80 was obtained for the prediction of flow distribution during the irrigation season, which varied between 0.71 and 0.92 depending on the EPP. When the PATs obtained from predicted and actual conditions were both tested under real conditions, an overall difference of 0.1 percent of energy recovery was achieved (see Fig. 13.19) [19]. Simulating the PATs obtained from the methodology in the hydraulic model, the pressure was reduced to the pressure required at hydrant level for almost the entire irrigation season (see Fig. 13.20).

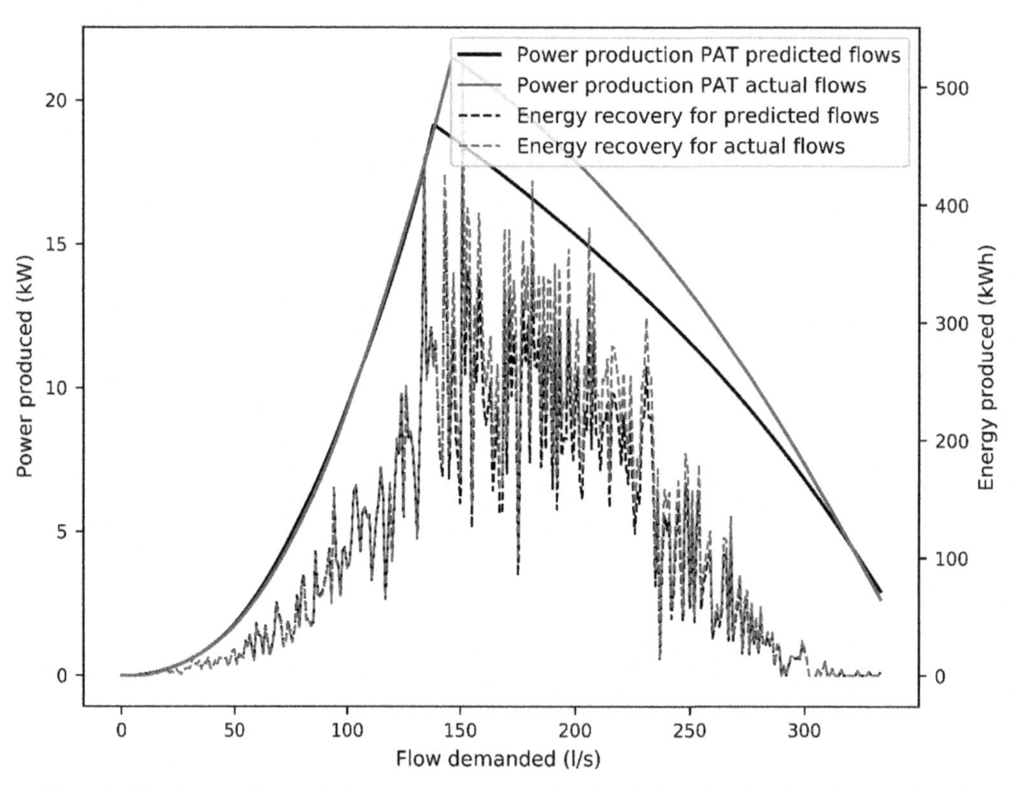

Fig. 13.19: Comparison of the energy recovery by the PATs obtained under predicted and actual conditions.
From: [17].

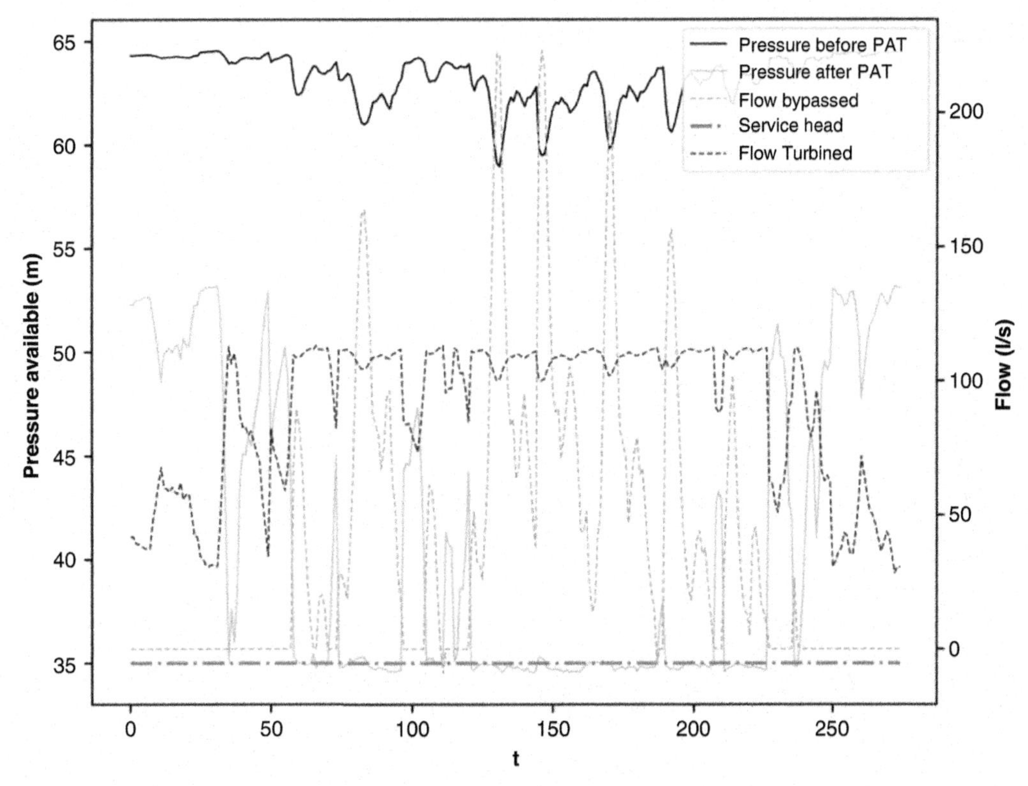

Fig. 13.20: Analysis of the actual pressure available and its variation when the predicted PAT was simulated, flow turbined and flow bypassed.
From: [17].

13.5 Conclusions

The optimization of the energy use in water transportation has become a key topic for water managers in recent years. Any energy recovery strategy has to be analysed in the context of the water energy food nexus for the strict relations existing between these three elements. A promising instrument in any energy recovery strategy is the use of mini or pico hydropower plants, converting electric or mechanical energy from the in-stream power drop of the pressure reducing valves.

The research of the last years produced great advances in the design of these plants on many aspects, like modelling, technology, electro-mechanical equipment and regulation. Additionally, the pay-back and potential use of the produced energy represent in the most of the cases, the main element in the assessment of the plant viability.

References

[1] M. Arriaga, Pump as turbine – a pico-hydro alternative in Lao People's Democratic Republic, Renew. Energy 35 (5) (2010) 1109–1115.

[2] A. Bolognesi, C. Bragalli, C. Lenzi, S. Artina, Energy efficiency optimization in water distribution systems, Procedia Eng. (2014).

[3] A. Carravetta, L. Antipodi, U. Golia, O. Fecarotta, Energy saving in a water supply network by coupling a pump and a Pump As Turbine (PAT) in a turbopump, Water (Switzerland) 9 (1) (2017).

[4] A. Carravetta, S. Derakhshan Houreh, H.M. Ramos, Introduction. Springer Tracts in Mechanical Engineering, 2018.

[5] A. Carravetta, S. Derakhshan Houreh, H.M. Ramos, Reverse pump theory. Springer Tracts in Mechanical Engineering, 2018.

[6] A. Carravetta, S. Derakhshan Houreh, H.M. Ramos, Industrial aspects of PAT design improvement. Springer Tracts in Mechanical Engineering, 2018.

[7] A. Carravetta, S. Derakhshan Houreh, H.M. Ramos, PAT selection. Springer Tracts in Mechanical Engineering, 2018.

[8] A. Carravetta, S. Derakhshan Houreh, H.M. Ramos, PAT control systems. Springer Tracts in Mechanical Engineering, 2018.

[9] A. Carravetta, O. Fecarotta, G. Del Giudice, H. Ramos, Energy recovery in water systems by PATs: a comparisons among the different installation schemes, Procedia Eng. (2014) 275–284.

[10] A. Carravetta, O. Fecarotta, R. Martino, L. Antipodi, PAT efficiency variation with design parameters, Procedia Eng. (2014) 285–291.

[11] A. Carravetta, O. Fecarotta, H.M. Ramos, A new low-cost installation scheme of PATs for pico-hydropower to recover energy in residential areas, Renew. Energy (2018f) .

[12] A. Carravetta, O. Fecarotta, M. Sinagra, T. Tucciarelli, Cost-benefit analysis for hydropower production in water distribution networks by a pump as turbine, J. Water Resour. Plan. Manag. 140 (6) (2014c) 04014002.

[13] A. Carravetta, G. del Giudice, O. Fecarotta, H. Ramos, PAT design strategy for energy recovery in water distribution networks by electrical regulation, Energies 6 (1) (2013) 411–424.

[14] A. Carravetta, G. Del Giudice, O. Fecarotta, H.M. Ramos, Energy production in water distribution networks: a PAT design strategy, Water Resour. Manag. 26 (13) (2012) 3947–3959.

[15] L. Corcoran, P. Coughlan, A. McNabola, Energy recovery potential using micro hydropower in water supply networks in the UK and Ireland, Water Sci. Technol. Water Supply (2013).

[16] L. Corcoran, A. Mcnabola, P. Coughlan, Optimization of water distribution networks for combined hydropower energy recovery and leakage reduction, J. Water Resour. Plan. Manag. A.S.C.E. 142 (2) (2015) 1–8.

[17] M. Crespo Chacón, "Pump-as-turbines for hydropower energy recovery from on-demand irrigation networks: flow fluctuation characterisation, energy potential extrapolation, and real-scale implementation, Trinity College Dublin (2020).

[18] C. Chacón, D. M., J.A. R., J.G. Morillo, A. McNabola, Pump-as-turbine selection methodology for energy recovery in irrigation networks: minimising the payback period, Water (Switzerland) (2019).

[19] M. Crespo Chacón, J.A. Rodríguez Díaz, J. García Morillo, A. McNabola, Hydropower energy recovery in irrigation networks: validation of a methodology for flow prediction and pump as turbine selection, Renew. Energy 147 (2020) 1728–1738.

[20] A.H. Elbatran, O.B. Yaakob, Y.M. Ahmed, H.M. Shabara, Operation, performance and economic analysis of low head micro-hydropower turbines for rural and remote areas: a review, Renew. Sust. Energ. Rev. (2015).

[21] O. Fecarotta, C. Aricò, A. Carravetta, R. Martino, H.M. Ramos, Hydropower potential in water distribution networks: pressure Control by PATs, Water Resources Management, Springer, Netherlands 29 (3), (2014), pp. 699–714.

[22] O. Fecarotta, A. Carravetta, M. Morani, R. Padulano, Optimal pump scheduling for urban drainage under variable flow conditions, Resources (2018).

[23] O. Fecarotta, R. Martino, M.C. Morani, Wastewater pump control under mechanical wear, Water (2019).
[24] O. Fecarotta, A. McNabola, Optimal location of pump as turbines (PATs) in water distribution networks to recover energy and reduce leakage, Water Resour. Manag. 31 (15) (2017) 5043–5059.
[25] O. Fecarotta, H.M. Ramos, S. Derakhshan, G. Del Giudice, A. Carravetta, Fine tuning of a PAT hydropower plant in a water supply network to improve the system effectiveness, J. Water Resour. Plan. Manag. 144 (8) (2018b) .
[26] G. Ferrarese, S. Malavasi, Perspectives of Water Distribution Networks with the GreenValve System, Water (Switzerland) 12 (6) (2020) 1579.
[27] N. Fontana, M. Giugni, D. Portolano, Losses reduction and energy production in water-distribution networks, J. Water Resour. Plan. Manag. 138 (3) (2012) 237–244.
[28] J. Gallagher, I.M. Harris, A.J. Packwood, A. McNabola, A.P. Williams, A strategic assessment of micro-hydropower in the UK and Irish water industry: identifying technical and economic constraints, Renewable Energy, Elsevier 81, (2015), 808–815.
[29] J. Gallagher, D. Styles, A. McNabola, A.P. Williams, Making green technology greener: achieving a balance between carbon and resource savings through ecodesign in hydropower systems, Resources, Conservation and Recycling, Elsevier B.V. 105 (2015), 11–17.
[30] J. García Morillo, A. McNabola, E. Camacho, P. Montesinos, J.A. Rodríguez Díaz, Hydro-power energy recovery in pressurized irrigation networks: a case study of an Irrigation District in the South of Spain, Agric. Water Manag. (2018).
[31] N.C. Ghosh, Integrated water resources management, Sustain. Utiliz. Nat. Resour. (2017).
[32] M. Giugni, N. Fontana, A. Ranucci, Optimal location of PRVs and turbines in water distribution systems, J. Water Resour. Plan. Manag. 140 (9) (2014) 06014004.
[33] I.E. Karadirek, S. Kara, G. Yilmaz, A. Muhammetoglu, H. Muhammetoglu, Implementation of hydraulic modelling for water-loss reduction through pressure management, Water Resour. Manag. 26 (9) (2012) 2555–2568.
[34] S. Malavasi, G. Ferrarese, GreenValve: regolazione e recupero di energia per un acquedotto intelligente, VIII Seminario Tecnologie e Strumenti Innovativi per le Infrastrutture Idrauliche "TeSI," Napoli, 2019.
[35] S. Malavasi, M.M.A. Rossi, G. Ferrarese, GreenValve: hydrodynamics and applications of the control valve for energy harvesting, Urban Water J. (2018).
[36] M.C. Morani, A. Carravetta, O. Fecarotta, A. McNabola, Energy transfer from the freshwater to the wastewater network using a PAT-equipped turbopump, Water (Switzerland) (2020).
[37] M.C. Morani, A. Carravetta, G. Del Giudice, A. McNabola, O. Fecarotta, A comparison of energy recovery by PATs against direct variable speed pumping in water distribution networks, fluids 3 (41) (2018).
[38] K.H. Motwani, S.V. Jain, R.N. Patel, Cost analysis of pump as turbine for pico hydropower plants - a case Study, Procedia Eng. (2013).
[39] A. Muhammetoglu, I.E. Karadirek, O. Ozen, H. Muhammetoglu, Full-scale PAT application for energy production and pressure reduction in a water distribution network, J. Water Resour. Plan. Manag. 143 (8) (2017) 04017040.
[40] H. Nautiyal, Varun, A. Kumar, Reverse running pumps analytical, experimental and computational study: a review, Renew. Sust. Energ. Rev. 14 (7) (2010) 2059–2067.
[41] N.T. Nazarov, Hydraulic machines. Turbines and pumps, Hydrotech. Constr. (1979).
[42] D. Novara, A. Carravetta, A. McNabola, H.M. Ramos, Cost model for pumps as turbines in run-of-river and in-pipe microhydropower applications, J. Water Resour. Plan. Manag. 145 (5) (2019) 04019012.
[43] F. De Paola, N. Fontana, E. Galdiero, M. Giugni, G.S. Degli Uberti, M. Vitaletti, Optimal design of district metered areas in water distribution networks, Procedia Eng. (2014).
[44] M.A. Perdigão, Implementation of Pump-as-Turbines as Energy Recovery Solutions within Water Distribution and Supply Systems - Case Study of Funchal Water Distribution System Pilot Zone, Instituto Superior Tecnico, Lisbon, 2018.
[45] M. Pérez-Sánchez, F.J. Sánchez-Romero, H.M. Ramos, P.A. López-Jiménez, Energy recovery in existing water networks: towards greater sustainability, Water (Switzerland) 9 (2) (2017).

[46] M. Ringel, Fostering the use of renewable energies in the European Union: the race between feed-in tariffs and green certificates, Renew. Energy (2006).
[47] I. Samora, "Optimization of low-head hydropower recovery in water supply networks, EPFL, Lausanne (2016).
[48] I. Samora, M.J. Franca, A.J. Schleiss, H.M. Ramos, Simulated annealing in optimization of energy production in a water supply network, Water Resources Management, Springer, 30 (2016a), pp. 1533–1547.
[49] I. Samora, V. Hasmatuchi, C. Münch-Alligné, M.J. Franca, A.J. Schleiss, H.M. Ramos, Experimental characterization of a five blade tubular propeller turbine for pipe inline installation, Renew. Energy (2016b).
[50] P.K. Swamee, A.K. Sharma, Design of water supply pipe networks, Des. Water Supply Pipe Netw. (2008).
[51] K. Vairavamoorthy, J. Lumbers, Leakage reduction in water distribution systems: optimal valve control, J. Hydraulic Eng. 124 (11) (1998) 1146–1154.
[52] M.R.N. Vilanova, J.A.P. Balestieri, Hydropower recovery in water supply systems: models and case study, Energ. Conv. Manag. 84 (2014).
[53] A.A. Williams, Pumps as turbines for low cost micro hydro power, Renew. Energy 9 (1–4) (1996) 1227–1234.
[54] E.B. Wylie, V.L. Streeter, Fluid Transient in Systems. Prentice-Hall, Prentice Hall, Englewood Cliffs, NJ, (1993), 7632.

CHAPTER 14

Soft water engineering design approaches for urban revitalization in post-soviet housing estates

Carme Machí Castañer[a], Daniel Jato-Espino[b,*]

[a]*University of São Paulo, Butantã, São PauloS,P, 05508-080, Brazil,* [b]*GITECO Research Group, Universidad de Cantabria, Av. de los Castros 44, Santander, 39005, Spain*

* *Corresponding author.*

14.1 Soviet-era and post-regime urban planning and construction processes

Soviet urban planning began in the early 20th century in Russia and spread to other Soviet Union countries, resulting in a large number of districts undergoing environmental decline [1]. Urban form was defined by social principles, establishing a regulatory basis of urban planning where the *neighbourhood* played a fundamental role [18,24]. The government invested in new industrial areas and housing estates located in peripheral areas of the city, which provided basic residential needs to enable the area to remain independent of historical urban centres [25]. There was a shift towards an industrialized construction process for the new housing, where the estate played a multiple role, i.e. as builder, investor and architect. This affected the final quality of the projects due to their inability to deal with local physical characteristics, such as topography, natural waterscapes, etc., leading to environmental issues and lack of adequate urban infrastructure [16,29].

There was a massive push in housing construction during the 1950s and 60s in the Soviet Union which significantly intensified in the 1980s, when many complex buildings of concrete-slab were constructed (see Fig. 14.1 using Petrzalka (Bratislava, Slovakia) as an example). During this development, Slovakia's concrete-slab technology represented 93.5 percent of the whole housing production, delivering 1,261,000 apartments between 1971 and 1980. The new dwellings were supposed to provide minimum housing standards with construction style determining the aesthetics of the block, i.e. their dimensions, the finishing and colour and shapes of their balconies, loggias and entrances etc.

In the late 1950s, there was a significant increase in housing stock, culminating in the emergence of model plans mainly based on prefabricated buildings [26]. Fig. 14.1A shows the large area that was occupied by a prefabricated development planned during the socialist

Sustainable Water Engineering. DOI: 10.1016/C2017-0-04301-X

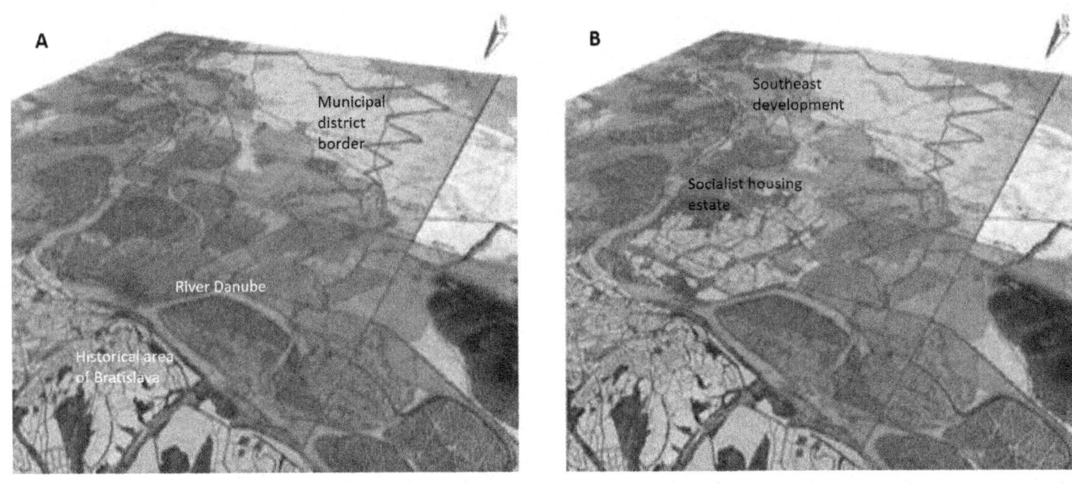

Fig. 14.1: Land use change and the spread of urbanisation in the area of Petrzalka (Bratislava, Slovakia) (a) 1910 (b) 2010.

regime for the south area of Bratislava (Slovakia). Today, there is a south-east development trend (Fig. 14.1B) associated with the north-south axes that links the Petrzalka's Socialist Housing Estate to the Bratislava's historical core. Residential blocks and clusters of blocks became the basic housing planning unit. Block height increased from the usual maximum of five stories, whilst public investment in urban facilities decreased with time [33]. As a result, some of the 'satellite neighbourhoods' that emerged in socialist cities, e.g. Moscow or St. Petersburg, were far from meeting social, cultural and everyday needs. Planning for these new districts differed from the traditional approach, since the spatial arrangement was defined by the neighbourhood unit instead of initially setting out a gridded road layout [12]. This inverted planning sequence consisted of first placing the buildings, and then allowing for the road geometry to accommodate communal housing. The road alignment was the result of the aggregation of neighbourhood units and generally had curved paths. As a consequence of this new planning approach, the traditional function of the street became traffic movement rather social interaction, with the interior of the housing cluster supporting people's everyday activities [23].

Additionally, cities experienced a decentralization process and the new socialist developments were made up of inner rings of suburban zones [2]. Several soviet housing estates in St. Petersburg were developed during the mid-20th century across surrounding green belt native forest. These neighbourhoods were planned around the idea of the *micro district* as a fundamental unit, popularly known as 'microraion', which served as primary structural elements. Collections of microraion, or clusters, made up whole districts. Within the microraion, each building was functionally and morphologically independent of the whole and each apartment was integrated into a neighbourhood community environment.

Socialist architecture was thus not an expression of an individual architect's conceptions, its focus was on social problems at the community level, where the block was an interdependent residential element. Architecture had to be economically efficient, dealing with minimum housing area but ensuring basic common standards. With the development of industrialized construction processes, residential blocks were built faster using prefabricated elements [32]. As shown in Fig. 14.2, the internal open space of the neighbourhood was enclosed by residential buildings, reflecting socialist living standards of a community living under principles of equality. A range of diverse microraion units were developed following the same concept of shared communal space in many socialist cities. Some examples of those are shown in the left images of Fig. 14.2 for the largest socialist housing estates erected in Bratislava's peripheral areas. Fig. 14.3 represents the spatial and functional distribution of urban-uses for Petrzalka's microraion units.

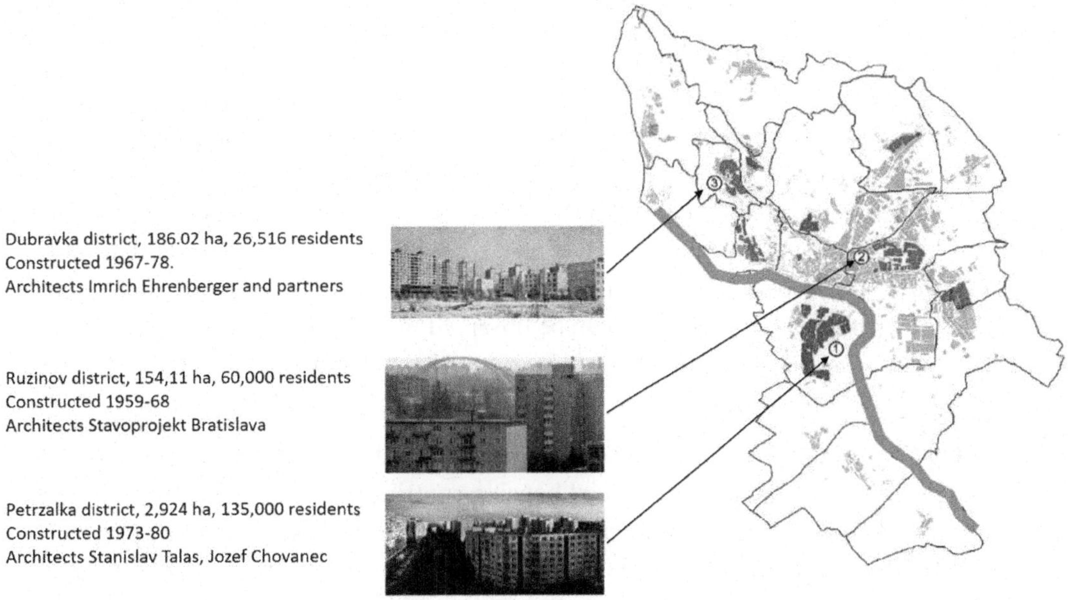

Dubravka district, 186.02 ha, 26,516 residents
Constructed 1967-78.
Architects Imrich Ehrenberger and partners

Ruzinov district, 154,11 ha, 60,000 residents
Constructed 1959-68
Architects Stavoprojekt Bratislava

Petrzalka district, 2,924 ha, 135,000 residents
Constructed 1973-80
Architects Stanislav Talas, Jozef Chovanec

Fig. 14.2: Post-socialist housing estates around Bratislava's city centre coloured by their construction period, highlighting the most populated ones, including Petrzalka's district.

These districts generally had a mono-functional character due to the lack of urban services. Fig. 14.3 shows the school units in the middle of the neighbourhoods, performing as the educational infrastructure at a community scale.

Under the socialist planning regime, all open space was owned by the government, therefore it was considered public and accessible [10], this helped control people's activities, leading to certain standards of social behaviour [9]. However, the publically-accessible green central areas associated with the microraions were small and isolated, reflecting the lack of public

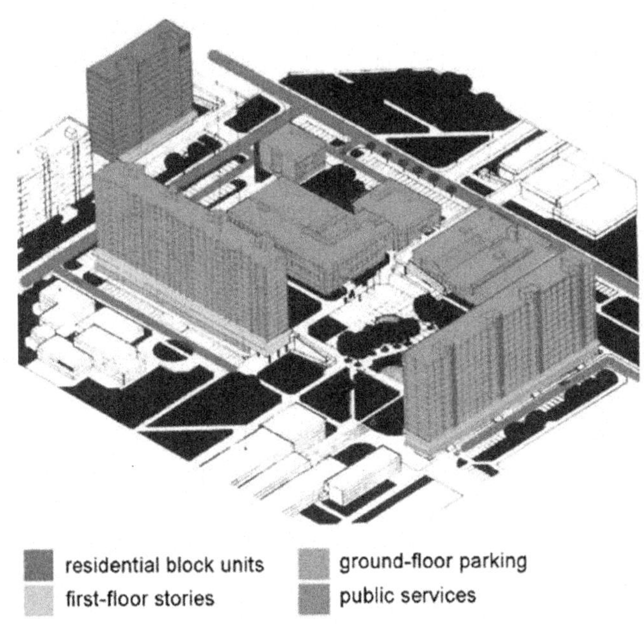

Fig. 14.3: Axionometric projection of a microraion in Petrzalka showing different uses of space.

investment, which limited the socialist housing estates' socio-environmental ambitions. Residential blocks were generally constructed first due to the urgent housing shortage, but other facilities, such as green space were not always carried out, or completed as proposed in the initial plan. An example of this situation can be observed in both St. Petersburg and Moscow, where satellite post-socialist districts turned into dormitory towns with a strong mono-functional characteristics [28].

After the fall of the Regime in former Soviet countries, a decline in affordable housing stock due to increasing costs and the lack of social housing contributed to a polarization of society. City centres became wealthy residential areas, whereas the middle-to-low income populations settled in peripheral districts, which generally corresponded to the post-socialist housing estates. This fact reinforced a polycentric system, and in addition to social polarization and decentralization, deindustrialization of former soviet countries affected relationships between where the workers lived and the old industrial areas [14,16,27]. As an example, St. Petersburg developed decentralization with a system of satellite suburbs (see Fig. 14.4), which lacked services and adequate transport to the city centre. As in the case of Bratislava, the polycentric model, highlighted in Fig. 14.2, changed urban growth and development, favouring a dramatic increase in Petrzalka's dependence on the city centre [14,16]. The most populated socialist housing estates around Bratislava's city centre are Petrzalka, Ruzinov and Dúbravka, with 135, 60 and 25 thousand inhabitants, respectively (Fig. 14.2). All these districts were built after the 1960s, coinciding with the period when massive pre-fabricated construction processes were intensified in soviet cities.

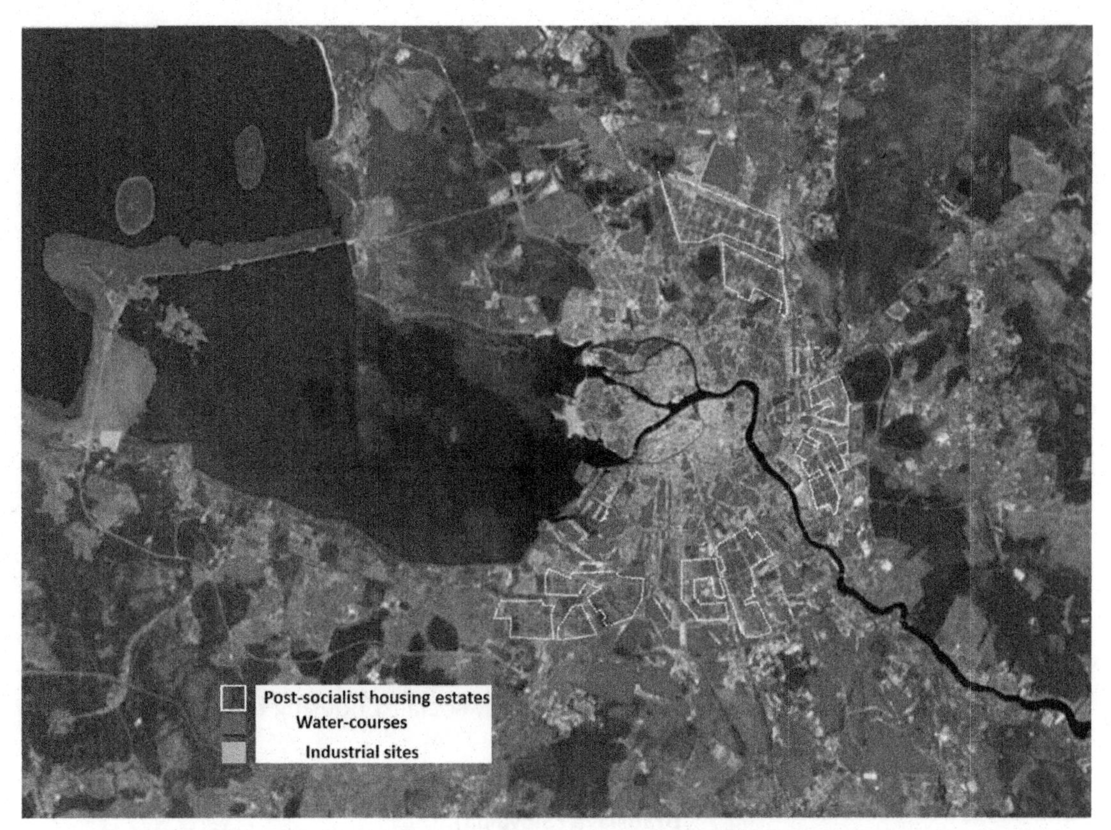

Fig. 14.4: Industrial sites and some of the main post-socialist housing estates around St. Petersburg's city centre (Russia).

Whilst residential buildings in Petrzalka are characterized by low provision of facilities, the original plan for the district included a central boulevard providing a large public space in the area, with pathways along existing water-courses, urban services and a new tram line to connect with the city centre. The proposed boulevard could have improved the living and environmental conditions for residents, but was never executed [11].

In Bratislava, Soviet housing estates were also considered very poor in terms of environmental quality, with lack of public transport systems, dubious safety and social cohesion. Interviews with local residents by Ira [16] assessed these post-socialist micro-districts, finding them to have the poorest aesthetic value and quality of housing. Overall, participants considered the historical core as being more suitable for sustainable development, whilst Petrzalka itself was regarded as unsustainable by 46 percent of the respondents. Fig. 14.5 represents Bratislava's sustainable development suitability for each of the city's areas. Soviet neighbourhoods were found to be less sustainable in terms of prospects for development, with their population representing more than 50 percent of the total in the city. Petrzalka's microdistricts

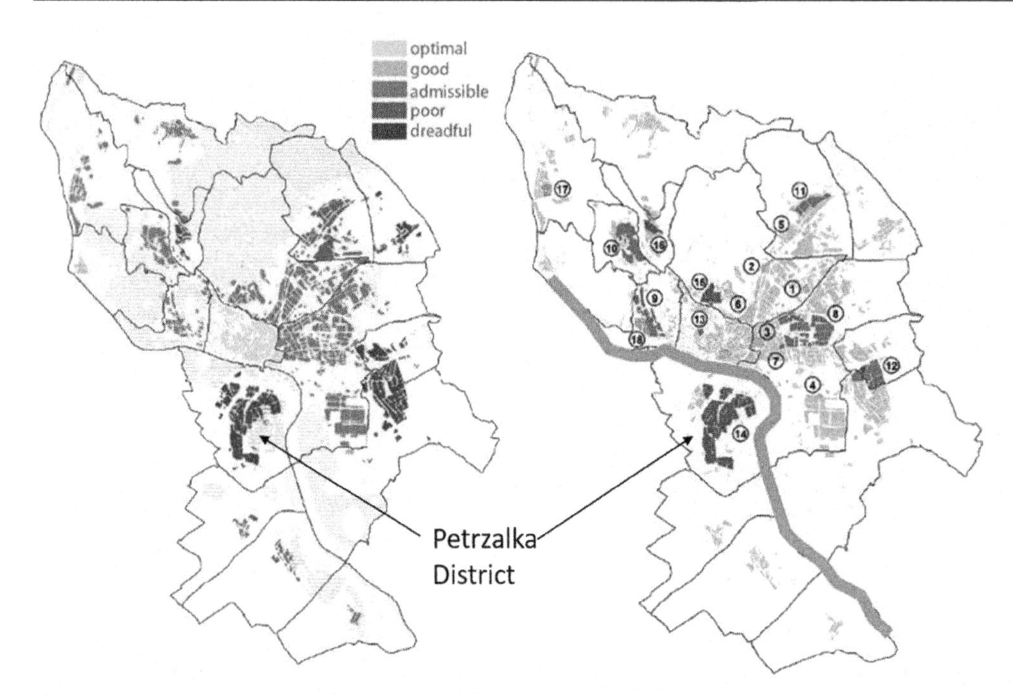

Perceptions of favourable development (the sustainability perspective)

Post-socialist housing estates in Bratislava

Fig. 14.5: Perception of favourable development -the sustainable perspective- in Bratislava (Slovakia).

have been rated as the most unsustainable zones in Bratislava (Fig. 14.5). Green areas within microraions were rarely used by locals and remained fragmented and isolated.

At the microraions scale, additional problems included the large distances between buildings which were not adequately managed, inner spaces with little functional role, resulting in little public amenity use [10]. Fig. 14.6 shows the open or green and built spatial configuration features. The full image of this configuration is the result of the aggregation of the different versions of the microraion nuclear unit, as shown at the top right diagram in Fig. 14.6. This leads to a most likely homogenous urban landscape. Each nuclear unit is of approximately 50 thousand square metres, projected with a small educational service within the residential building, as depicted in the bottom diagram in Fig. 14.6. Every unit is assembled alongside two others, composing residential pockets surrounded by the street systems. The lack of urban centrality influences some aspects of the urbanity and liveability of the place. Urban centralities play a key role in providing a wider range of public services and enhancing the interaction among neighbours from different parts of a given urbanized area [13,15,17]. Also, the urban landscape of Petrzalka is homogeneous due to the lack of hierarchy between buildings, which may undermine residents' sense of belonging to the place.

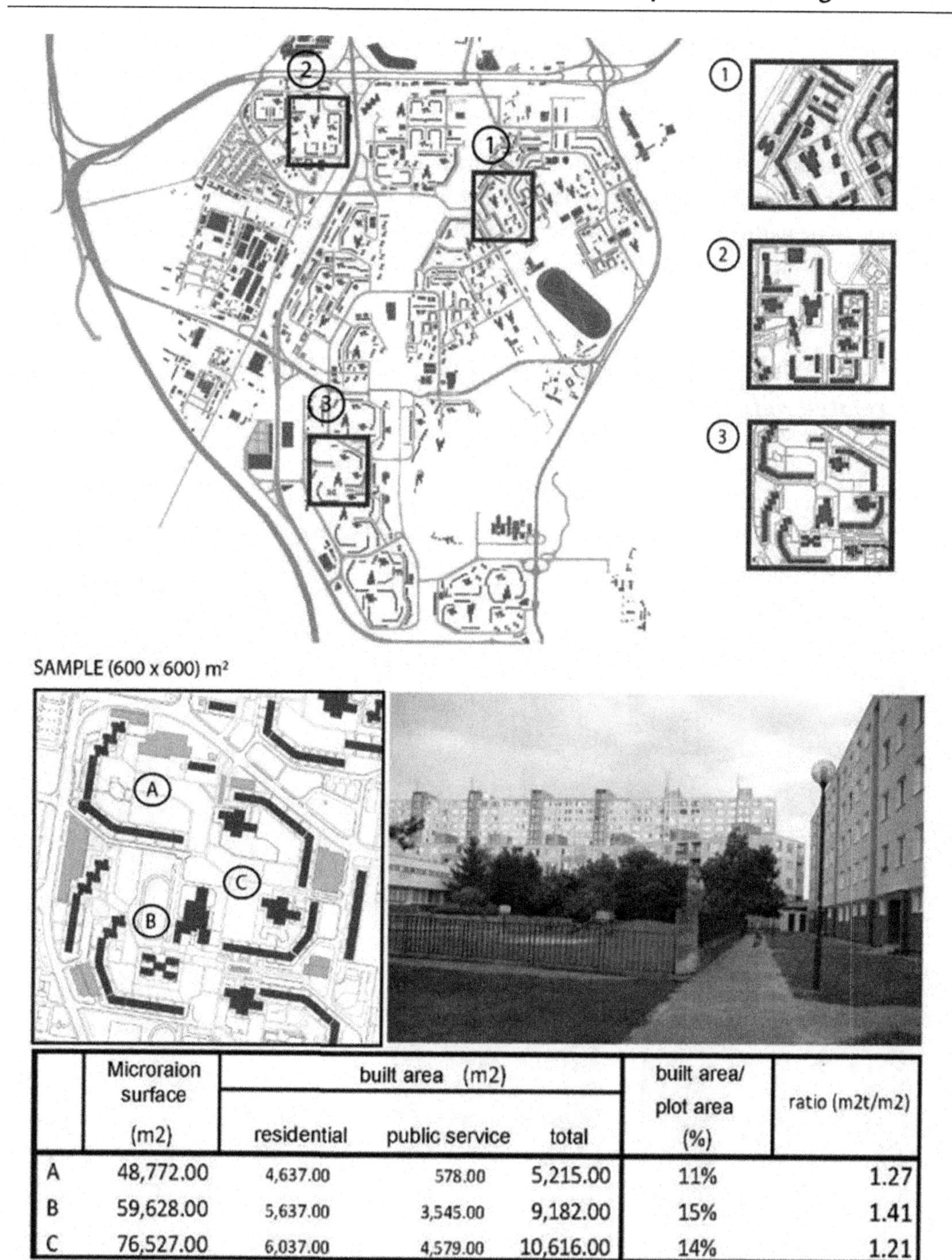

	Microraion surface (m2)	built area (m2)			built area/ plot area (%)	ratio (m2t/m2)
		residential	public service	total		
A	48,772.00	4,637.00	578.00	5,215.00	11%	1.27
B	59,628.00	5,637.00	3,545.00	9,182.00	15%	1.41
C	76,527.00	6,037.00	4,579.00	10,616.00	14%	1.21

Fig. 14.6: An example of a microraion in a housing state in Petrzalka, including data related to building configuration.

Overall, post-socialist neighbourhoods were primarily mono-functional, dominantly residential with few amenities. In addition, they were commonly isolated from commerce, in contrast to the increasing economic wealth and activity in old-urban centres of many post-soviet cities.

The specifics of the urban planning philosophy in these post-soviet cities led to a variety of environmental deficiencies, amongst which proper water quantity and quality management were found to be particularly affected. Hence, after identifying environmental and social shortcomings in these areas, the following section will specifically concentrate on water-related issues associated with post-Soviet construction.

14.2 Intra-urban transformations of post-socialist housing estates: water-related deficiencies

Many post-socialist areas experienced environmental and social decay (see Fig. 14.7) related to air pollution (as a consequence of the increase in the use of private cars showed in Fig. 14.7C), insecurity (such as vandalism and crime) and increasing difficulties in accessing satisfactory public transportation, non-contaminated water sources and proper drainage systems [7,16].

Fig. 14.7: Social and environmental decay in post-socialist Petrzalka, Bratislava (Slovakia).

Neighbourhood public space was intended to extend private life into public communal living but, as noted earlier, this was not integrated into the existing natural landscape features, such as creeks, water bodies or the original topography [10]. St. Petersburg's microraion neighbourhoods are a clear example of a hydrological landscape transformed by socialist planning, since they were constructed over the original swamp, leading to a fragmented hydrologic network, with stagnant and increasingly contaminated water [19,20]. There are many examples of neglected fluvial environments in St. Petersburg including the progressive occupation of the River Neva's tributaries and other minor rivers embankments with industrial activities (Fig. 14.8) [19,20]. As a result, the outskirts of the city flood regularly, leading to both water and soil pollution; these areas require increased wastewater purification services and strategies to mitigate flooding problems. In addition, difficulties associated with proper sewage treatment remains a problem in St. Petersburg and ultimately affects water quality in urban channels and rivers [20].

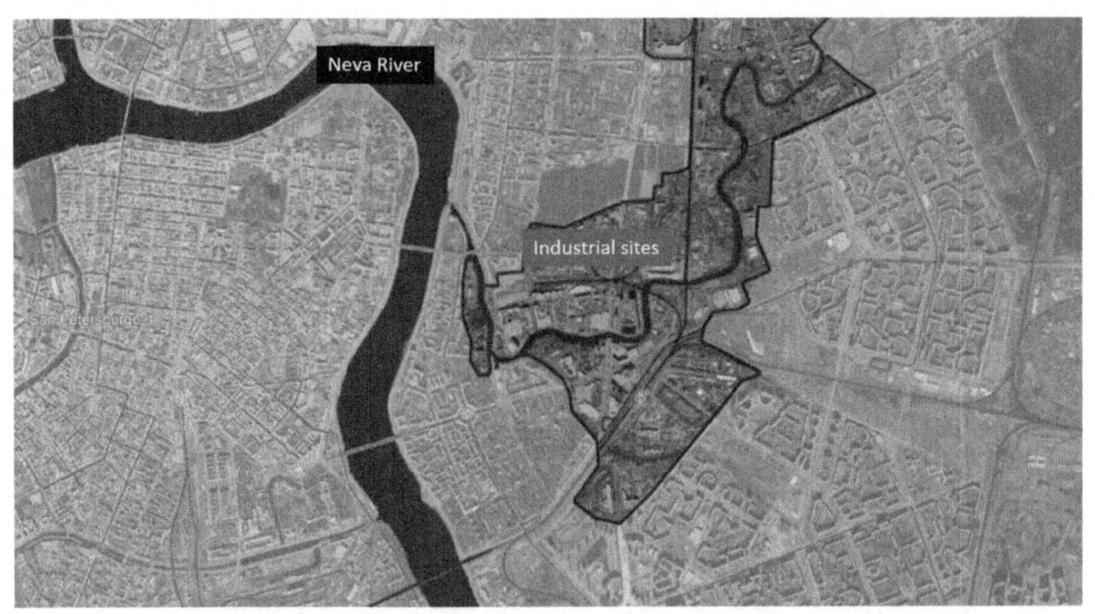

Fig. 14.8: Industrial development near tributaries of the River Neva, St. Petersburg (Russia).

The neighbourhood of Petrzalka in Bratislava, is another example of post-socialist districts having to deal with ground and surface water quality and quantity issues. In this region, the storm drain network was incorporated into a pre-established grid of massive prefabricated residential buildings. The traces of the original hydrological landscapes that connected to the Danube River were erased, favouring the impacts of urban contaminants from runoff collected by conventional storm-drain on surface and sub-surface local water [5,11].

St. Petersburg is presently dealing with increasing water contamination and frequent flooding due to the effects of climate change. Additionally, demands for water can hardly be provided during peak periods, even though the city grew over a swamp historically characterized by having abundant water resources. Deactivated industrial sites became environmentally degraded zones, with the embankments and green areas of St. Petersburg's where the tributaries of the River Neva are impacted by old factories. The city's micro-districts sometimes flood, with associated water and soil pollution, requiring increased wastewater treatment. As previously mentioned, adequate sewage treatment was an issue, due to both low-density urbanization in the post-socialist areas of St. Petersburg, and the small number of Water Treatment Plants. Added to the flat topography of the location, these issues made it necessary to deepen the sewers, which improved the conveyance of runoff once captured but did not increase the capacity of filtrating water at the source.

A similar problem developed in Petrzalka due to its development in an area with a high-water table level. Several streams crossed the area, connected to the River Danube in the north of the district (see Fig. 14.9). In order to build the new residential complex, a small village

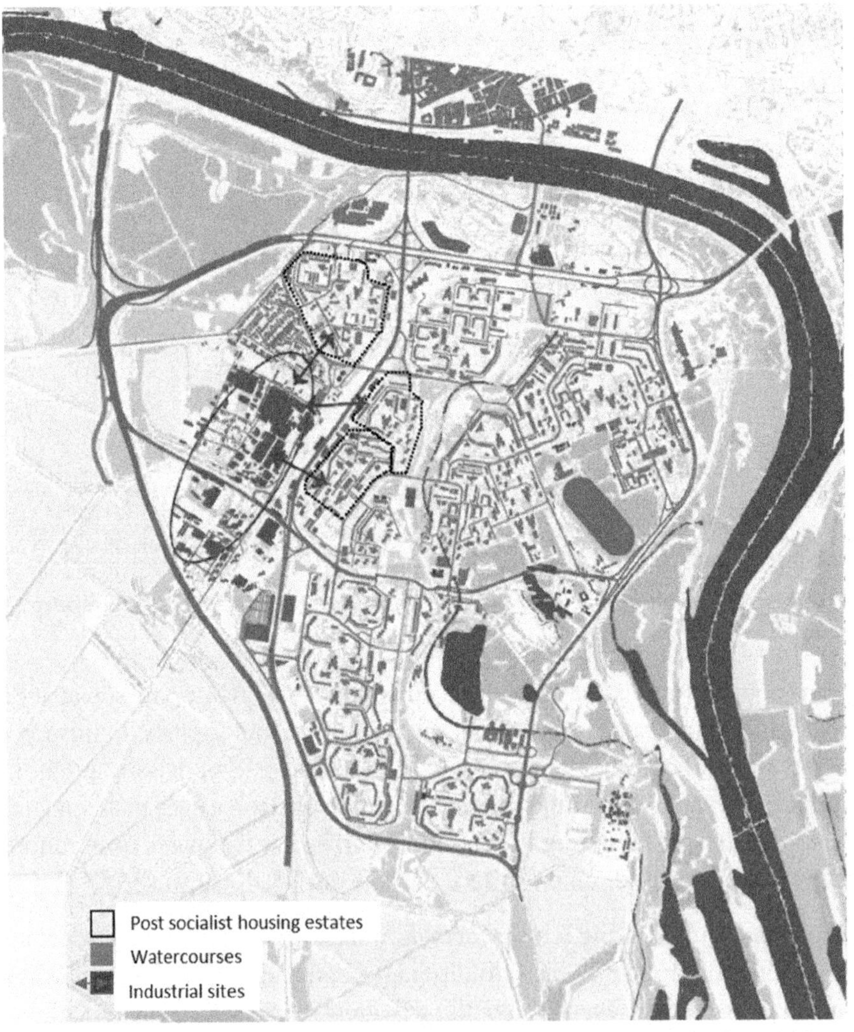

Fig. 14.9: Petrzalka's water-course system and the location of the Petrzalka's industrial site, indicating its relation with the closest microraion units.

was destroyed; large portions of the swampy areas were unfilled, but the high ground water table still restricted the final design of buildings. The ground water is connected to the River Danube and is influenced by its rising levels, carrying potential problems for building foundations. Retaining walls anchored to waterproof substratum are a highly expensive solution, so the parking lots of residential buildings are on the ground floor, with commercial stores on the first floor (Figs. 14.7B and 14.7D), which disconnects them from the street and public space. Finally, as already mentioned for the case of St. Petersburg, industrial activities also developed near to the Petrzalka's residential districts that may have contributed to contaminated soil and underground water in this area (Figs. 14.8 and 14.9)

Microraions are usually bounded by roads carrying vehicular traffic, reflecting the historical gradual shift in emphasis in major Soviet cities from being based on walking to trams, trolleybuses, buses or even the metro; however, the use of private vehicles has intensified dramatically recently impacting major roads in terms of environmental pollution and its potential to affect human health. Chronic congestion, decay in the environmental value of green spaces and the quality problems affecting local water sources in many post-socialist districts have had important negative impact on their future sustainability, which justifies the need for designing and implementing alternative water engineering techniques to revitalize these cities. A restoration strategy based on using soft water engineering practices to deal with peak runoff volume and pollution will be discussed in the following sections

14.3 Developing sustainable water management through soft water engineering in post-socialist housing estates

Proposals to integrate soft water engineering solutions to regenerate post-socialist housing estates is investigated in this section. The proposals are based on intervening at three different scales and urban contexts. The conventional hard engineering approach of extending pipelines and drilling deeper responded to the constant growing demand for water [3,31]. The alternative to hard engineering is to promote water efficiency and conservation based on relatively small-scale and more efficient solutions using so-called "soft" alternatives [3,31]. These measures are also called "soft path" [4] and include Nature Based Solutions (NBS; [21]) and Sustainable Drainage Systems (SuDS; [6]). These approaches rely on taking account of pre-development topography and hydrology [22] using on-site water resources and methods of management that are analogues to natural processes [3,31]. SuDS uses Green Infrastructure (GI) to encourage infiltration, detention and slow conveyance to address the equal roles of water quantity reduction, water quality improvement, provision of amenity of residents and space for biodiversity. Thus, using GI, the following sections consider three strategies:

1. Regeneration of old industrial sites in post-socialist districts, commonly located near residential neighbourhoods, which are progressively abandoned and degraded.

2. Water body recovery in Petrzalka and post-socialist St. Petersburg based on the implementation of green corridors with their associated soft water engineering techniques.
3. Soft water engineering systems integrated into microraions in St. Petersburg and Petrzalka to mitigate current water management problems and to provide additional environmental and social benefits to revitalize these neighbourhoods.

14.3.1 Regenerating abandoned post-socialist industrial sites using GI and accessibility-based planning approaches

St. Petersburg has a large industrial belt that impacts soil and air quality, with old industrial sites along the River Ohkta and its tributary, the Okkervil, negatively influencing water quality. In addition, there is a lack of green and public spaces and the presence of high speed roads affects the pedestrian and bicycle accessibility of these areas. Environmental impacts from such industrial activity are often associated with economic decline. The main strategy to establish links between regenerated industrial sites and green areas is based on the introduction of soft water engineering systems, such as SuDS, and GI, to make better use of the land and improve the cityscape and finally, to integrate a more accessibility-based planning of the mobility system. Urban voids and abandoned buildings can be strategically re-purposed using GI and SuDS for more sustainable water management, making them more attractive and usable for residents of the surrounding area. The same concepts are proposed for an old industrial site, in the west of Petrzalka, which is also characterized by environmental degradation and abandonment.

New green areas with SuDS for water retention and treatment in the revitalized industrial area can be encouraged, together with an increase in interconnectivity between existing streets (Fig. 14.10).

Improving pedestrian and bicycle accessibility between the site and surrounding microraions would provide amenity and convenience (Fig. 14.11). Additionally, transforming existing roads into boulevards, with multiple uses, and increasing the provision of pedestrian circulation, new trees, SuDS and public transport, as depicted in Fig. 14.12, provides a means of sustainably managing water, as well as providing amenity for residents, space for biodiversity and potential mitigation and adaptation measure for climate change.

It is widely accepted that, in order to improve accessibility in degraded urban areas, multimodal transport, with walkable distances between transportation nodes is a requirement [8,30]. Reclaiming streets for people through traffic strategies promotes sustainability, whilst the reduction of road areas help increase the amount of suitable spaces for technologies to perform as source control measures. SuDS such as bio-swales, bio-retention basins and porous pavements are among the possible technologies that can be utilized for the retention of runoff diffuse contaminants in the Petrzalka and St. Petersburg case studies. At the same time, both walking and cycling is constrained by motorized modals in both Petrzalka and

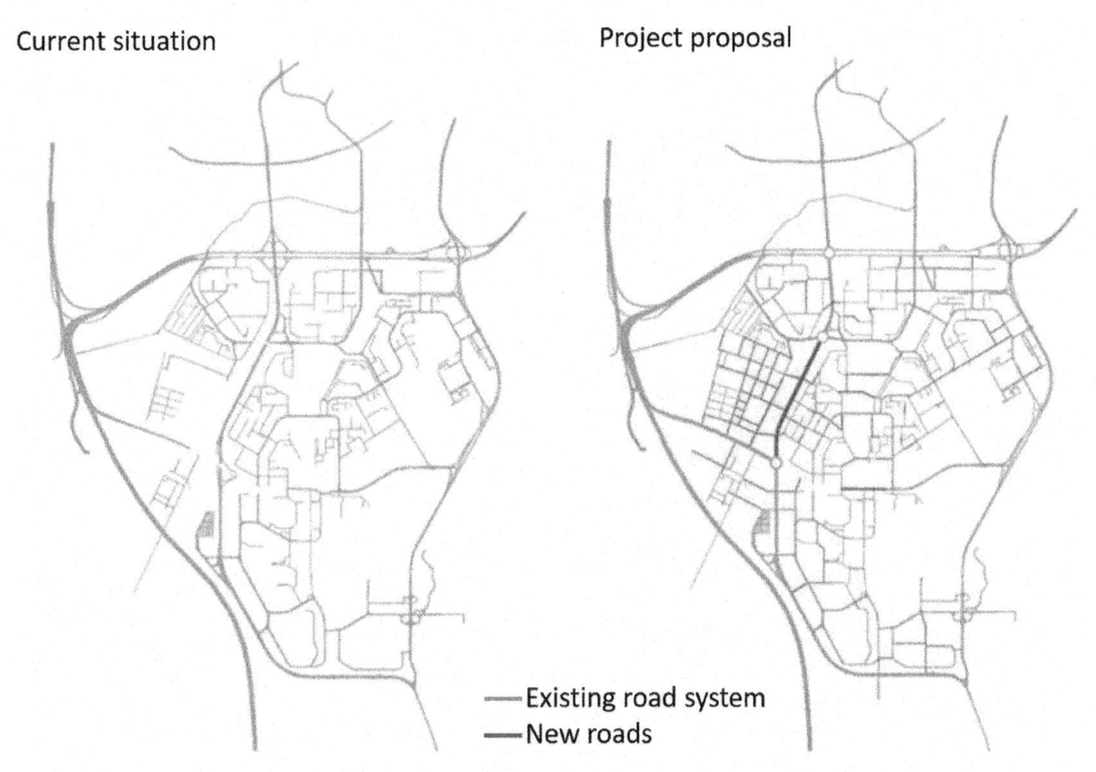

Fig. 14.10: Improvement of internal road connectivity within the industrial site and neighbouring areas of Petrzalka in Bratislava (Slovakia).

St. Petersburg post-socialist neighbourhoods, such that there is a need for promoting environmentally efficient streets with space for SuDS and vegetation. A south-north highway system in Petrzalka that separates the industrial area from residential units has consequential impacts on connectivity between both areas for users, while reducing the capacity for the integration of green-blue corridors with water retention and detention uses. The accessibility-based approach could therefore improve east-west connectivity for establishing GI systems and including both vehicular transport and non-motorized slower mobility (Fig. 14.12).

14.3.2 Integrating SuDS into post-socialist microraion neighbourhoods to address water management issues

This section discusses some opportunities for improving the current water engineering system of post-soviet housing estates. The main question is whether the integration of soft water engineering systems for microraions in St. Petersburg and Petrzalka, as well as other post-soviet housing states, can mitigate water management problems and provide additional environmental and social benefits to revitalize these neighbourhoods.

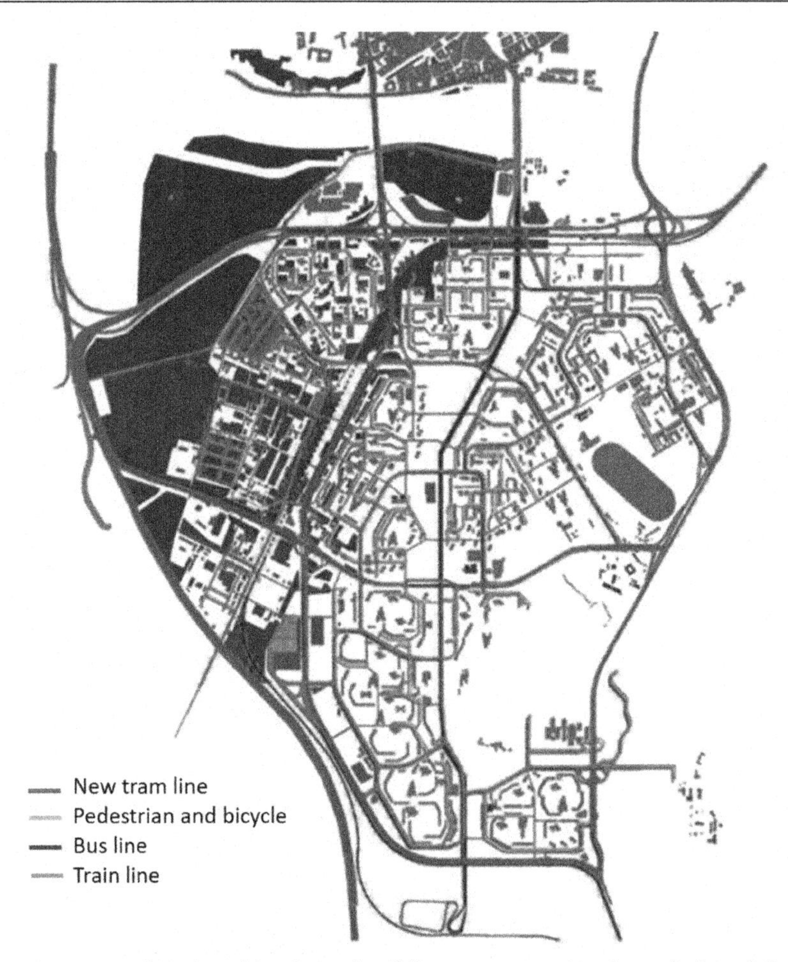

Fig. 14.11: New pedestrian, bicycle and public transportation in an industrial site and neighbouring areas, Petrzalka, Bratislava, Slovakia.

The integration of NBS into housing estates means less reliance on large, monofunctional and centralized design approaches. This means designs need to be flexible enough to combine conventional storm-drain projects with nature-based facilities that have the ability to work in a diffused and decentralized manner in the urban environment. The efficiency of SuDS/ GI systems to restore connectivity between hydrological elements, keep stormwater clean at the source and reduce water discharge volumes downstream has attracted interest globally [6]. In terms of urban revitalization, the integration of SuDS can transform people's perception of the place, including their everyday lives, allowing recycling and re-use of this resource. Some ideas are shown in Fig. 14.13 and Fig. 14.14 for Petrzalka, where there was a need for increasing water peak storage as well as to address the current demand for parking areas. Some ideas have been developed for Petrzalka's microraions to cope with the abovementioned problems. Fig. 14.14 shows the new proposed residential typologies to be combined

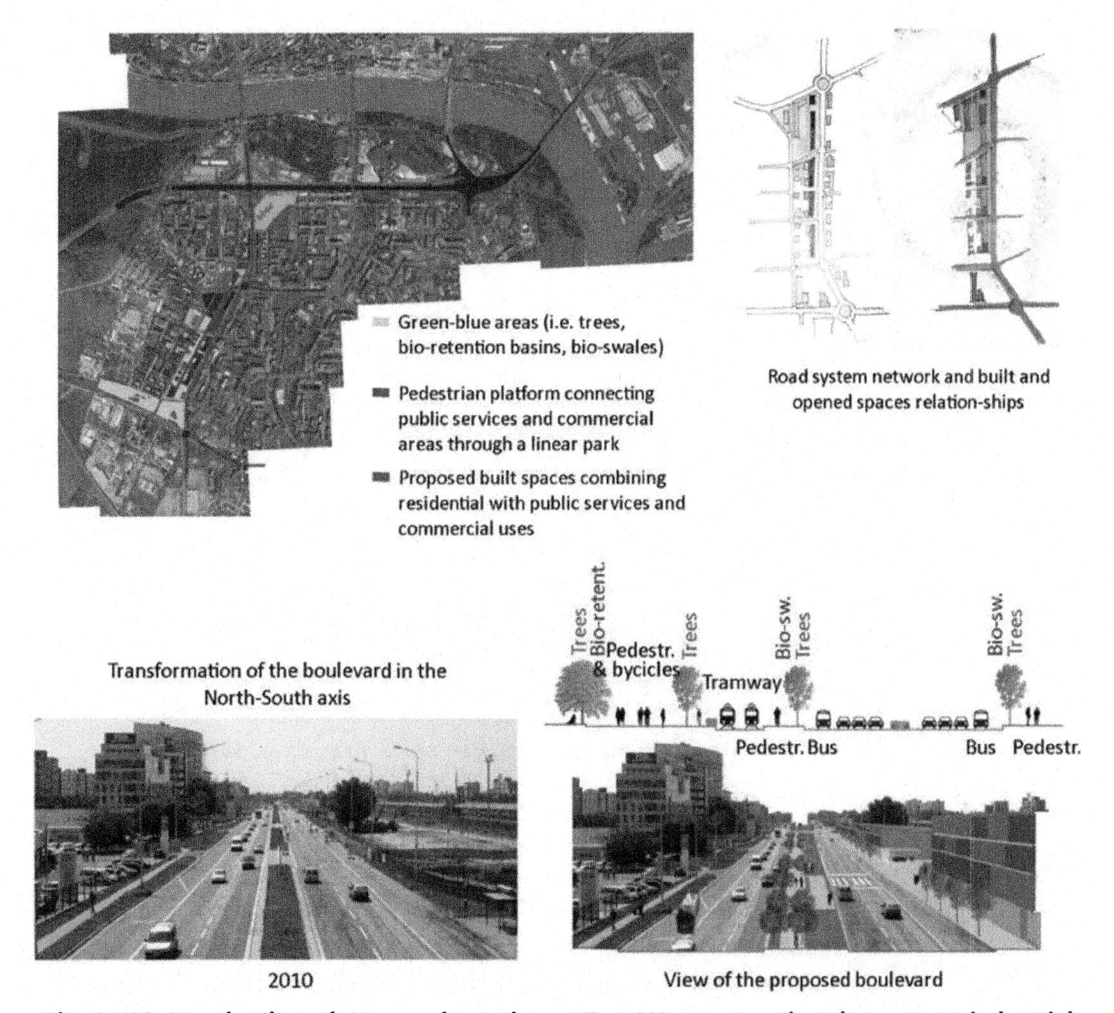

Fig. 14.12: New boulevard proposed to enhance East-West connections between an industrial site and neighbouring areas of Petrzalka, Bratislava, Slovakia.

with parking stories that contribute to reducing the presence of vehicles in the neighbourhood. Also, water quality of the receiving bodies is increasingly damaged by diffuse contaminants carried in stormwater runoff. Such problems can be addressed by incorporating sensitive and diffuse solutions. Fig. 14.15 corresponds to a view for a proposed sustainable stormwater system based on bio-swales and bio-retention systems to contribute to reducing diffuse contamination in a residential cluster in Petrzalka. In addition to the projected SuDS, rainwater harvesting devices can be placed in the roof areas of the new proposed buildings, as depicted in Fig. 14.14.

There are traces of the original hydrology still visible, although it has changed in response to urbanisation. One of the main aims, therefore, would be to treat the water before it

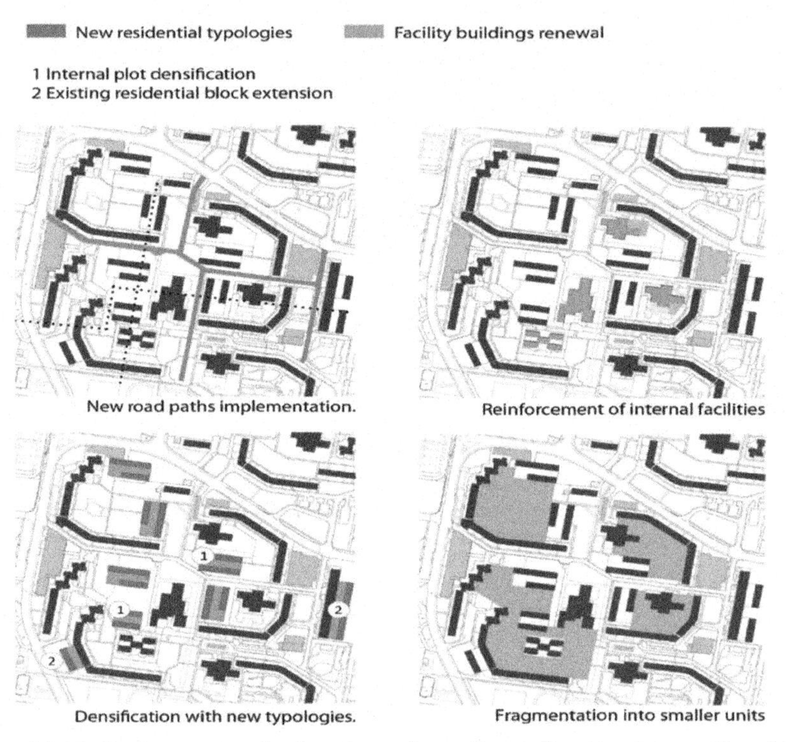

Fig. 14.13: Project strategies in microraions, Petrzalka, Bratislava, Slovakia.

reached the receiving body, by integrating SuDS in the microraions, potentially by making use of the empty spaces between buildings. Designs do need to be playful and reflect natural runoff management, using artificial creeks combined with bio-retention ponds and other GI elements, to treat and retain water contributing to sustainable water management in the area. Such approaches can also be integrated into responses to the potential impacts of global climate change. Bratislava's preparation for climate change includes providing resilience against the expected increases in long periods of drought and more extreme rainfall events, which may result in an increased risk of local flooding. This includes "green and soft adaptation measures to maximise the use of rainwater and GI" in its "Action Plan for Adaptation Climate Change in Bratislava 2017–2020", which was published in 2016. Among the adaptation options, green spaces and corridors in urban areas, as well as water sensitive urban and building designs are being considered. Small projects to support SuDS are also contemplated in these kind of interventions for Bratislava. At the same time, new housing volumes proposed within the courtyards as an urban planning tool towards increasing social diversity and social interaction can pioneer SuDS integrated at the building scale.

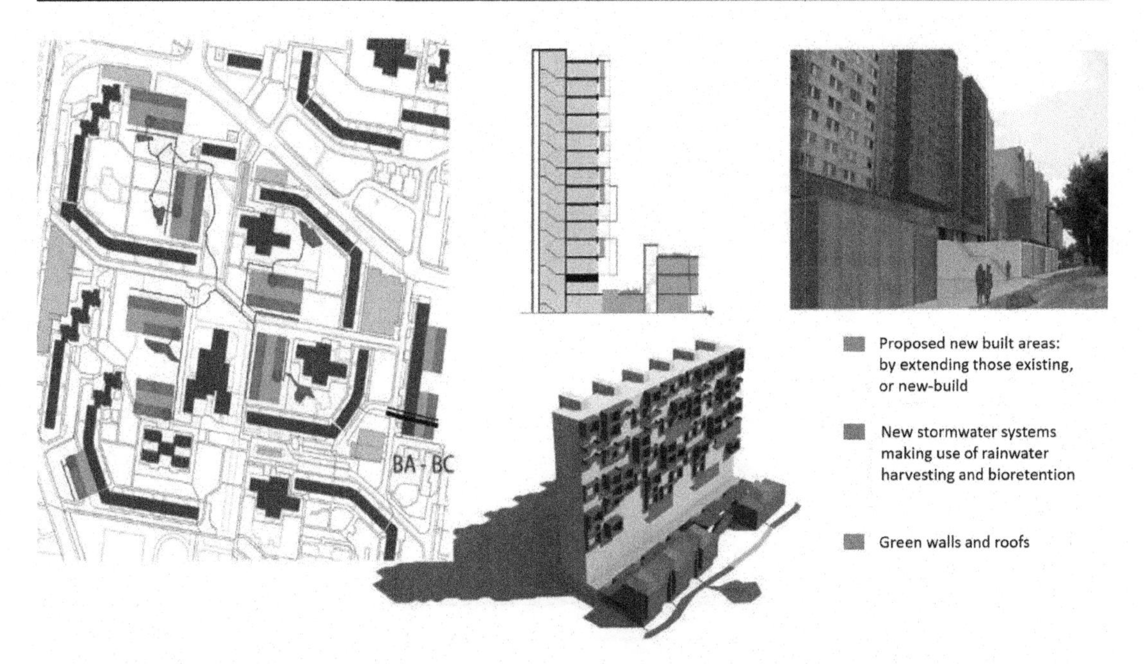

Fig. 14.14: Project strategies for integrating SuDS at the microraions scale, Petrzalka, Bratislava, Slovakia.

Once initiated at the small scale, there may be opportunities for SuDS to expand and connect via blue–green networks with existing watercourses, providing new hydrological functions in the landscape and more NBS thinking in terms of planning water infrastructure in the urban environment, as proposed in Fig. 14.14. In this context, the different landscapes can be seen at their unit scale within the broader picture of the district; thus, the design of water infrastructure systems becomes closer to the ideal of an integrated network.

The major challenges previously discussed in St Petersburg lead to the question of how to combine approaches to address the environmental, social and aesthetic issues in the micro neighbourhood context and whether there are suitable areas in which to exploring innovative water management solutions. One obvious solution would be to investigate the potential of the microraion internal spaces in which to install approaches such as bioretention, small-scale wetlands and other green-and-grey networks to encourage the filtering and infiltration of stormwater while making water available for secondary use. This approach can be integrated into any existing drainage system, enhancing both the multifunctionality of water infrastructures and the generation of more inclusive environments. Together with the integration of GI into microraion space, encouraging various construction styles which combine housing with water retention and storage systems, current conditions can be gradually transformed (Fig. 14.15) in Petrzalka.

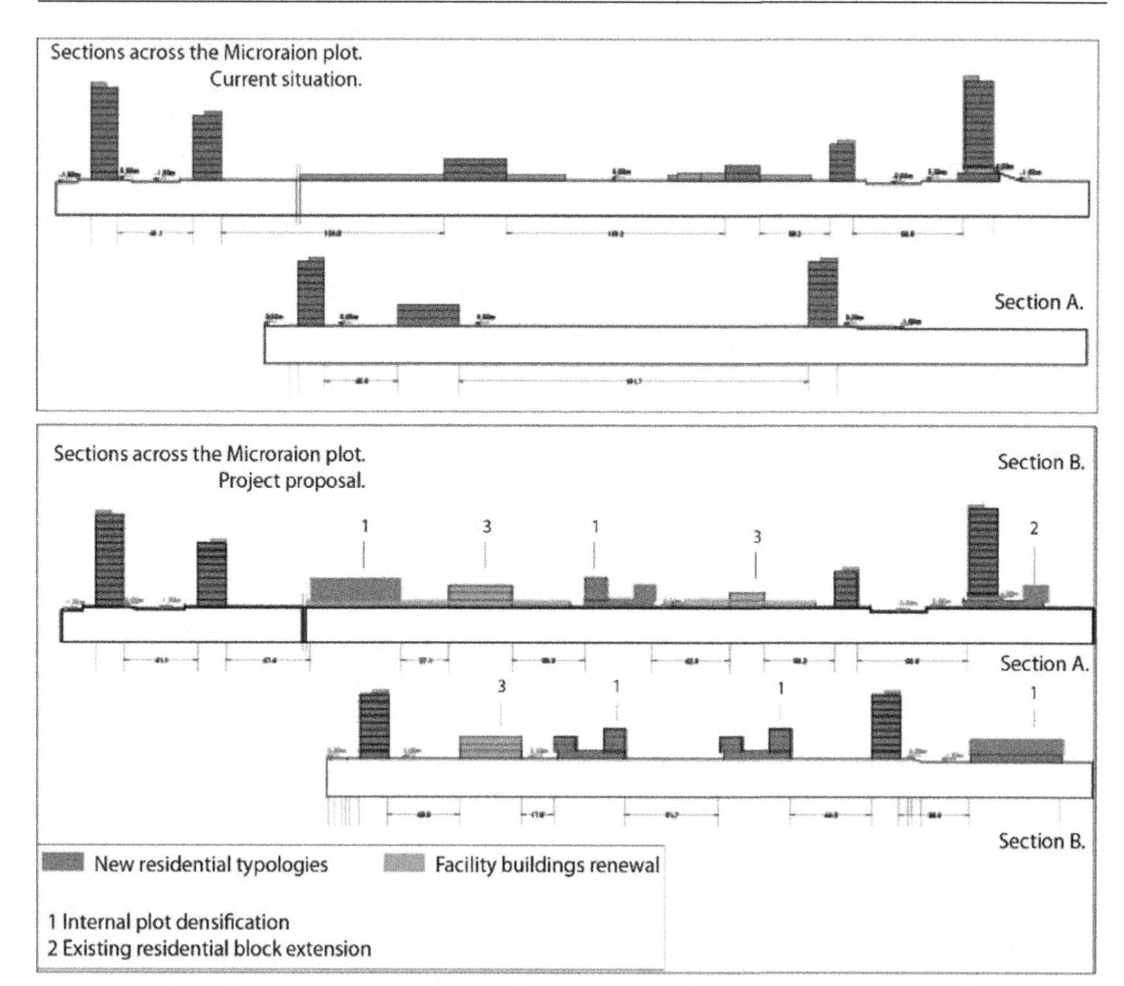

Fig. 14.15: Proposal to integrate new building typologies in the microraions, Petrzalka, Bratislava, Slovakia.

The microraion internal densification with new combined uses (i.e. residential, parking, commercial and public services) is depicted in Fig. 14.16. Also, the renewal of existing facilities within the plots is fundamental to enhance a more liveable neighbourhood. Yellow blocks represent existing educational buildings that could be more intensely used by local residents by including a renewal project for them. The combination of the abovementioned urban strategies together with the integration of SuDS may guarantee a more environmentally efficient and liveable Petrzalka.

In its strategy to address the issues detailed in this chapter, Bratislava City lists the many benefits of the NBS approach, including adaptation to climate change by sequestering and storing

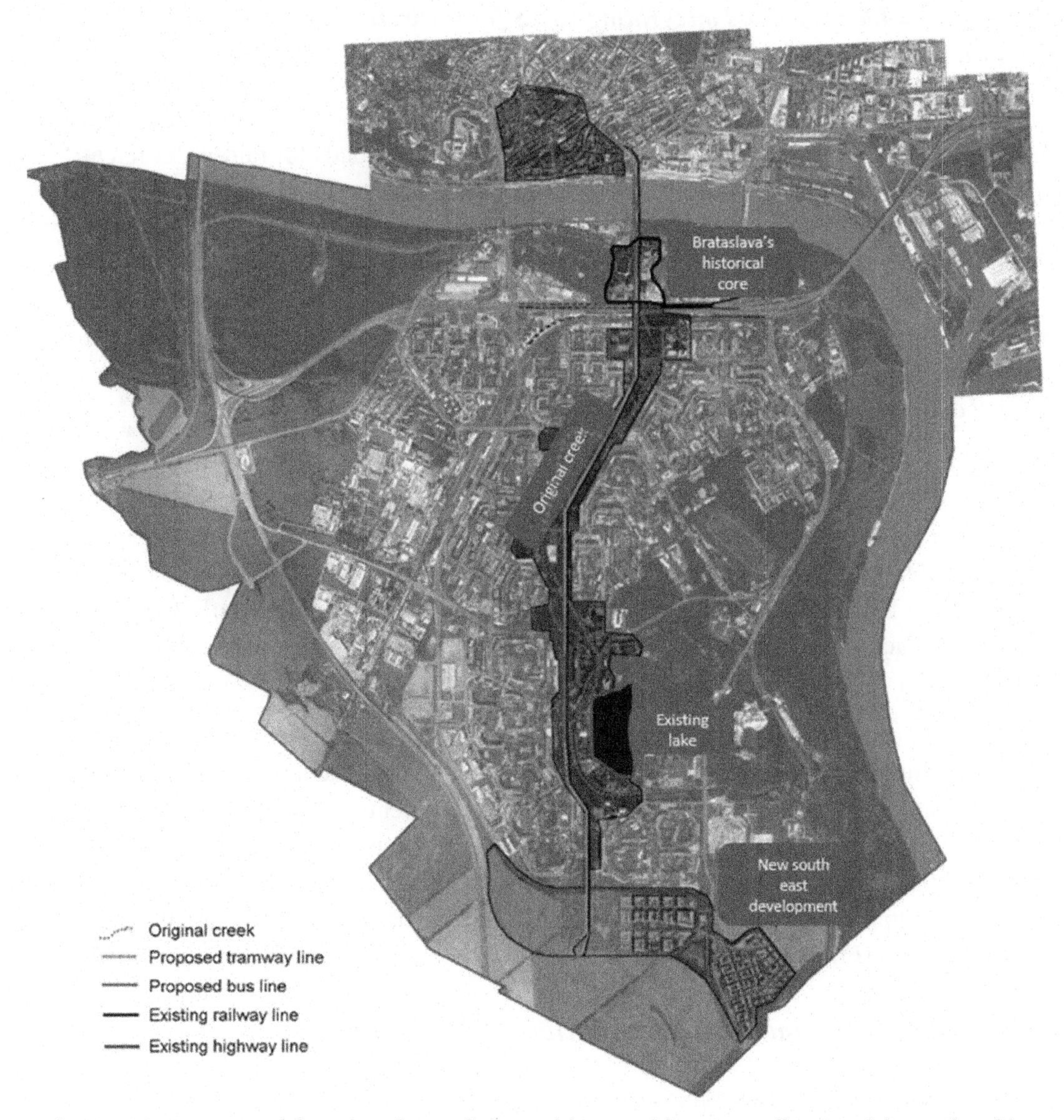

Fig. 14.16: A proposal for a south-north green-blue corridor, Petrzalka, Bratislava, Slovakia.

carbon; management of flood risk and resilience by reduction of flood peak and water volume which will reduce sewer system loads. Blue and GI need to be better connected through the city, by integrating corridors which will encourage biodiversity, but also provide amenity for residents and thus improve social interaction and inclusion. The following section considers the potential to design and install such blue-green corridors into the city, and the benefits of using such an approach.

14.3.3 Green-blue corridors integrating SuDS for water treatment in post-socialist quarters

For the integration of all the above mentioned GI technologies, the concept of blue–green corridors can be introduced at a larger scale. Within this idea, ponds for rain-water retention can work as secondary systems that can store excess water volume, particularly during storm events, whereas bio-swales and bio-retention facilities retain water from low-intensity rain storms. This strategy may combine small-to-medium size water retention and detention devices that incorporate hydrological functions such as filtration, infiltration and bio-retention for minor flooding mitigation and water quality improvement and, at the same time, contribute to create high quality landscapes for people recreation with urban amenities and services associated to them.

With careful planning and design, each microraion can be connected with one another (Fig. 14.14). Overall, a blue–green corridor can perform as the main green area to be linked to the whole GI services. This central park connects the microraion units with the existing lakes and, later, to the Danube River, thereby reinstalling the connectivity of the hydrological landscapes both superficially and sub-superficially. Thus, within this park, water ponds and other GI facilities can be created to connect stormwater systems with the existing hydrology.

Fig. 14.16 represents this linear park and highlights the connection of Petrzalka with the city core facilitated by the implementation of a green boulevard and a new tramway line in the south–north direction. Thanks to this, the new developments, currently implemented at south–east areas near to Petrzalka, can be connected to the linear park as well, as depicted in Fig. 14.16. In addition, the new tramway is implemented to provide non-motorized public transportation to the local residents from Petrzalka and the new developed areas through this south–north axes (Fig. 14.16).

14.4 Conclusions and recommendations

The urban transformation process experienced by post-socialist estates in Eastern Europe and Russia involved severe environmental impacts, whereby pollution of both surface and ground water increased over the last decades. The particular urban configuration of these areas also hindered control of high runoff rates derived from intense storm events, leading to flooding. Inadequate water management resulted in decreased quality of life for the inhabitants of these post-soviet cities.

Taking Petrzalka and St. Petersburg as two examples of the resulting problems in these areas in terms of water management, a strategy based on the implementation of soft engineering techniques, such as SuDS utilising GI and NBS, has the ability to restore them from an environmental and social point of view. This approach can reduce the use of traditional hard engineering drainage solutions, thereby reducing the degree to which ecosystems are affected and can preserve the natural hydrological cycle as much as possible. In particular, the specific measures proposed include bio-retention areas, wetlands and green–blue corridors to revitalize residential microdistricts and marginal industrial sites, providing diverse benefits such as water treatment, harvesting and reuse, air purification and aesthetic benefits. In the end, the inclusion of these elements results in the improvement of living conditions in the area at different levels.

Implementation of soft water engineering techniques must not focus on the complete replacement of traditional hard solutions. Both approaches can coexist to maximise the benefits associated with each of them. In terms of water management, soft practices can be especially effective to deal with stormwater at source by reducing peak flows and pollutant loads, such that the amount and condition of water to be captured and conveyed by hard drainage systems is reduced. This integrated effect is especially necessary in the context of climate change and urban sprawl, whereby the magnitude and frequency of storms is projected to rise, as well as the intensity of production and consumption habits that pollute the environment.

References

[1] C. Becker, S.J. Mendelsohn, K. Benderskaya, Russian urbanization in the Soviet and post-Soviet eras, Urban. Emerg. Popul. Issues Working Paper 9 (2012) 1–128.

[2] I. Brade, C. Smigiel, Z. Kovacs, Suburban residential development in post-socialist urban regions: the case of Moscow, Sofia, and Budapest, in: H. Kilper (Ed.), German Annual of Spatial Research and Policy 2009: New Disparities in Spatial Development in Europe, Springer, BerlinGermany, 2009, pp. 79–104.

[3] D.B. Brooks, Beyond greater efficiency: the concept of water soft path, Can. Water Resour. J. 30 (2005) 83–92, doi:10.4296/cwrj300183.

[4] D.B. Brooks, O.M. Brandes, S. Gurman, Making the Most of the Water We Have - The Soft Path Approach to Water Management, Earthscan, ed.,Oxfordshire (U.K.), 2009.

[5] C. Castañer, The Post-Socialist Transformations: Petrzalka as case study, Universitat Politècnica de Catalunya (UPC). Barcelona (Spain), 2012, pp. 1–63.

[6] S.M. Charlesworth, C. Booth (Eds.), Sustainable Surface Water Management; a handbook for SuDS, Wiley Blackwell publishing, 2017.

[7] M. Czepczyński, Cultural landscapes of post-socialist cities: representation of powers and needs. Re-Materialising Cultural Geography, Ashgate, 2008.

[8] I. de Solà-Morales, M. Rubió, Las formas de crecimiento urbano, Edicions UPC, Barcelona (Spain), 1997, pp. 1–196.

[9] D. Dushkova, D. Haase, A. Haase, Urban green space in transition: historical parks and Soviet heritage in Arkhangelsk, Russia. Crit. Hous. Anal. 3 (2016) 1, doi:10.13060/23362839.2016.3.2.300.

[10] B. Engel, Public space in the Blue Cities in Russia, Prog. Plann. 66 (2006) 147–239, doi:10.1016/j.progress.2006.08.004.

[11] K. Gross, The International Urbanistic Competition: Bratislava-Petržalka, Vydavateĺstvo SlovenskéNo fondu výtvarných umení, Bratislava (Slokavia), 1969, pp. 1–102.

[12] G. Herbert, The neighbourhood unit principle and organic theory, Sociol. Rev. 11 (1963) 165–213, doi:10.1111/j.1467-954X.1963.tb01231.x.

[13] B. Hillier, et al., Natural movement: or configuration and attraction in urban pedestrian movement, Environment and Planning B: Planning and Design 20 (1) (1993) 29–66, doi:10.1068/b200029.
[14] S. Hirt, K. Stanilov, Planning Sustainable Cities: Global Report on Human Settlements, Nairobi, Kenya, 2009, pp. 1–108.
[15] F. Holanda, et al., Arquitetura e Urbanidade, ProEditores Associados, São Paulo (Brazil), 2003, pp. 1–192.
[16] V. Ira, The changing intra-urban structure of the Bratislava city and its perception, Geogr. Časopis 55 (2003) 91–107.
[17] J. Jacobs, The Death and Life of Great American Cities, Random House USA Inc., New York (U.S.), 1961, pp. 1–458.
[18] H.F. Jahn, The housing revolution in Petrograd 1917-1920, Jahrb. Gesch. Osteur. 38 (1990) 212–227.
[19] A. Korzun, N. Nagornova, N. Oblomkova, J. Saukkoriipi, E. Salminen, Nutrient Loads to the Baltic Sea from Leningrad Oblast and Transboundary Rivers, Baltic Marine Environment Protection Commission, Vantaa (Finland), 2014, pp. 1–128.
[20] K.I. Krasnoborodko, A.M. Alexeev, L.I. Tsvetkova, L.I. Zhukova, The development of water supply and sewerage systems in St. Petersburg, Eur. Water Manag. 2 (1999) 51–61.
[21] T. Lavers, S. Charlesworth, Opportunity mapping of natural flood management measures: a case study from the headwaters of the Warwickshire-Avon, Environ. Sci. Pollut. Res. 25 (2018) 19313–19322. doi:10.1007/s11356-017-0418-z.
[22] A.B. Lovins, Soft Energy Paths Toward A Durable Peace, Harper & Row, New York (U.S.), 1977.
[23] J. Maxim, Mass housing and collective experience: on the notion of microraion in Romania in the 1950s and 1960s, J. Archit. 14 (2009) 7–26, doi:10.1080/13602360802705155.
[24] M.W. Mehaffy, S. Porta, O. Romice, The "neighborhood unit" on trial: a case study in the impacts of urban morphology, J. Urban. Int. Res. Placemaking Urban Sustain. 8 (2015) 199–217, doi:10.1080/17549175.2014.908786.
[25] H.W. Morton, Housing in the Soviet Union, Proc. Acad. Polit. Sci. 35 (1984) 69–80, doi:10.2307/1174118.
[26] S.E. Reid, Khrushchev modern: agency and modernization in the Soviet home, Cah. Monde Russe 47 (2006) 227–268.
[27] J. Robinson, The post-Soviet city: identity and community development. In: City Futures in a Globalising World: An International Conference on Globalism and Urban Change, Chicago, U.S., 2009, pp. 1–17.
[28] R. Rudolph, I. Brade, Moscow: processes of restructuring in the post-Soviet metropolitan periphery, Cities 22 (2005) 135–150, doi:10.1016/j.cities.2005.01.005.
[29] M. Whitehead, Hollow sustainabilities? Perspectives on sustainable development in the post-socialist world, Geogr. Compass 4 (2010) 1618–1634, doi:10.1111/j.1749-8198.2010.00398.x.
[30] P. Wirth, J. Chang, R.U. Syrbe, W. Wende, T. Hu, Green infrastructure: a planning concept for the urban transformation of former coal-mining cities, Int. J. Coal Sci. Technol. 5 (2018) 78–91, doi:10.1007/s40789-018-0200-y.
[31] G. Wolff, P. Gleick, The soft path for water. In: The World's Water 2002–2003, Island Press, Washington, D.C.(U.S.), 2002, pp. 1–32.
[32] K.E. Zarecor, Architecture in Eastern Europe and the Former Soviet Union, in: E.G. Haddad, D. Rifkind (Eds.), A Critical History of Contemporary Architecture, 1960–2010, Iowa State University, Iowa (U.S.), 2014, pp. 255–274.
[33] Z.L. Zile, Programs and problems of city planning in the Soviet Union, Washingt. Univ. Law Rev. 1 (1963) 19–59.

CHAPTER 15

Canals—The past, the present and potential futures

Carly B. Rose
Visiting Fellow, CABER, UWE, Bristol, UK

15.1 Introduction

Our species has been modifying the natural world, to better meet our own needs, for millennia. The development of agriculture entailed digging irrigation and drainage channels, while the need for reliable transport arteries led first to river modifications (such as "canalisation" by straightening) and subsequently to the creation of wholly artificial watercourses. This reached a peak in the UK during the industrial revolution and was responsible for many of the technological advances that would later contribute to railway and road construction. It also led to the establishment of a new profession: Civil Engineering (as distinct from its military forerunner).

The transport function of the UK canal system was largely superseded by the introduction of railways and motor vehicles. The remaining canals were then given a new lease of life, however, as a new leisure-boating industry arose, which also prompted the restoration of some derelict waterways. By contrast, many of the European waterways continue to be used for freight transport, which has the advantage of offering a lower carbon solution than road haulage. As we move towards a future of climate changes the canal network offers some innovative solutions to a variety of problems, including droughts, floods and heatwaves. The story of the rise, fall and resurgence of the UKs inland waterways looks set to continue for many years to come.

15.2 The past

The first known "artificial river" was the Grand Canal in China, which was designed for navigation purposes, as opposed to water supply. Constructed around the 5th century BCE as a military supply route, it also featured the first recorded use of the pound lock in the 10th century [25]. The summit section did, however, suffer from water shortages for many centuries, which underlines the need for rigorous planning to obtain and retain adequate water for effective functioning of artificial waterways.

Sustainable Water Engineering. DOI: 10.1016/C2017-0-04301-X

The Romans, although better known for their water supply infrastructure, also created some canals for navigation in the first century CE. They are believed to have been responsible for at least one of the earliest artificial waterways in Britain, the Foss Dyke, which joins the Rivers Trent and Witham [9]. In the UK, navigable rivers were the major trading routes for many centuries; in winter the unpaved roads were typically impassable due to deep mud, and in the summer months the deep wheel ruts would be baked hard. Even after the introduction of long–distance turnpike roads, water transport was still more efficient for heavy or fragile goods [53]. Artificial waterways, therefore, offered a more reliable alternative to both natural rivers, which could suffer from low water levels in the summer, and the fledgling road network.

15.2.1 Early canals

The first documented canal in Europe was a section of the Stecknitz Canal in Germany. It was built in the late 14th century to serve the salt trade by connecting the catchment basins of the Trave and Elbe rivers [67]. Later examples of similar endeavours include the Brussels–Scheldt Maritime Canal (Belgium) in the mid-sixteenth century, and the Canal de Briaré (France) in the mid 17th century [19]. The European country perhaps most strongly linked with canals, the Netherlands, also embarked upon its canal-construction programme around the same time; although earlier works had been constructed, these were drainage-related, as opposed to dedicated navigation channels.

These early waterways derived their water supplies from the rivers they connected. By contrast, the Canal du Midi in France provided a shortcut for sea-going vessels. It connected the Bay of Biscay, via the navigable River Garonne at Toulouse, and the Mediterranean, at Sète on the southern coast [62]. This was the first canal to use a sequence of locks to allow craft to climb up and down hills in order to cross a watershed [61]. This method, however, meant that water would be lost from the summit level in both directions each time a vessel made the passage. Unlike its predecessors this canal lacked a ready-made water supply, and also passed through an area prone to low rainfall. The engineer responsible for the project, therefore, designed a "feeder" canal, which conveyed water from the mountains near Carcassonne, as well as creating the first artificial reservoir dedicated to supplying a canal, the "Bassin de Saint–Ferréol". Subsequent increases in traffic meant a second reservoir had to be built, highlighting that even well-designed canals can face water shortages, a problem that will be discussed later in this chapter.

15.2.2 Canal growth in the industrial revolution

During the second half of the 18th century good transport links became increasingly vital. Raw materials needed to be moved from their sources to the new factories, and finished products needed to be delivered to the consumers, without incurring unacceptable damages in the case of fragile goods. Despite the existence of navigable canals in Europe, the idea was not taken up in England until a wealthy landowner recognised the value of building a canal, which would create an uninterrupted navigable route for coal from his mines direct to the major industrial city of Manchester. The Bridgewater Canal provided a catalyst for a century of canal building known as "canal mania", and by1850 the system was close to 4000 miles in length [5]. The technology and associated expertise also became exports in themselves: Thomas Telford designed the Göta Canal in Sweden [52] and James Brindley's eponymous nephew was responsible for many early canals in the USA [40].

For two centuries the UK inland waterway system was the direct equivalent of the modern motorway (highway) network for long–distance transport of goods. This has been estimated to have exceed 30 million tonnes per year at its peak [28]. In the 20th century, however, the canals proved unable to compete with other transport methods due to a failure to modernize, in sharp contrast with the situation in mainland Europe.

15.3 The present—canals in the 21st century

15.3.1 The UK experience

The expansion of the railways was the initial trigger for the demise of UK canals as a commercial proposition. The waterways could not compete with rail's speed or carrying capacity, particularly for passenger transport, and the least profitable routes were soon abandoned. A substantial amount of freight business, however, continued until the 1950s when road haulage eclipsed both its predecessors [28].

A lack of investment in modernising waterborne transport exacerbated the situation in the UK. The infrastructure was, for the most part, designed for "narrowboats", less than two metres wide and with a draught of just over one metre. A rare exception was a scheme to deepen and widen the northern sections of the Grand Union Canal, funded by a government grant in the 1930s. There was insufficient funding for the work to be applied to the entire canal, which compromised the effectiveness of the initial grant [49]. Construction of larger lock chambers, capable of accommodating vessels of a size that would have been more cost-effective, was hampered by water supply problems. The nationalisation programme of the post-war years included the waterway network, but attempts to increase commercial traffic were unsuccessful and commercial canal carrying formally ended in 1963, with only a small number of independent operators continuing to trade [49].

15.3.2 The European experience

The UK experience contrasts markedly with much of Europe, where canal systems were modernized and developed via public funding. Upgrading to the standard Freycinet gauge (300 t) had commenced on the French network in 1879 [28]. The interconnected nature of the European waterways was, of course, an advantage; in order to create a coherent network, minimum standards for waterways of international importance were established in 1954 [39]. Continuing investment included, for example, the canalization of 275 km of the Moselle River linking Germany, Luxembourg and France. More recently, the Strépy–Thieu boat lift on the Canal du Centre in Belgium was built in 2002 to accommodate craft of 1350 t [51].

15.3.3 Resurgence, restoration and regeneration

Just as the UKs waterborne freight industry was declining, the 1960s saw increased interest in pleasure-boating on the canals. This eventually gave rise to new industries, including the building of craft designed specifically for leisure use, and the appearance of holiday operators whose "selling points" included the peace and remoteness of many of the extant canals. The income from water-based tourism in rural areas in need of economic regeneration can be of great importance, as well as providing a much-needed source of employment. Some previously abandoned waterways have been brought back into use with the aid of European Union funding, for example the restoration of Droitwich canal in the UK, as well as some French canals [2,12]. An added benefit of such initiatives can be the enhancement and/or creation of green "wildlife corridors" which can improve biodiversity in urban areas.

The shift from freight to leisure use in UK canals has, however, had impacts upon both water level management and maintenance regimes, as will now be discussed.

15.3.4 Water level management

Those waterways still used by freight traffic require a greater depth of water to be maintained in the channel, as laden vessels sit lower in the water than leisure craft. This necessitates regular dredging to remove silt, as well as preventing bankside vegetation from encroaching, to ensure the design profile is not compromised. As a consequence, waterways used for commercial boating require higher maintenance expenditure than the canals used predominantly for leisure. The responsible body for most waterways in England and Wales is currently the Canal & River Trust (C&RT), which employs a risk-based approach to dredging requirements. The "minimum open channel" dimensions for each navigation is defined, and dredging works are prioritised accordingly [32]. C&RT monitor and control water in the canals within their remit via a sophisticated telemetry system (MEICA SCADA), which governs flow (lock bypasses and reservoir outflows) and levels in reservoirs and pounds [10]. Pump and sluice operations are automated, with any excess water being drained off via sluice gates, overflow weirs and spillways, allowing around 20 cm of freeboard to the top of the banks. Any initiatives designed to re-introduce water-borne freight, therefore, necessitate review and amendment of the current water level management and maintenance regimes. An example is provided by the operation to transfer waste and construction materials by water during the creation of the Olympic Park for the London 2012 Games, and the scope for continuing waterborne transport thereafter [37,64].

Canals rarely pose a flood risk in themselves, provided structures and equipment function correctly. The chief exceptions are: where breaches occur due to structural failure, for example embankments weakened by burrowing animals; where extreme rainfall events lead to overtopping and associated damage (e.g. [31]); or overtopping in the event of damage or vandalism, typically where mechanisms are triggered accidentally or maliciously (e.g. [41,48]). Where such escapes do occur, or where access is required for planned maintenance, the water can be controlled by inserting "stop planks" (effectively a temporary dam) using pre-formed vertical channels in the canal walls. Some canals may themselves be flooded from adjacent rivers or streams, particularly during extreme weather events [50].

Waterborne transport can be environmentally beneficial, efficient and cost-effective for appropriate cargoes. In view of the climate emergency, governments worldwide are increasingly keen to examine methods that can offer reduced environmental impacts (e.g. [1,14]), and these issues will now be discussed in more detail.

15.3.5 Carbon emissions and climate change issues

Analysis undertaken for the UK Government's Commission for Integrated Transport of CO_2 emissions [45] demonstrated that movement of freight on inland waterways is relatively energy efficient. It generates around 35 g of CO_2/t km, compared to the weighted averages of 14.5 g for rail freight, 200 g for road freight and in excess of 1500 g for air freight. A report by the Inland Waterways Advisory Council [34] stated:

> "...waterborne freight transport could make a useful contribution towards meeting the UK Government's commitment to reducing carbon emissions by 60% by 2050"

The UK government has supported initiatives through financial incentives such as the Waterborne Freight Grant scheme [18], which aids companies with the operating costs associated with running water freight transport, where the latter was more expensive than the road-based alternative. As an example, Lille is one of the most important inland ports in France, with over a million tonnes of waterborne freight passing through the port annually. Given adequate infrastructure, transport of containers by barge over relatively short distances can be successful and could be more widely adopted in the UK. The European Union funded "Creating" project looked at optimal technical solutions and innovative ship designs to strengthen the position of inland shipping within the logistics chain [22,1]. To this end C&RT have already applied to create an "inland port" in the Yorkshire city of Leeds. This involves building two new wharves on a brown-field site on the River Aire, which would be capable of handling up to 200,000 t a year of bulk cargoes such as gravel, aggregates, steel, timber and shipping containers [46,66].

Water transport is suitable for a variety of non-perishable cargoes: steel and other metal products, forest products, and bulk cargoes such as grain, aggregates, coal, petroleum products, chemicals, waste and cement [36], as well as indivisible abnormal loads on some of the larger canals and rivers [15]. Whilst unable to rival the European model, within the limitations of the existing UK network there are niche opportunities to be exploited, where the origin and destination of goods are both situated on waterways. An example is the supermarket chain Tesco, which began to use barges on the Manchester Ship Canal to transport wine from Liverpool in 2007; it is estimated that this has taken 50 lorries off the roads each week and has reduced associated carbon emissions by 80 percent [63].

Some unusual non-transport uses for the existing canal system and its infrastructure have also been developed, such as the installation of a 400-mile fibre-optic network under canal towpaths [56]. Not only did this offer the advantage of a reduced risk of the cables being damaged by other utilities, compared to highway installation, but also conveyance of construction materials and subsequent maintenance activities could be conducted using waterborne transport, again reducing emissions [65]. Another example of mitigation opportunities afforded by the use of waterways is the scope to use them as heat sources or sinks, for heating or cooling adjacent buildings [35]. It is reported that one multinational company with headquarters in London saved over £100k and 276 tons of carbon per year by switching to the new method [59].

Whether providing waterborne transport, leisure activities or these more novel uses, canal systems being artificial waterways require adequate water supplies on a continual basis. A wide variety of methods has been developed to achieve this, and these will be examined in detail in the next section.

15.3.6 Sourcing and conserving water supplies

15.3.6.1 Sources of supply

Canals have been designed with, or undergone post-construction adaptations to utilise, many different sources of water. The effects of climate change will, however, require ongoing reviews to ensure adaptions are put into place in a timely fashion [35,6].

a) Natural sources, at or above the summit level
    - From natural lakes, such as Lough Shark on the Newry Canal (Northern Ireland) [49];
    - From natural rivers, such as the River Dee (UK), which supplies both the Llangollen Canal, which is 5 km/3 miles distant and, fed by gravity alone, the Shropshire Union Canal at Hurleston Junction, which is 66 km/41 miles distant [23];
    - Feeders, such as that from the mountains near Carcassonne (20 km/12.5 miles distant) supplying the Canal du Midi in France [24].
b) Dedicated reservoirs, at or near the summit level
    - The Bassin de Saint–Ferréol on the Canal du Midi (as referred to earlier), created by damming the Laudot River, and Bosley reservoir supplying the Macclesfield Canal in the UK, which is fed by a total of 13 streams that drain from the surrounding hills [44].

c) Water pumped from natural sources below summit level
   - Crofton pumping station on the Kennet and Avon Canal (UK) lifts water from a lake, Wilton Water [5].

d) Input from mine pumping
   - The Bridgewater Canal (UK) originally used water drained from the Worsley mines [49], and the Birmingham Canal Navigations (UK) benefitted from numerous mine water supplies, including Bradley, Deepfields and Stow Heath [55].

e) Input from treated effluent
   - Barnhurst Sewage Treatment Works, Wolverhampton (UK), supplies both the Shropshire Union and the Staffordshire & Worcestershire Canals, although such sources need to be monitored for quality issues to comply with the Water Framework Directive 2000 [38]. A similar arrangement applies to the restored Canal de Roubaix, which uses water from the Wattrelos wastewater treatment plant, after it has passed through a reed-bed filtration system [3].

f) Other sources, such as surface water inputs and groundwater
   - The Water Act 2003 removed the historical exemption for surface water transfers into canals, with the change coming into effect in January 2018 [6,8]. This may result in reduced water supply to a number of UK canals that have previously been fed from land drainage and surface water inputs. Similar impacts may also arise from changes to groundwater abstractions (e.g. those supplying the Grand Union Canal). As the licence applications are undergoing determination at the time of publication, however, the outcome remains uncertain.

15.3.6.2 Water conservation

Water can be lost from a canal system via a number of routes. These include seepage/leakage, losses consequent upon lock usage, and evaporation. The extent of these losses depends upon factors such as the construction methods employed, surface area/ambient temperature, and whether any mitigating features have been incorporated, originally or retrospectively.

The topmost 20 cm of the canal lining is at the greatest risk of leaks, as it is subjected to continual wetting and drying from both water level fluctuations and the wash from boat traffic, as well as damage caused by burrowing animals, all of which can affect the structural integrity of the lining [7]. To reduce physical damage from boat wash as far as is practicable, there is a speed limit of 4 mph (6.4 km/h) on the navigations managed by the Canal & River Trust in England and Wales, or 5 knots on Scottish canals.

Waterways with traditional lock designs are inevitably subject to water loss, as each descending boat takes a lockful of water from the summit level (upper pound) down to the sump level (trough pound), where any excess water is discharged to avoid flooding. Typically, a narrow lock in the UK uses around 140 m^3 of water each time a boat descends. Volumes are, of course, greater for the wider lock chambers on the "broad" canal network [6]. Even when no boats are using a lock there is usually a residual flow from the upper level, which is either routed around the chamber via a bypass channel termed a "bywash", or allowed to pass through the chamber, exiting via leakage through the lock gates themselves.

Not all canals benefit from reliable supplies from major rivers, hence various methods have been developed to conserve water, ranging from structural and mechanical solutions to changes in custom and practice for those using the system.

a) Structural remedies for use with locks

- Storage within modified canal pounds, such as those constructed on the lock flight at Devizes (UK), which rises over 70 m in just over 3 km, with the intervals between adjacent locks being, of necessity, very short. Each of the 16 locks was built with, in effect, its own reservoir, with the intervening pounds being extended sideways, to store the water needed for its operation (Fig. 15.1).

Fig. 15.1: Water storage area (upper right) situated between two lock chambers at Caen Hill, Devizes. Licensed for use: CC BY-SA/2.0 - © Chris Talbot - geograph.org.uk/p/2837654.

- Side ponds and related techniques, which employ connecting culverts to direct the water emptied from a lock into a separate storage area, or "water saving basin" (e.g. [60]). The basin is then drained to partly fill the chamber for use by a boat climbing the flight, a method that is also useful where the pounds between lock chambers are exceptionally short [16]. A variation uses paired, or twinned, locks with a connection between the two, again to allow water being drained from one chamber to partly fill the other (Fig. 15.2).

Fig. 15.2: Twinned lock chambers – Camden, London (UK) [copyright Myrabella/Wikimedia Commons/CC BY-SA 3.0 & GFDL].

- It should be noted that "side ponds" are distinct from "side pounds", such as those found on the Caen Hill flight at Devizes, which are "*Sections of canal used to increase the water storage between locks in a flight* e.g. *with a steep gradient* (definition from: [6]).

b) Mechanical remedies

- Back-pumping schemes, whereby water is recycled back to the top of a flight of locks using a pump or a sequence of pumps at each lock, an example being that on the Canal de Roubaix (Blue Links, no date). This is often necessary on restored canals where the original water sources are no longer available, or inadequate, but can prove costly; the scheme installed at Foxhangers Wharf to serve the 29-lock Caen Hill flight as part of the restoration of the Kennet and Avon canal cost £1million in 1996 [43]. C&RT are currently conducting feasibility studies on new back–pumping schemes on a number of waterways, including the Leeds-Liverpool and Rochdale canals [6,8].
- Boat-lifts as an alternative to locks, whereby boats can be raised and lowered. These have the advantage of saving water by using a water-filled tank, or caisson, to transfer boats between levels. An example is the unique Falkirk Wheel (Fig. 15.3), which replaced a series of 11 locks with a height difference of 35 m. Each rotation takes around 20 min, compared to the 2 h that would have been required to negotiate the original lock flight. As well as being a functional structure on a canal restored only in 2002, this was intentionally designed to be a major tourist attraction and has successfully provided economic regeneration in the local area [4].

Fig. 15.3: The Falkirk Wheel, Scotland (UK) with the visitor centre on the left.

- Inclined planes are another method, which make use of water-filled tanks, often in pairs acting as counterweights; others lift boats out of the water entirely, transporting them on wheeled cradles. These alternatives offer a speedier solution than transiting a sequence of locks. An example is the Strepy–Thieu lift on the Canal du Centre in Belgium, which was constructed in the 1960s as part of the work necessary to bring the Belgian network up to the European 1350 t barge standard. There are no functional inclined planes in the UK at present.

c) Custom and practice
- Standard practices applied to the UK canal system when commercial freight was the norm, and waterway staffing levels were far higher, have now been modified to reflect the predominant usage by the leisure boating community, who are perceived as being less adept at managing complex mechanisms.
- Side pond use has largely been discontinued, or retained only where experienced staff-members are available to operate the mechanisms, as anecdotal evidence indicates that water was wasted in the past by incorrect use. In the light of climate change, however, C&RT are already reviewing this matter, and have stated that:

"An automated system that would only allow paddles and sluices to be operated correctly, or the use of volunteer lock keepers at sites with side ponds … would help reduce this risk" [6].

- As a matter of "good practice", rather than a legally enforceable measure, it is customary for boats to share broad locks, where these exist. As the chambers can take two standard narrowboats side by side, waiting for another vessel travelling in the same direction allows water to be saved.
- Anti-vandal devices. Locks are sometimes deliberately misused in order to drain the water from canals (particularly in some urban areas). To prevent unauthorised use of lock mechanisms anti-vandal devices may be fitted, requiring the use of a special key that can be purchased only from the relevant authorities.

Canal systems of both the UK and Europe have seen major changes since their inception, and this shows no sign of abating. As we move forward with the challenges of climate change in mind, canals may be able to offer yet more additional functions and services over and above those envisaged by their original architects. Some of the projected developments in this area will now be examined.

15.4 Potential futures—climate proofing

To develop climate resilient infrastructure in the most cost-effective manner, designs can incorporate two or more functions. Road and rail embankments can act as flood defences, and reservoirs can be used for flood control and as water storage [54]. Canals can also have additional uses over and above or in some cases as an alternative to transport use: for example, the Tavistock Canal (UK) is used to convey water for a hydropower scheme at Morwellham Quay [42]. Other climate resilience uses will now be considered.

15.4.1 Flood alleviation

Canals can, in some instances, provide flood flow routes or storage [20]. An example is the Gloucester & Sharpness Canal (UK), which provides some flood storage to attenuate peak flows on the River Severn at Gloucester [26]. Indirectly, flood storage is afforded by the embankment of the Basingstoke Canal, as it cuts across the floodplain of a watercourse [30]. The Dahme Flood Relief Canal (Germany) diverts water from the upper reaches of the River Spree, but is also a navigable channel.

One of the key challenges for the UK in the coming decades will be the management of both water supplies and flood risk. Water shortages are expected to be exacerbated by climate change [17,29] but it is anticipated these effects will impact more upon the south, particularly the south–east, of the country in the summer months. In contrast, the north–west is likely to experience additional rainfall in winter, with concomitant increases in flood risk. This combination of circumstances has led to renewed interest in the concept of water transfer between river basins.

15.4.2 Water transfers

Proposals for a piped national water grid in the UK have been put forward several times, commencing in the 1940s [21,33], but this has been ruled out on each occasion on the grounds of cost and practicality. There are precedents for the use of regional water grids; however, pipeline transfers already exist between Wales and England, with the Elan Valley dams supplying Birmingham, and the Vyrnwy reservoir supplying Liverpool. Pipelines are not the only means of transferring water over distance, and making use of existing infrastructure, such as the canal network, may provide a more cost-effective solution in some instances.

Currently, there are three such bulk transfers in operation. Firstly, the Bridgwater and Taunton Canal has conveyed water to Taunton via Durleigh Reservoir since 1962, by transferring water from the River Tone to the River Parrett [57]. Secondly, the Llangollen Canal has long been used to transfer supplies from the River Dee to the Nantwich area, via Hurleston Reservoir [11]. A third example is provided by the Gloucester & Sharpness Canal, which conveys water to Bristol Water for public supply use, thus, transferring water originating from the River Severn over a distance of 26 km [11].

A review of the possibility of extending this principle [11] concluded that large scale water transfer is feasible, but at considerable economic cost; the organisation is, however: *"...working to progress canal water transfer with water companies"*. An example is the recurrent suggestion for transfer of water from the River Severn to the River Thames, using a variety of conveyance methods, which have been examined on a number of occasions (e.g. [13,47]). Ongoing consideration of this option by Thames Water has been requested by both Ofwat (regulator of the water sector in England and Wales) and the Environment Agency [58], as an alternative to the construction of one or more large reservoirs to serve the water-stressed South East area. It has been suggested there may be a role for the utilisation of the Cotswold Canals, once fully restored, in conjunction with piped conveyance [58], although water quality and other environmental considerations, such as invasive non-native species risks, would need to be addressed.

In the light of continuing climate change, however, surmounting such barriers may prove to be unavoidable. For example, during a period of severe drought in early 2012, a statement from the head of water resources at the Environment Agency [27] stated:

> "Water companies are starting to plan to see if there are any prudent actions that can be taken, which include further transfers of water between companies and even from river basin to river basin. Water takes a lot of energy to transfer, so it won't be something that happens every day, but it can be very useful for meeting drought demands. There are risks, however, such as transferring invasive species and changing river chemistry, so we have to weigh this up."

It remains to be seen how this situation will develop in the future, with the needs of consumers and the requirement to protect the natural environment potentially in opposition.

The canal systems across Europe have already undergone many changes since their original construction as transport arteries. Their future may lie in their ability to provide assistance in attaining climate resilience.

15.5 Conclusion

Artificial waterways have been constructed for centuries, whether to convey water itself, people or cargoes. From our earliest agriculturally-driven needs to the creation of extensive water-borne transport networks, humankind has modified the natural world, finding a variety of methods to overcome the geographic challenges. The Civil Engineering profession can thank the industrial revolution for providing the catalyst that drove expertise in large scale construction projects, motivated by aims other than those of the military. The inventiveness of the human mind in the face of problems is clearly illustrated by the variety of methods developed for water supply and conservation purposes.

Compared to natural watercourses, canals possess significant advantages as a transport system, but have not always been able to compete with rail and road haulage. Their success varies from country to country, however, as demonstrated by the continued use of the European waterways for freight. The leisure boating industry is still the principal alternative use within the UK, but other opportunities have been developed in recent years, notably the use of towpaths for accommodating the fibre-optic network and, in some instances, providing a means of flood alleviation. Other innovative uses continue to arise, many of them driven by the need to adapt to climate change, and some of these may provide an incentive for further restoration initiatives.

Canals have been of benefit to humankind both in the past and in the present day, and have the potential to prove useful for the foreseeable future.

References

[1] H.G. Blaauw, CREATING Project NR. FP6–506542 (Concepts to Reduce Environmental impact and Attain optimal Transport performance by Inland NaviGation), (2008).

[2] Blue Links, Blue Links, a Leading EU–funded Project for Sustainable Canal Restoration, (2009) [Online]. Available: http://www.bluelinks2008.org/default.html.

[3] Blue Links, (no date). New Water Supply System, [Online]. Available: http://www.bluelinks2008.org/en/flashInfo/news5.html.

[4] British Waterways Scotland & Hayward, D., The Falkirk Wheel, British Waterways Scotland, Glasgow, 2004.

[5] A. Burton, D. Pratt, The Anatomy of Canals – The Mania Years, Tempus Publishing Ltd, Stroud, 2002.

[6] Canal & River Trust, Putting the Water into Waterways – Water Resources Strategy 2015–2020, Canal & River Trust, 2015.

[7] Canal & River Trust, Canal and River Maintenance, Canal & River Trust, 2017.

[8] Canal & River Trust, Putting the Water into Waterways – Water Resources Strategy 2015–2020", Canal & River Trust, 2018.

[9] Canal & River Trust, (no date-a). Fossdyke Navigation [Online]. Available: https://canalrivertrust.org.uk/enjoy–the–waterways/canal–and–river–network/fossdyke–navigation 2019].
[10] Canal & River Trust, (no date-b). Using Technology to Manage our Waterways, [Online]. Available: https://canalrivertrust.org.uk/specialist–teams/managing–our–water/meica–scada–using–technology–to–manage–our–waterways.
[11] Canal & River Trust and Black & Veatch, Water Transfer for Public Water Supply via the CRT Canal Network, Canal & River Trust, 2016.
[12] CanalLink, Final Report – Interreg IIIB North Sea Region – New Opportunities for Inland Waterways Across the North Sea, 2006.
[13] Cascade Consulting, Assessment of Alternative Schemes for Draft Final Water Resources Management Plan – Severn–Thames transfer: Longdon Marsh A. Reading, Thames Water Utilities Ltd, UK, 1992.
[14] Central Europe Programme, Project Stories from the CENTRAL EUROPE Programme – Sustainable Public Transport and Logistics, CENTRAL EUROPE Programme – Joint Technical Secretariat, Vienna, Austria, 2014.
[15] Commercial Boat Operators Association, Inland Waterway Freight Routes for Abnormal Indivisible Loads, 2018. [Online]. Available: http://www.cboa.org.uk/inland–waterway–freight–routes.html.
[16] H.R. de Salis, Bradshaw's Canals and Navigable Rivers of England and Wales, David & Charles, Newton Abbot, 1969.
[17] Defra, Water for Life, H M Government, London, 2011.
[18] Department for Transport, Waterborne Freight Grant (WFG) Scheme 2015 to 2020, H M Government, London, 2015. [Online]. Available: https://www.gov.uk/government/publications/waterborne–freight–grant–scheme–guide–2015–to–2020 (Accessed 2019).
[19] F. Derrick, A Voyage through Time on the Canal du Midi. France Today, 2014. [Online]. Available: https://www.francetoday.com/culture/events/voyage_through_time_on_the_canal_du_midi/.
[20] R.W. Dun, Reducing uncertainty in the hydraulic analysis of canals, Proceed. Instit. Civil Eng.-Wat. Manag. 159 (2006) 211–224.
[21] Environment Agency, Do We Need Large–Scale Water Transfers for South East England?, Environment Agency, Bristol, 2006.
[22] European Commission Research & Innovation (Transport Projects), "CREATING" Shifting Cargo from Road to Water, 2006. [Online]. Available: http://www.welcomeeurope.com/news–europe/creating–deplacer–cargaison–route–voies–d–eau–8791+8691.html#replierTexte.
[23] S. Fisher, Canals of Britain – A Comprehensive Guide, Adlard Coles Nautical, London, 2009.
[24] French Waterways. Canal du Midi [Online]. Available: https://www.french–waterways.com/waterways/south/canal–midi/2019.
[25] B. Gascoigne, History of Canals, 2001. [Online]. Available: http://www.historyworld.net/wrldhis/PlainTextHistories.asp?groupid=141&HistoryID=aa19>rack=pthc#ixzz1u6tc3Nwc.
[26] Gloucester Docks and Sharpness Canal, History, 2012. [Online]. Available: http://www.gloucesterdocks.me.uk/studies/twofloods.htm.
[27] R. Gray, Rivers to Be Linked to Ease Drought Under Water Company Plans *The Daily Telegraph*, 2012. [Online]. Available: http://www.telegraph.co.uk/earth/drought/9178003/Rivers–to–be–linked–to–ease–drought–under–water–company–plans.html (Accessed 4 April 2012).
[28] C. Hadfield, Canals of the World, Basil Blackwell, Oxford, 1964.
[29] J.W. Hall, G. Watts, M. Keil, L. de Vial, R. Street, K. Conlan, P.E. O'Connell, K.J. Beven, C.G. Kilsby, Towards risk-based water resources planning in England and Wales under a changing climate, Water and Environ. J. 26 (2012) 118–129.
[30] Hart District Council, Strategic Flood Risk Assessment – December 2016, Hants: Hart District Council, Fleet, 2016.
[31] P. Higson, Llangollen Canal Closed by Storm Damage, 2019. [Online]. (Accessed 26/06/2019).
[32] G. Holland, Dredging Prioritisation Paper (WUSIG Meeting 13 April 2011), British Waterways, Watford, 2011.
[33] Howard Humphreys Consulting Engineers, National Water Resources Strategy: Comparative Environmental Appraisal of Strategc Options, National Rivers Authority, Worthing, 1994.

[34] Inland Waterways Advisory Council (IWAC), Decreasing our Carbon Footprint – Moving More Freight onto the Inland Waterways of England and Wales, IWAC, London, 2008.
[35] Inland Waterways Advisory Council (IWAC), Climate Change Mitigation and Adaptation – Implications for Inland Waterways in England and Wales, IWAC, London, 2009.
[36] Inland Waterways Association, Advantages of Waterways Freight, (2010). [Online]. Available: https://www.waterways.org.uk/activities/freight/advantages_of_freight.
[37] International Olympic Committee, Factsheet – London 2012 Facts & Figures, (2013). (update – July 2013).
[38] Jacobs, Birmingham Resilience Project – Technical Appendix 14.1: Water Framework Directive Compliance Assessment, Severn Trent Water Ltd., 2015.
[39] T.G. Jordan, The European Cultural Area – A Systematic Geography, 2nd ed., Harper & Row Publishers Inc., New York, 1988.
[40] R.J. Kapsch, Y.E. Long, James Brindley, American canal engineer, Int. J. History of Eng. Technol. 81 (2011) 22–59.
[41] R. Kell, Canal Overflows Near Blackburn Businesses after Vandalism at Water Lock, Lancashire Telegraph, 2018, 20th December.
[42] KSB Ltd & South West Water, Capturing untapped Energy: pump as Turbine – Morwellham Quay, (2014).
[43] P. Lindley–Jones, Restoring the Kennet & Avon Canal, Stroud Tempus, UK, 2002.
[44] Macclesfield Canal Society, (no date). The Macclesfield Canal – Reservoirs and Feeders [Online]. (Accessed 7 May 2012 2012).
[45] A. McKinnon, CO2 Emissions from Freight Transport: An Analysis of UK Data, 2010.
[46] J. Morton, New port to be built in Stourton, South Leeds Life (2019) 1 January.
[47] Mott MacDonald, Thames Water WRMP19 – Resource Options – Raw Water Transfers Feasibility Report. Reading: Thames Water Utilities Limited, (2016).
[48] T. Oakley, Why Would Someone do This? Canal Locks Targeted Over and Over Again by Vandals, Express and Star, 2018, 27 Jun 27.
[49] E. Paget–Tomlinson, The Illustrated History of Canal & River Navigations, Ashbourne Publishing Ltd, Landmark, 2006.
[50] Pennine Waterways News, Rochdale Canal Remains Blocked After Floods, 2012. [Online]. Available: http://waterwaynews.blogspot.com/2012/06/rochdale–canal–remains–blocked–after.html (Accessed 2019).
[51] Province de Hainault, (no date). The Boat Lift of Strépy–Thieu [Online]. Available: https://voiesdeau.hainaut.be/en/placeofinterest/the–boat–lift–of–strepy–thieu/.
[52] L.T.C. Rolt, Thomas Telford, Sutton Publishing Ltd., Stroud, 2007.
[53] A. Rosevear, Turnpike Roads in England, 2008. [Online]. Available: http://www.turnpikes.org.uk/The%20Turnpike%20Roads.htm.
[54] Secretary of State for Environment Food and Rural Affairs, Climate Resilient Infrastructure: Preparing for a Changing Climate, H M Government, London, 2011.
[55] R. Shill, Water Sources on the Birmingham Canal Navigations, 2006. [Online]. Available: http://allensregister.com/bp188_chasewater.php.
[56] S. Sim, Regenerating the Waterways, Watford: British Waterways, 2003.
[57] Somerset Rivers, (no date). Bridgwater and Taunton Canal [Online]. Available: http://www.somersetrivers.co.uk/index.php?module=Content&func=view&pid=73.
[58] Thames Water, Draft Water Resources Management Plan 2019 – Statement of Response – Technical Appendices – Appendix J: Severn Thames Transfer – Further Work, Reading: Thames Water, 2018.
[59] The Engineer, Canal Cooling, 2008. [Online]. Available: https://www.theengineer.co.uk/canal–cooling/.
[60] The Macclesfield Canal, (no date). Bosley Side Ponds [Online]. Available: https://macclesfieldcanal.org.uk/1831–1834.
[61] Tresor–Languedoc, (no date). Canal du Midi (source: UNESCO) [Online]. Available: https://www.tresor–languedoc.com/canal–du–midi–source–unesco/.
[62] United Nations Educational Scientific and Cultural Organization, (no date). Canal du Midi [Online]. Available: http://whc.unesco.org/en/list/770/.
[63] D. Ward, Wine on the water as Tesco turns to barges to cut emissions, The Guardian (2007) Fri 19 Oct 2007 (14.47).

[64] R. Wilkinson, River Lee—bow back rivers, Notes and News (2009) [Online].
[65] Wood Hall and Heward Engineering, (no date). Infrastructure Maintenance – Cables [Online]. Available: http://www.whhbarges.co.uk/infrastructure_maintenance.htm.
[66] Yorkshire Evening Post, "Inland Port" Could be Operational in Leeds by Spring 2020, Yorkshire Evening Post, 2018, Monday 25 June 2018.
[67] J. Zumerchik, S.L. Danver, Seas and Waterways of the World: An Encyclopedia of History, Uses, and Issues – Vol. 1. ABC – Clio, Santa Barbara, California, 2008.

CHAPTER 16

Towards sustainable water engineering: Insights and inferences for the future

Susanne M. Charlesworth[a,*], Colin A. Booth[b], Kemi Adeyeye[c]
[a]*CAWR, Coventry University, Coventry, UK* [b]*University of the West of England, Bristol, UK*
[c]*Department of Architecture and Civil Engineering, University of Bath, Bath, UK*
**Corresponding author.*

16.1 Introduction

The role of engineers in the management of water has mainly been in the design and delivery of usually grey "hard" infrastructure which makes use of concrete, metal and plastic products. This has included concrete channels and plastic pipes whose function is to deliver potable water supplies or, conversely, to remove excess surface water as quickly as possible. Latterly, this is not sustainable in terms of performance or environmental impact as the many instances of flooding worldwide demonstrate. Floodwaters carry pollutants, cause erosion, and negatively impact human and environmental health.

Sustainable water engineering (SWE) helps to address these challenges, in addition to delivering cyclical and efficient use of water (and energy) resources at the city, neighbourhood or building scale. In addition to natural water management, SWE solutions can deliver complimentary benefits such as enabling the reuse, recycling and repurposing of wastewater, rather than simply disposing of it. This is particularly the case where cities have reached their peak in meeting water demand, and are essentially "running out of water", as well as time to address the issue, as is detailed in Chapter 6 of this book and Richter et al. [8].

The focus of this book is on the many ways in which engineering solutions can incorporate aspects of sustainability to ensure that they are resilient to change, convey multiple benefits to the planet and to people rather than just the one role of a pipe or a drain. Engineers have an important role to play in ensuring programmes, such as the UN sustainable development goals (SDG), discussed in Chapter 1, are achieved. However, as the chapters in this book illustrate, SWE cannot be achieved without engaging stakeholders, policymakers, landscape designers, environmental organisations, local authorities, and communities; much more can be accomplished by working collaboratively (i.e. following the meaning of Aristotle's phrase, the whole is greater than the sum of its parts).

Sustainable Water Engineering. DOI: 10.1016/C2017-0-04301-X

16.2 Historical perspectives

The usual definition of sustainability is the one from the Brundtland's WCED [12] report, whereby the importance for the whole of humanity is conveyed in the title "*Our Common Future*" and encapsulated in the subtitle: "*World Commission on Environment and Development*". The complexities surrounding the concept are discussed in Chapter 2, which presents the ancient city of Constantinople as an historic example where engineered water solutions assured 1200 years of its existence. To a certain extent, Constantinople could be considered *sustainable*, in terms of managing its water supply, and was *resilient*, a word commonly used in conjunction with sustainability. Resilience is generally used to describe the ability to bounce back from impactful change. An example is a drainage system that continues to function efficiently in the face of change, such as flooding, so that a city continues to effectively manage excess surface water, recover and resume its usual functions [2]. Chapter 2 also illustrates that, far from being new or innovative, many of these methods of sustainably managing water are far from being novel. Therefore, rather than reinventing the (water) wheel, it is proposed that engineers revisit and relearn what has been forgotten in the drive for technological macro–fixes.

16.3 Transport infrastructure

The use of canals (Chapter 15) is a case in point, whereby the European canal system continues to serve multiple functions, for example, as an important transport medium. The use of canals in antiquity to transport people and goods developed to address burgeoning urbanisation and industry. Their use was largely superseded by the increased use of the railways, followed by greater use of vehicular transport and the accompanying construction of roads and highways necessary to accommodate the increased volumes of traffic. Yet, Chapter 15 suggests that the UK and other context–relevant countries could learn from the continued use of European canals in the transport of goods, which could reduce the overreliance on road transport.

Drainage from highways (Chapter 10) offers great opportunities to sustainably address the safety of road–users and protection of the paved structure by using recycled materials, such as recycled asphalt pavement, infiltrating structures, such as pervious paving systems (PPS), management of excess surface water using highway filter drains, and slow conveyance to the receiving environment using swales.

16.4 Water supply and demand

Chapter 2 showed that in antiquity, the city of Constantinople was able to successfully deliver water to its population using engineered strategies for 1200 years. However, the modern megacity of Istanbul with a population of 15.2 million in 2020 (https://worldpopulationreview.com/

world-cities/istanbul-population/), struggled with water supply deficiencies in the 1990s [1]. By 2025, about half of the global population will live in areas classified as "water-stressed" [11]. With 785 million people worldwide lacking access to basic drinking water services, it is no surprise then, that the total number of diarrhoeal deaths in 2016 was 1.4 million (Chapter 7). Of those, after adjusting for the likely effect of non-blinding bias, 485,000 deaths were attributable to inadequate water, 432,000 to inadequate sanitation and 165,000 to inadequate hygiene behaviours. Inadequate WASH together caused 829,000 diarrhoeal deaths, which corresponds to ~60 percent of total diarrhoeal deaths in 2016 that would have been preventable through improved drinking water and/or sanitation services and/or simply handwashing with soap [11,7]. Despite the Coronavirus Pandemic in 2020, the WHO/UNICEF joint monitoring program estimates 4 in 10 households still do not have soap and water in their dwellings, which affects their ability to stay safe or simply wash their hands (Chapter 7). To this end, Chapter 3 covers water supply and the concepts of wholesomeness in terms of water quality standards, efficient water distribution systems, and the role of the engineer in maintaining a sustainable water supply.

Examples of water–related emergencies worldwide include Cape Town (South Africa) and Saõ Paolo (Brazil), where (in 2018) the former nearly ran out of water for the populace. The cause, effect and management responses are detailed in Chapter 6; here, the author lists cities around the world where this situation will become increasingly common due to mounting pressures around urbanisation and changes imposed on the hydrological cycle due to increasing development and the profligate use of potable water supplies.

Rainwater harvesting (Chapter 8) and greywater recycling (Chapter 4) make use of water that would otherwise be wasted. However, conserving water in this way is pointless if it is then wasted in inefficient delivery systems, and inefficient water use actions and behaviours. At the front end of the water supply spectrum, Chapter 13 recognises the additional value of water flowing in engineered networks as a useful source of energy. It details the process, hardware and water–energy nexus benefits of micro hydropower recovered in water networks in terms of reducing the environmental, economic, and social impact of such networks. On the demand side, Chapter 5 considers the efficient use of domestic water, holistically accounting for the engineering of appliances to reduce water use on the one hand, and the socio–economic aspects, which depend on user engagement, better marketing strategies and improved policy on the other.

16.5 Soft infrastructure

Innovative strategies are needed to transition towns and cities to sustainable water management, whilst accounting for climate change and population growth [6]. These do not need to be mega–structures but can include the use of blue (water) and green (vegetation) infrastructure, in combination with the use of native planting to achieve a multiple benefit solution.

There are many ways in which engineers can address sustainability and resilience in water management systems; Fenner et al. [4] describes these as approaches with the ability to adapt to change, are flexible and also able to incorporate blue and green infrastructure (GBI) with grey, concrete-based water management techniques. Encapsulated in the concept of ecosystem services and the SuDS Square (as discussed in Chapter 11) is the information needed to design such systems, as well as the multiple benefits that such an approach confers. Chapter 11 also illustrates how "engineered" interventions, such as constructed wetlands, can be integrated into the overall design. Two case studies are presented in this chapter, firstly, the design of a SuDS management train in the City of Coventry, UK, which details the necessary information for the decision-making process to ensure that the proposed interventions are appropriate for the site conditions. The second case study illustrates the use of SuDS in a challenging environment, whereby the system was designed, constructed and operated by the community in a refugee camp in the Kurdistan Region of Iraq. A further interesting case study is presented in Chapter 14 of the design of sustainable drainage, nature based, and GBI in Post-Soviet housing complexes to provide more pleasant living conditions, but also resilience to flooding, climate change and pollution. The integration of open water into residential areas as suggested in Chapter 14 provides multiple benefits including the potential for increased biodiversity and amenity for residents.

Focusing on the use of GBI, Chapter 9 details the use of phytoremediation to address the treatment of polluted runoff from domestic greywater, municipal wastewater, leachates and industrial effluents. These engineered or constructed wetlands offer low–tech, cost–effective means of dealing with metals and organics, trapping suspended solids, and contribute to removing nutrients such as nitrogen and phosphorus. The harvested biomass can be used as mulches and fertilisers but can also provide much-needed feedstock for biogas generation. The chapter concludes with excellent Nigerian examples to illustrate the multiple benefits of this sustainable engineering solution.

Soft infrastructure needs to be flexible in design in order to manage future challenges, one of which is global climate change. The phrase "Think globally, act locally" was initiated in the early 1900s by Scottish town planner Patrick Geddes, but became an environmental catch phrase in the 1970s [3]. It remains relevant as the phrase is still in use, for example by the UNDP [10] with respect to achievement of the SDGs. Therefore, although GBI are utilised locally, they have a cumulative global effect, for example in mitigating and adapting to climate change.

16.6 Retrofitting sustainable devices

There is not a specific chapter devoted to this particular subject, but it is a thread running throughout and is sometimes used as a barrier to the development of sustainable approaches to water infrastructure and delivery. Whilst it is relatively easy to design sustainable options into new build, retrofit is often perceived as being difficult and expensive. However, Chapter 11 provides an international perspective on the retrofit of SuDS devices on-street and also around the home, showing the possibilities, and arguing that barriers such as land-take, expense and issues around installing interventions around, onto and into existing structures can be overcome fairly readily and cost effectively if site specific characteristics are accounted for.

At the building scale, the retrofitting of water saving devices is becoming standard practice but, as Chapter 5 clearly shows, without accounting for the eventual user or the effective use of policy, take-up will not be sufficient to be impactful. Community engagement, it is argued, can be achieved via raising awareness, appropriate market/marketing strategies and implementing effective policy to encourage their use.

Greywater and RWH (Chapters 4 and 8, respectively) can supplement water supplies but both need careful design to optimise financial returns so, in Chapter 8, tank size is important since over-sizing offers minimal benefit. Chapter 4 introduces the complexities around water company attitudes to reducing their return on supplying water of potable standards when GW replaces some of their provision.

As is described in Chapter 10, the retrofit of some SuDS devices during highway renovation is quite possible, and the widespread use of combined kerb and drainage components can be used to replace solid kerbs, with stormwater potentially directed into a SuDS management train of PPS or bioremediation.

16.7 Addressing climate change through sustainable water engineering

There is enormous potential in providing engineered solutions, which can mitigate environmental changes that have already taken place, and adapting to those which may occur in the future. Examples in Chapter 6 show that engineers can be involved in the production of local models of water supply and management under climate change scenarios in order to predict future water use and distribution.

Highways need to be specifically designed with climate change in mind, since changing rainfall patterns, high temperatures and alterations in drainage will impact road safety and the longevity of the trafficked pavement. In Chapter 10, various suggestions are made in terms of maintenance regimes, road structures, including the use of pervious surfaces, and also GBI, such as filter drains and swales placed parallel with the road to manage excess runoff and issues with pollution. This is further explored in Chapter 11 where the use of GBI in cities in the design of SuDS can tackle multiple issues, such as changes to the climate, improving biodiversity and perceptions of human health. In the future, with burgeoning populations and the impacts of climate change to account for, a one–benefit solution will not be useful, but with proactive collaborative working, solutions with multiple benefits can be co-designed and managed for the benefit of all. Chapter 12 moves away from the city, into upstream, rural environments to upscale to a catchment-based approach in tackling the issue of flooding of downstream communities. Natural flood risk management combines the benefits of GBI, SuDS and Nature Based Solutions by "working with nature" in large scale, but in dispersed interventions, carefully designed into the catchment to de-synchronise and smooth flood peaks.

16.8 The future of sustainable water engineering

Whilst acknowledging the statement: "*Sustainable water management is a major challenge*" [5], the chapters in this book have shown the possibilities of addressing issues associated with the water supply and management infrastructure in a sustainable and resilient manner. The future for SWE will probably include smart technological solutions to issues around data collection that will enable more efficient strategies to be implemented, particularly in the developing world, where access to water is a constant issue. For instance, Swan et al. [9] argue that the use of smart pumps attached to improved sources of water, such as boreholes, would enable the rapid identification of inefficient or faulty infrastructure, reducing the necessity for site visits, improve monitoring and data collection, and subsequently reduce the time taken to repair defective pumps. In this way, the use of smart pumps may significantly improve the likelihood of delivery of SDG 6's target of universal access to safe and affordable drinking water by 2030. Furthermore, in addressing the energy–water nexus, You et al. [13] harness the wasted power of human urine to generate electricity in combining waste-to-power in order to enhance the benefits of WASH in a remote region of Uganda. If implemented, these novel approaches have the potential to give the engineer valuable tools with which to enhance the sustainability and resilience of interventions in the management of water.

The authors in this book expound the benefits of sustainable engineering solutions, but also emphasise the importance of collaborative working between engineers and designers, strong policies and standards, and engaging proactively with end users. The importance of community and stakeholder engagement, based on education, knowledge exchange and the dissemination of relevant, robust information should also not be overlooked. Lastly, this book concurs the need for more research, practical investigations and demonstrators to further establish efficiency in performance, and efficacy of solutions to deliver multifaceted sustainable benefits.

References

[1] D. Altinbilek, Water management in Istanbul, Water Resour. Develop. 22 (2) (2006) 241–253, doi:10.1080/07900620600709563.

[2] K. Bertilsson, K. Wiklund, I. de Moura Tebaldi, O. Moura Rezende, A. Pires Veról, M. Gomes Miguez, Urban flood resilience–a multi-criteria index to integrate flood resilience into urban planning, J. Hydrol. 573 (2019) 970–982.

[3] É. Darier, R. Schüle, "Think globally, act locally"? climate change and public participation in Manchester and Frankfurt, Local Environ. 4 (3) (1999) 1–14.

[4] R. Fenner, E. O'Donnell, S. Ahilan, D. Dawson, Achieving urban flood resilience in an uncertain future, Water 11 (2019) 1082, 1–9, doi:10.3390/w11051082 1–9.
[5] S.H.A. Koop, C.J. van Leeuwen, The challenges of water, waste and climate change in cities, Environ. Dev. Sustain. 19 (2017) 385–418, doi:10.1007/s10668-016-9760-4.
[6] K. van Leeuwen, R. Sjerps, Istanbul: the challenges of integrated water resources management in Europa's megacity, Environ. Dev. Sustain. 18 (2016) 1–17, doi:10.1007/s10668-015-9636-z.
[7] A. Prüss-Ustün, J. Wolf, J. Bartram, T. Clasen, O. Cumming, M.C. Freeman, R. Johnston, Burden of disease from inadequate water, sanitation and hygiene for selected adverse health outcomes: an updated analysis with a focus on low-and middle-income countries, Int. J. Hyg. Environ. Health 222 (5) (2019) 765–777.
[8] B.D. Richter, D. Abell, E. Bacha, K. Brauman, S. Calos, A. Cohn, C. Disla, S.F. O'Brien, D. Hodges, S. Kaiser, M. Loughran, C. Mestre, M. Reardon, E. Siegfried, Tapped out: how can cities secure their water future? Water Policy 15 (2013) 335–363.
[9] A. Swan, N. Cooper, W. Gamble, M. Pritchard, Using smart pumps to help deliver universal access to safe and affordable drinking water, P. Inst. Civil Eng-Eng. Su. 171 (6) (2018) 277–285.
[10] UNDP (United Nations Development Programme), Think Globally, Act Locally, 2018. Available at: https://www.undp.org/content/undp/en/home/blog/2018/think-globally-act-locally.html.
[11] WHO, Drinking Water: Key Facts, 2019. Available at: https://www.who.int/news-room/fact-sheets/detail/drinking-water.
[12] World Commission on Environment and Development, Our Common Future, Oxford University Press, Oxford, 1987.
[13] J. You, C. Staddon, A. Cook, J. Walker, J. Boulton, W. Powell, I. Ieropoulos, Multidimensional benefits of improved sanitation: evaluating 'PEE POWER®' in Kisoro, Uganda, Int. J. Environ. Res. Public Health 17 (2020) 2175, doi:10.3390/ijerph17072175.

Index

Page numbers followed by '*t*' indicate tables and *f* indicate figures.

H

I

J

K

L

M

N

S

T

U

V

W

Y

Printed in the United States
By Bookmasters